Two Manuals Combined into One Book

Military First Aid

and

Military Mountaineering

Department of Defense

Pentagon
Publishing

*FIELD MANUAL
NO. 4-25.11
NAVY TACTICAL
REFERENCE
PUBLICATION
NO. 4-02.1
AIR FORCE MANUAL
NO. 44-163(I)

HEADQUARTERS
DEPARTMENT OF THE ARMY,
THE NAVY, AND THE AIR FORCE
Washington, DC, 23 December 2002

FIRST AID

TABLE OF CONTENTS

*This publication supersedes FM 21-11, 27 October 1988

PREFACE

This manual meets the first aid training needs of individual service members. Because medical personnel will not always be readily available, the nonmedical service members must rely heavily on their own skills and knowledge of life-sustaining methods to survive on the integrated battlefield. This publication outlines both self-aid and aid to other service members (buddy aid). More importantly, it emphasizes prompt and effective action in sustaining life and preventing or minimizing further suffering and disability. First aid is the emergency care given to the sick, injured, or wounded before being treated by medical personnel. The term *first aid* can be defined as "urgent and immediate lifesaving and other measures, which can be performed for casualties by nonmedical personnel when medical personnel are not immediately available." Nonmedical service members have received basic first aid training and should remain skilled in the correct procedures for giving first aid. This manual is directed to *all* service members. The procedures discussed apply to all types of casualties and the measures described are for use by both male and female service members.

This publication is in consonance with the following North Atlantic Treaty Organization (NATO) International Standardization Agreements (STANAGs) and American, British. Canadian, and Australian Quadripartite Standardization Agreements (QSTAGs).

TITLE	STANAG	QSTAG
Medical Training in First Aid, Basic Hygiene and Emergency Care	2122	535
First Aid Kits and Emergency Medical Care Kits	2126	
Medical First Aid and Hygiene Training in NBC Operations	2358	
First Aid Material for Chemical Injuries	2871	

These agreements are available on request, using Department of Defense (DD) Form 1425 from the Standardization Documents Order Desk, 700 Robins Avenue, Building 4, Section D, Philadelphia, Pennsylvania 19111-5094.

Unless this publication states otherwise, masculine nouns and pronouns do not refer exclusively to men.

Use of trade or brand names in this publication is for illustrative purposes only and does not imply endorsement by the Department of Defense (DOD).

The proponent for this publication is the US Army Medical Department Center and School. Submit comments and recommendations for the improvement of this publication directly to the **Commander, US Army Medical Department Center and School, ATTN: MCCS-FCD-L, 1400 East Grayson Street, Fort Sam Houston, Texas 78234-5052**.

CHAPTER 1

FUNDAMENTAL CRITERIA FOR FIRST AID

"The fate of the wounded rests in the hands of the ones who apply the first dressing."

Nicholas Senn (1898) (49th President of the American Medical Association)

1-1. General

When a nonmedical service member comes upon an unconscious or injured service member, he must accurately evaluate the casualty to determine the first aid measures needed to prevent further injury or death. He should seek medical assistance as soon as possible, but he should not interrupt the performance of first aid measures. To interrupt the first aid measures may cause more harm than good to the casualty. Remember that in a chemical environment, the service member should not evaluate the casualty until the casualty has been masked. After performing first aid, the service member must proceed with the evaluation and continue to monitor the casualty for development of conditions which may require the performance of necessary basic lifesaving measures, such as clearing the airway, rescue breathing, preventing shock, and controlling bleeding. He should continue to monitor the casualty until relieved by medical personnel.

Service members may have to depend upon their first aid knowledge and skills to save themselves (self-aid) or other service members (buddy aid/ combat lifesaver). They may be able to save a life, prevent permanent disability, or reduce long periods of hospitalization by knowing **WHAT** to do, **WHAT NOT** to do, and **WHEN** to seek medical assistance.

NOTE

The prevalence of various body armor systems currently fielded to US service members, and those in development for future fielding, may present a temporary obstacle to effective evaluation of an injured service member. You may have to *carefully remove* the body armor from the injured service member to complete the evaluation or administer first aid. Begin by removing the outer–most hard or soft body armor components (open, unfasten or cut the closures, fasteners, or straps), then remove any successive layers in the same manner. Be sure to follow other notes, cautions and warnings regarding procedures in contaminated situations and when a broken back or neck is suspected. Continue to evaluate.

1-2. Terminology

To enhance the understanding of the material contained in this publication, the following terms are used—

- *Combat lifesaver.* This is a US Army program governed by Army Regulation (AR) 350-41. The combat lifesaver is a member of a nonmedical unit selected by the unit commander for additional training beyond basic first aid procedures (referred to as *enhanced first aid*). A minimum of one individual per squad, crew, team, or equivalent-sized unit should be trained. The primary duty of this individual does not change. The additional duty of combat lifesaver is to provide enhanced first aid for injuries based on his training before the trauma specialist (military occupational specialty [MOS] 91W) arrives. The combat lifesaver's training is normally provided by medical personnel assigned, attached, or in direct support (DS) of the unit. The senior medical person designated by the commander manages the training program.

- *Trauma Specialist (US Army) or Hospital Corpsman (HM).* A medical specialist trained in emergency medical treatment (EMT) procedures and assigned or attached in support of a combat or combat support unit or marine forces.

- *Casualty evacuation.* Casualty evacuation (CASEVAC) is a term used by nonmedical units to refer to the movement of casualties aboard nonmedical vehicles or aircraft. See also the term *transported* below. Refer to FM 8-10-6 for additional information.

> **CAUTION**
>
> Casualties transported in this manner do not receive en route medical care.

- *Enhanced first aid (US Army).* Enhanced first aid is administered by the combat lifesaver. It includes measures, which require an additional level of training above self-aid and buddy aid, such as the initiation of intravenous (IV) fluids.

- *Medical evacuation.* Medical evacuation is the timely, efficient movement of the wounded, injured, or ill service members from the battlefield and other locations to medical treatment facilities (MTFs). Medical personnel provide en route medical care during the evacuation. Once the casualty has entered the medical stream (trauma specialist, hospital corpsman, evacuation crew, or MTF), the role of first aid in the care of the casualty ceases and the casualty becomes the responsibility of the health service support (HSS) chain. Once he has entered the HSS chain he is referred to as a *patient*.

- *First aid measures.* Urgent and immediate lifesaving and other measures, which can be performed for casualties (or performed by the casualty himself) by nonmedical personnel when medical personnel are not immediately available.

- *Medical treatment.* Medical treatment is the care and management of wounded, injured, or ill service members by medically trained (MOS-trained) HM, and area of concentration (AOC) personnel. It may include EMT, advanced trauma management (ATM), and resuscitative and surgical intervention.

- *Medical treatment facility.* Any facility established for the purpose of providing medical treatment. This includes battalion aid stations, Level II facilities, dispensaries, clinics, and hospitals.

- *Self-aid/buddy aid.* Each individual service member is trained to be proficient in a variety of specific first aid procedures. This training enables the service member or a buddy to apply immediate first aid measures to alleviate a life-threatening situation.

- *Transported.* A casualty is moved to an MTF in a nonmedical vehicle without en route care provided by a medically-trained service member (such as a Trauma Specialist or HM). First aid measures should be continually performed while the casualty is being transported. If the casualty is acquired by a dedicated medical vehicle with a medically-trained crew, the role of first aid ceases and the casualty becomes the responsibility of the HSS chain, and is then referred to as a *patient*. This method of transporting a casualty is also referred to as *CASEVAC*.

1-3. Understanding Vital Body Functions for First Aid

In order for the service member to learn to perform first aid procedures, he must have a basic understanding of what the vital body functions are and what the result will be if they are damaged or not functioning.

a. Breathing Process. All humans must have oxygen to live. Through the breathing process, the lungs draw oxygen from the air and put it into the blood. The heart pumps the blood through the body to be used by the cells that require a constant supply of oxygen. Some cells are more dependent on a constant supply of oxygen than others. For example, cells of the brain may die within 4 to 6 minutes without oxygen. Once these cells die, they are lost forever since they do not regenerate. This could result in permanent brain damage, paralysis, or death.

b. *Respiration.* Respiration occurs when a person inhales (oxygen is taken into the body) and then exhales (carbon dioxide [CO_2] is expelled from the body). Respiration involves the—

- *Airway.* The airway consists of the nose, mouth, throat, voice box, and windpipe. It is the canal through which air passes to and from the lungs.

- *Lungs.* The lungs are two elastic organs made up of thousands of tiny air spaces and covered by an airtight membrane. The *bronchial tree* is a part of the lungs.

- *Rib cage.* The rib cage is formed by the muscle-connected ribs, which join the spine in back, and the breastbone in front. The top part of the rib cage is closed by the structure of the neck, and the bottom part is separated from the abdominal cavity by a large dome-shaped muscle called the *diaphragm* (Figure 1-1). The diaphragm and rib muscles, which are under the control of the respiratory center in the brain, automatically *contract* and *relax*. *Contraction* increases and *relaxation* decreases the size of the rib cage. When the rib cage increases and then decreases, the air pressure in the lungs is first less and then more than the atmospheric pressure, thus causing the air to rush into and out of the lungs to equalize the pressure. This cycle of inhaling and exhaling is repeated about 12 to 18 times per minute.

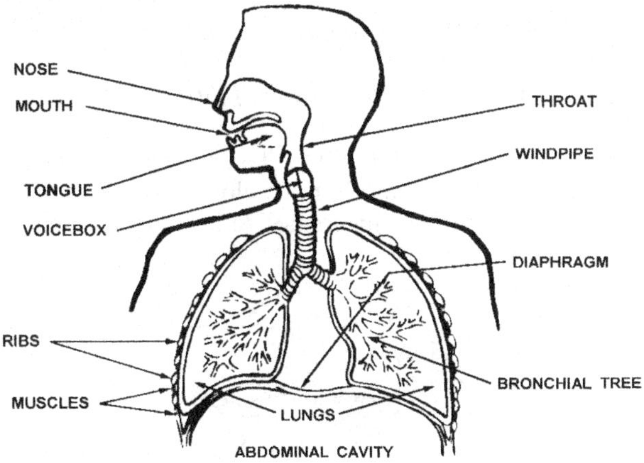

Figure 1-1. Airway, lungs, and rib cage.

c. Blood Circulation. The heart and the blood vessels (arteries, veins, and capillaries) circulate blood through the body tissues. The heart is divided into two separate halves, each acting as a pump. The left side pumps oxygenated blood (bright red) through the arteries into the capillaries; nutrients and oxygen pass from the blood through the walls of the capillaries into the cells. At the same time waste products and CO_2 enter the capillaries. From the capillaries the oxygen poor blood is carried through the veins to the right side of the heart and then into the lungs where it expels the CO_2 and picks up oxygen. Blood in the veins is dark red because of its low oxygen content. Blood does not flow through the veins in spurts as it does through the arteries. The entire system of the heart, blood vessels, and lymphatics is called the *circulatory system.*

(1) *Heartbeat.* The heart functions as a pump to circulate the blood continuously through the blood vessels to all parts of the body. It contracts, forcing the blood from its chambers; then it relaxes, permitting its chambers to refill with blood. The rhythmical cycle of contraction and relaxation is called the *heartbeat.* The normal heartbeat is from 60 to 80 beats per minute.

(2) *Pulse.* The heartbeat causes a rhythmical expansion and contraction of the arteries as it forces blood through them. This cycle of expansion and contraction can be felt (monitored) at various points in the body and is called the *pulse.* The common points for checking the pulse are at the—

- Side of the neck (*carotid*).
- Groin (*femoral*).
- Wrist (*radial*).
- Ankle (*posterior tibial*).

(*a*) *Carotid pulse.* To check the carotid pulse, feel for a pulse on the side of the casualty's neck closest to you. This is done by placing the tips of your first two fingers beside his Adam's apple (Figure 1-2).

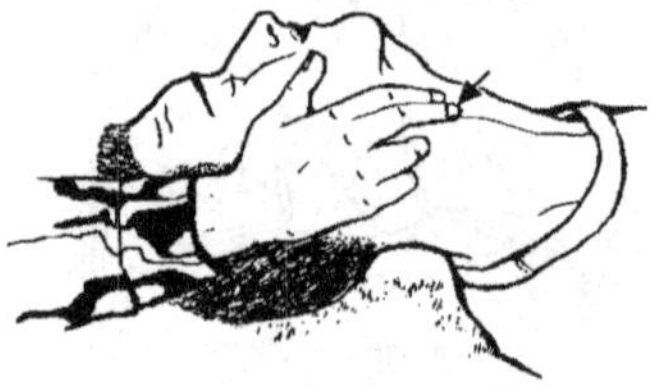

Figure 1-2. Carotid pulse.

(*b*) *Femoral pulse*. To check the femoral pulse, press the tips of your first two fingers into the middle of the groin (Figure 1-3).

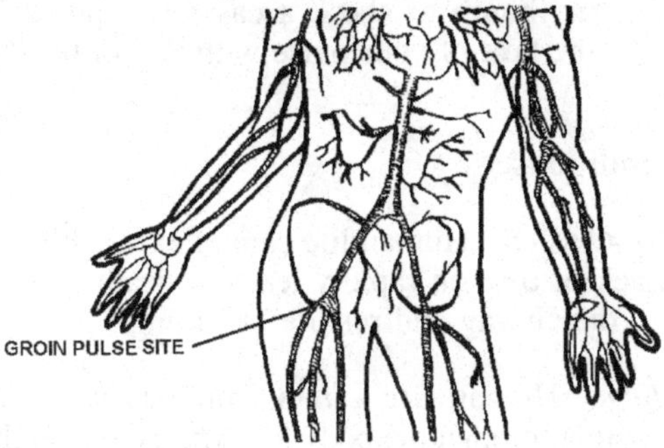

Figure 1-3. Femoral pulse.

(*c*) *Radial pulse*. To check the radial pulse, place your first two fingers on the thumb side of the casualty's wrist (Figure 1-4).

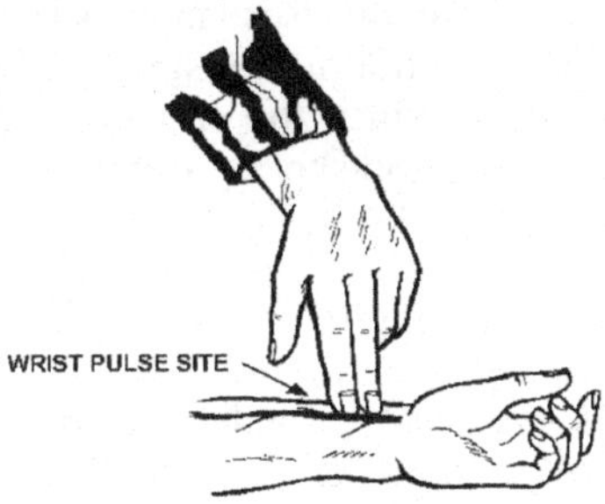

Figure 1-4. Radial pulse.

(*d*) *Posterior tibial pulse*. To check the posterior tibial pulse, place your first two fingers on the inside of the ankle (Figure 1-5).

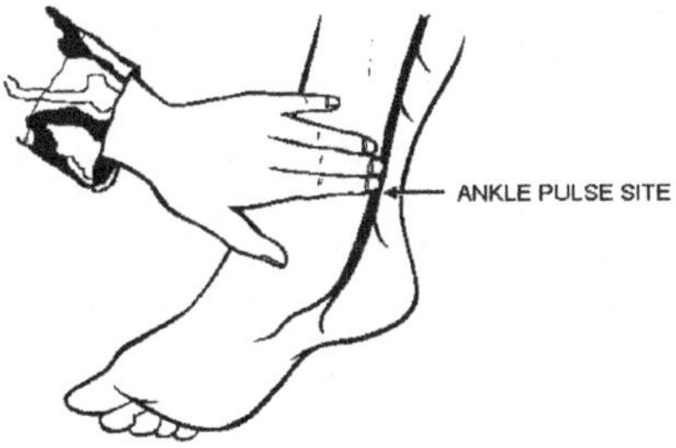

Figure 1-5. Posterior tibial pulse.

NOTE

DO NOT use your thumb to check a casualty's pulse because you may confuse the beat of your pulse with that of the casualty.

1-4. Adverse Conditions

a. Lack of Oxygen. Human life cannot exist without a continuous intake of oxygen. Lack of oxygen rapidly leads to death. First aid involves knowing how to open the airway and restore breathing.

b. Bleeding. Human life cannot continue without an adequate volume of blood circulating through the body to carry oxygen to the tissues. An important first aid measure is to stop the bleeding to prevent the loss of blood.

c. Shock. Shock means there is an inadequate blood flow to the vital tissues and organs. Shock that remains uncorrected may result in death even though the injury or condition causing the shock would not otherwise be fatal. Shock can result from many causes, such as loss of blood, loss of fluid from deep burns, pain, and reaction to the sight of a wound or blood. First aid includes preventing shock, since the casualty's chances of survival are much greater if he does not develop shock. Refer to paragraphs 2-21 through 2-24 for a further discussion of shock.

d. Infection. Recovery from a severe injury or a wound depends largely upon how well the injury or wound was initially protected. Infections result from the multiplication and growth (spread) of harmful microscopic organisms (sometimes referred to as germs). These harmful microscopic organisms are in the air, water, and soil, and on the skin and clothing. Some of these organisms will immediately invade (contaminate) a break in the skin or an open wound. The objective is to keep wounds clean and free of these organisms. A good working knowledge of basic first aid measures also includes knowing how to dress a wound to avoid infection or additional contamination.

1-5. Basics of First Aid

Most injured or ill service members are able to return to their units to fight or support primarily because they are given appropriate and timely first aid followed by the best medical care possible. Therefore, all service members must remember the basics.

- Check for **BREATHING**: Lack of oxygen intake (through a compromised airway or inadequate breathing) can lead to brain damage or death in very few minutes.

- Check for **BLEEDING**: Life cannot continue without an adequate volume of blood to carry oxygen to tissues.

- Check for **SHOCK**: Unless shock is prevented, first aid performed, and medical treatment provided, death may result even though the injury would not otherwise be fatal.

1-6. Evaluating a Casualty

a. The time may come when you must instantly apply your knowledge of first aid measures. This could occur during combat operations, in training situations, or while in a nonduty status. Any service member observing an unconscious and/or ill, injured, or wounded person must carefully and skillfully evaluate him to determine the first aid measures required to prevent further injury or death. He should seek help from medical personnel as soon as possible, but must not interrupt his evaluation of the casualty or fail to administer first aid measures. A second service member may be sent to find medical help. One of the cardinal principles for assisting a casualty is that you (the initial rescuer) must continue the evaluation and first aid measures, as the tactical situation permits, until another individual relieves you. If, during any part of the evaluation, the casualty exhibits the conditions (such as shock) for which the service member is checking, the service member must stop the evaluation and immediately administer first aid. In a chemical environment, the service member should not evaluate the casualty until both the individual and the casualty have been masked. If it is suspected that a nerve agent was used, administer the casualty's own nerve agent antidote autoinjector. After providing first aid, the service member must proceed with the evaluation and continue to monitor the casualty for further complications until relieved by medical personnel.

WARNING

Do not use your own nerve agent antidote autoinjector on the casualty.

NOTE

Remember, when evaluating and/or administering first aid to a casualty, you should seek medical aid as soon as possible. **DO NOT** stop first aid measures, but if the situation allows, send another service member to find medical aid.

b. To evaluate a casualty, perform the following steps:

(1) *Check the casualty for responsiveness.* This is done by gently shaking or tapping him while calmly asking, "Are you OK?" Watch for a response. If the casualty does not respond, go to step (2). If the casualty responds, continue with the evaluation.

(*a*) If the casualty is conscious, ask him where he feels different than usual or where it hurts. Ask him to identify the location of pain if he can, or to identify the area in which there is no feeling.

(*b*) If the casualty is conscious but is choking and cannot talk, stop the evaluation and begin first aid measures. Refer to paragraphs 2-10 and 2-11 for specific information on opening the airway.

WARNING

If a broken back or neck is suspected, do not move the casualty unless his life is in immediate danger (such as close to a burning vehicle). Movement may cause permanent paralysis or death.

(2) *Check for breathing.* (Refer to paragraph 2-6 for this procedure.)

(*a*) If the casualty is breathing, proceed to step (3).

(*b*) If the casualty is not breathing, stop the evaluation and begin first aid measures to attempt to ventilate the casualty. Attempt to open the airway, if an airway obstruction is apparent, clear the airway obstruction, then ventilate (see paragraphs 2-10 and 2-11).

(*c*) After successfully ventilating the casualty, proceed to step (3).

(3) *Check for pulse.* (Refer to paragraph 1-3*c*(2) for specific methods.) If a pulse is present and the casualty is breathing, proceed to step (4).

(*a*) If a pulse is present, but the casualty is still not breathing, start rescue breathing.

(*b*) If a pulse is not present, seek medical personnel for help.

(4) *Check for bleeding.* Look for spurts of blood or blood-soaked clothes. Also check for *both* entry and exit wounds. If the casualty is bleeding from an open wound, stop the evaluation and begin first aid procedures as follows for a—

(*a*) Wound of the arm or leg (refer to paragraphs 2-16 through 2-18 for information on putting on a field or pressure dressing).

(*b*) Partial or complete amputation, apply dressing (refer to paragraph 2-16 to 2-18) and then apply tourniquet if bleeding is not stopped (refer to paragraph 2-20 for information on putting on a tourniquet).

(*c*) Open head wound (refer to paragraph 3-10 for information on applying a dressing to an open head wound).

(*d*) Open chest wound (refer to paragraph 3-5 for information on applying a dressing to an open chest wound).

(*e*) Open abdominal wound (refer to paragraph 3-7 for information on applying a dressing to an open abdominal wound).

WARNING

In a chemically contaminated area, do not expose the wounds. Apply field dressing and then pressure dressing over wound area as needed.

(5) *Check for shock.* (Refer to paragraph 2-24 for first aid measures for shock.) If the signs and symptoms of shock are present, stop the evaluation, and begin first aid measures immediately. The following are the nine signs and symptoms of shock.

(*a*) Sweaty but cool skin (clammy skin).

(*b*) Paleness of skin. (In dark-skinned service members look for a grayish cast to the skin.)

(*c*) Restlessness or nervousness.

(*d*) Thirst.

(*e*) Loss of blood (bleeding).

(*f*) Confusion (does not seem aware of surroundings).

(*g*) Faster than normal breathing rate.

(*h*) Blotchy or bluish skin, especially around the mouth.

(*i*) Nausea or vomiting.

WARNING

Leg fractures must be splinted before elevating the legs as a first aid measure for shock.

(6) *Check for fractures.*

(*a*) Check for the following signs and symptoms of a back or neck injury and perform first aid procedures as necessary.

- Pain or tenderness of the back or neck area.
- Cuts or bruises on the back or neck area.
- Inability of a casualty to move or decreased sensation to extremities (paralysis or numbness).
  - Ask about ability to move (paralysis).
  - Touch the casualty's arms and legs and ask whether he can feel your hand (numbness).
- Unusual body or limb position.

(*b*) Immobilize any casualty suspected of having a back or neck injury by doing the following:

- Tell the casualty not to move.
- If a back injury is suspected, place padding (rolled or folded to conform to the shape of the arch) under the natural arch of the casualty's back. (For example, a blanket/poncho may be used as padding.)

WARNING

Do not move casualty to place padding.

- If a neck injury is suspected, immediately immobilize (manually) the head and neck. Place a roll of cloth under the casualty's neck, and put weighted boots (filled with dirt or sand) or rocks on both sides of his head.

(*c*) Check the casualty's arms and legs for open or closed fractures.

- Check for *open* fractures by looking for—
  - Bleeding.
  - Bones sticking through the skin.
  - Check for pulse.
- Check for *closed* fractures by looking for—
  - Swelling.
  - Discoloration.
  - Deformity.
  - Unusual body position.
  - Check for pulse.

(*d*) Stop the evaluation and begin first aid measures if a fracture to an arm or leg is suspected. Refer to Chapter 4 for information on splinting a suspected fracture.

(*e*) Check for signs/symptoms of fractures of other body areas (for example, shoulder or hip) and provide first aid as necessary.

(7) *Check for burns.* Look carefully for reddened, blistered, or charred skin; also check for singed clothing. If burns are found, stop the evaluation and begin first aid procedures. Refer to paragraph 3-9 for information on giving first aid for burns.

NOTE

Burns to the upper torso and face may cause respiratory complications. When evaluating the casualty, look for singed nose hair, soot around the nostrils, and listen for abnormal breath sounds or difficulty breathing.

(8) *Check for possible head injury.*

(*a*) Look for the following signs and symptoms:

- Unequal pupils.
- Fluid from the ear(s), nose, mouth, or injury site.
- Slurred speech.
- Confusion.
- Sleepiness.
- Loss of memory or consciousness.
- Staggering in walking.
- Headache.
- Dizziness.
- Nausea or vomiting.
- Paralysis.
- Convulsions or twitches.
- Bruising around the eyes and behind the ears.

(*b*) If a head injury is suspected, continue to watch for signs which would require performance of rescue breathing, first aid measures for shock, or control of bleeding; seek medical aid. Refer to paragraph 3-10 for information on first aid measures for head injuries.

CHAPTER 2

BASIC MEASURES FOR FIRST AID

2-1. General

Several conditions that require immediate attention are an inadequate airway, lack of breathing, and excessive loss of blood (circulation). A casualty without a clear airway or who is not breathing may die from lack of oxygen. Excessive loss of blood may lead to shock, and shock can lead to death; therefore, you must act immediately to control the loss of blood. All wounds are considered to be contaminated, since infection-producing organisms (germs) are always present on the skin and clothing, and in the soil, water, and air. Any missile or instrument (such as a bullet, shrapnel, knife, or bayonet) causing a wound pushes or carries the germs into that wound. Infection results as these organisms multiply. That a wound is contaminated does not lessen the importance of protecting it from further contamination. You must dress and bandage a wound as soon as possible to prevent further contamination.

NOTE

It is also important that you attend to any airway, breathing, or bleeding problems **IMMEDIATELY** because these problems, if left unattended, may become life threatening.

Section I. OPEN THE AIRWAY AND RESTORE BREATHING

2-2. Breathing Process

All humans must have oxygen to live. Through the breathing process, the lungs draw oxygen from the air and put it into the blood. The heart pumps the blood through the body to be used by the cells that require a constant supply of oxygen. Some cells are more dependent on a constant supply of oxygen than others. For example, cells of the brain may die within 4 to 6 minutes without oxygen. Once these cells die, they are lost forever since they do not regenerate. This could result in permanent brain damage, paralysis, or death.

2-3. Assessment of and Positioning the Casualty

a. **CHECK** for responsiveness (Figure 2-1A)—establish whether the casualty is conscious by gently shaking him and asking, "Are you OK?"

b. **CALL** for help (Figure 2-1B).

c. **POSITION** the unconscious casualty so that he is lying on his back and on a firm surface (Figure 2-1C).

WARNING

If the casualty is lying on his chest (prone position), cautiously roll the casualty as a unit so that his body does not twist (which may further complicate a back, neck, or spinal injury).

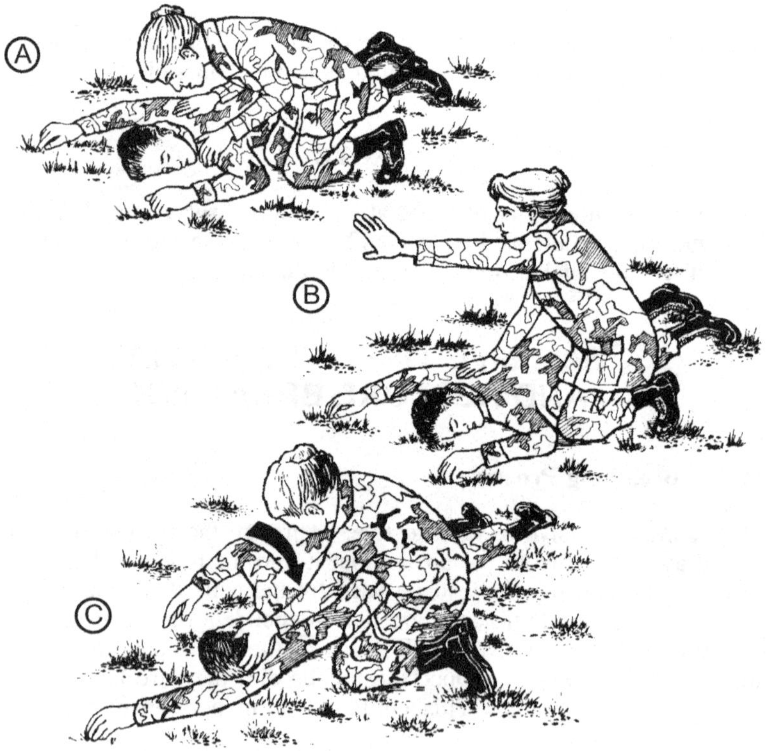

Figure 2-1. Assessment (Illustrated A—C).

(1) Straighten the casualty's legs. Take the casualty's arm that is nearest to you and move it so that it is straight and above his head. Repeat the procedure for the other arm.

(2) Kneel beside the casualty with your knees near his shoulders (leave space to roll his body) (Figure 2-1B). Place one hand behind his head and neck for support. With your other hand, grasp the casualty under his far arm (Figure 2-1C).

(3) Roll the casualty towards you using a steady, even pull. His head and neck should stay in line with his back.

(4) Return the casualty's arms to his side. Straighten his legs. Reposition yourself so that you are now kneeling at the level of the casualty's shoulders. However, if a neck injury is suspected and the jaw-thrust technique will be used, kneel at the casualty's head, looking towards his feet.

2-4. Opening the Airway of an Unconscious or Not Breathing Casualty

The tongue is the single most common cause of an airway obstruction (Figure 2-2). In most cases, simply using the head-tilt/chin-lift technique can clear the airway. This action pulls the tongue away from the air passage in the throat (Figure 2-3).

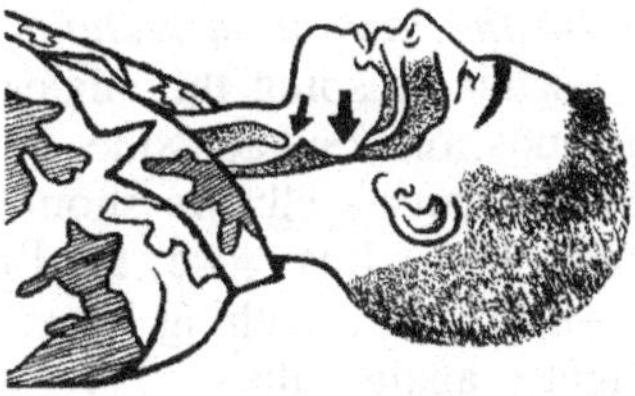

Figure 2-2. Airway blocked by tongue.

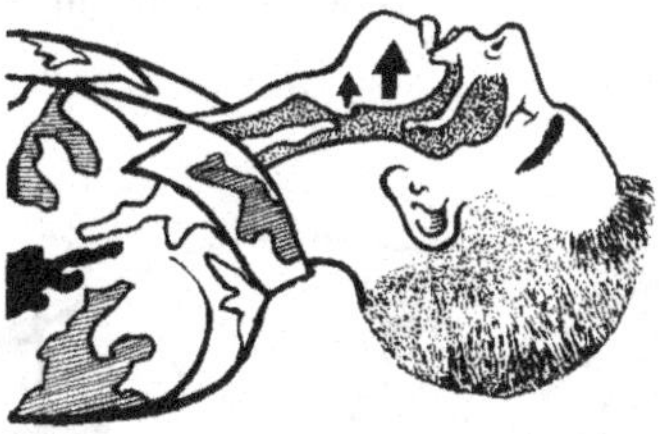

Figure 2-3. Airway opened by extending neck.

a. Call for help and then position the casualty. Move (roll) the casualty onto his back (Figure 2-1C). (Refer to paragraph 2-3*c* for information on positioning the casualty.)

NOTE

Perform finger sweep. If foreign material or vomitus is visible in the mouth, it should be removed, but do not spend an excessive amount of time doing so.

b. Open the airway using the jaw-thrust or head-tilt/chin-lift technique.

CAUTION

The head-tilt/chin-lift technique is an important procedure in opening the airway; however, use extreme care because excess force in performing this maneuver may cause further spinal injury. In a casualty with a suspected neck injury or severe head trauma, the safest approach to opening the airway is the jaw-thrust technique because in most cases it can be accomplished without extending the neck.

(1) *Perform the jaw-thrust technique.* The jaw-thrust may be accomplished by the rescuer grasping the angles of the casualty's lower jaw and lifting with both hands, one on each side, displacing the jaw forward and up (Figure 2-4). The rescuer's elbows should rest on the surface on which the casualty is lying. If the lips close, the lower lip can be retracted with the thumb. If mouth-to-mouth breathing is necessary, close the nostrils by placing your cheek tightly against them. The head should be carefully supported without tilting it backwards or turning it from side to side. If this is unsuccessful, the head should be tilted back very slightly. The jaw-thrust is the safest first approach to opening the airway of a casualty who has a suspected neck injury because in most cases it can be accomplished without extending the neck.

Figure 2-4. Jaw-thrust technique of opening airway.

(2) *Perform the head-tilt/chin-lift technique.* Place one hand on the casualty's forehead and apply firm, backward pressure with the palm to tilt the head back. Place the fingertips of the other hand under the bony part of the lower jaw and lift, bringing the chin forward. The thumb should not be used to lift the chin (Figure 2-5).

NOTE

The fingers should not press deeply into the soft tissue under the chin because the airway may be obstructed.

Figure 2-5. Head-tilt/chin-lift technique of opening airway.

(3) *Check for breathing (while maintaining an airway).* After establishing an open airway, it is important to maintain that airway in an open position. Often the act of just opening and maintaining the airway will allow the casualty to breathe properly. Once the rescuer uses one of the techniques to open the airway (jaw-thrust or head-tilt/chin-lift), he should maintain that head position to keep the airway open. Failure to maintain the open airway will prevent the casualty from receiving an adequate supply of oxygen. Therefore, while maintaining an open airway the rescuer should check for breathing by observing the casualty's chest and performing the following actions within 3 to 5 seconds:

(*a*) **LOOK** for the chest to rise and fall.

(*b*) **LISTEN** for air escaping during exhalation by placing your ear near the casualty's mouth.

(*c*) **FEEL** for the flow of air on your cheek (see Figure 2-6).

(*d*) **PERFORM** rescue breathing if the casualty does not resume breathing spontaneously.

NOTE

If the casualty resumes breathing, monitor and maintain the open airway. He should be transported to an MTF, as soon as practical.

2-5. Rescue Breathing (Artificial Respiration)

a. If the casualty does not promptly resume adequate spontaneous breathing after the airway is open, rescue breathing (artificial respiration) must be started. Be calm! Think and act quickly! The sooner you begin rescue breathing, the more likely you are to restore the casualty's breathing. If you are in doubt whether the casualty is breathing, give artificial respiration, since it can do no harm to a person who is breathing. If the casualty is breathing, you can feel and see his chest move. If the casualty is breathing, you can feel and hear air being expelled by putting your hand or ear close to his mouth and nose.

b. There are several methods of administering rescue breathing. The mouth-to-mouth method is preferred; however, it cannot be used in all situations. If the casualty has a severe jaw fracture or mouth wound or his jaws are tightly closed by spasms, use the mouth-to-nose method.

2-6. Preliminary Steps—All Rescue Breathing Methods

a. Establish unresponsiveness. Call for help. Turn or position the casualty.

b. Open the airway.

c. Check for breathing by placing your ear over the casualty's mouth and nose, and looking toward his chest.

(1) **LOOK** for rise and fall of the casualty's chest (Figure 2-6).

(2) **LISTEN** for sounds of breathing.

(3) **FEEL** for breath on the side of your face. If the chest does not rise and fall and no air is exhaled, then the casualty is not breathing.

(4) **PERFORM** rescue breathing if the casualty is not breathing.

NOTE

Although the rescuer may notice that the casualty is making respiratory efforts, the airway may still be obstructed and opening the airway may be all that is needed. If the casualty resumes breathing, the rescuer should continue to maintain an open airway.

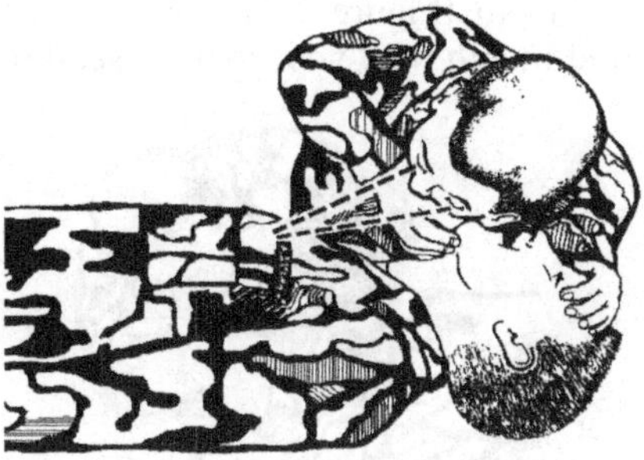

Figure 2-6. Check for breathing.

2-7. Mouth-to-Mouth Method

In this method of rescue breathing, you inflate the casualty's lungs with air from your lungs. This can be accomplished by blowing air into the person's mouth. The mouth-to-mouth rescue breathing method is performed as follows:

a. If the casualty is not breathing, place your hand on his forehead, and pinch his nostrils together with the thumb and index finger of this hand. Let this same hand exert pressure on his forehead to maintain the backward head tilt and maintain an open airway. With your other hand, keep your fingertips on the bony part of the lower jaw near the chin and lift (Figure 2-7).

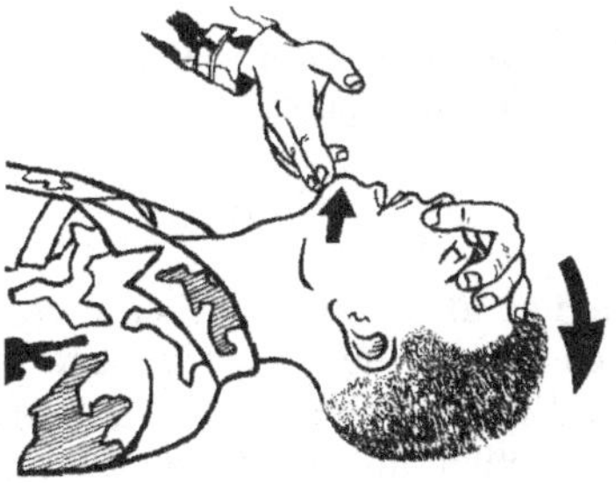

Figure 2-7. Head tilt/chin lift.

NOTE

If you suspect the casualty has a neck injury and you are using the jaw-thrust technique, close the nostrils by placing your cheek tightly against them.

b. Take a deep breath and place your mouth (in an airtight seal) around the casualty's mouth (Figure 2-8). (If the injured person is small, cover both his nose and mouth with your mouth, sealing your lips against the skin of his face.)

Figure 2-8. Rescue breathing.

c. Blow two full breaths into the casualty's mouth (1 to 1 1/2 seconds per breath), taking a breath of fresh air each time before you blow. Watch out of the corner of your eye for the casualty's chest to rise. If the chest rises, sufficient air is getting into the casualty's lungs. Therefore, proceed as described in step (1). If the chest does not rise, do the following (*a*, *b*, and *c* below) and then attempt to ventilate again.

(1) Take corrective action immediately by reestablishing the airway. Make sure that air is not leaking from around your mouth or out of the casualty's pinched nose.

(2) Reattempt to ventilate.

(3) If the chest still does not rise, take the necessary action to open an obstructed airway (paragraph 2-10).

NOTE

If the initial attempt to ventilate the casualty is unsuccessful, reposition the casualty's head and repeat rescue breathing. Improper chin and head positioning is the most common cause of difficulty with ventilation. If the casualty cannot be ventilated after repositioning the head, proceed with foreign-body airway obstruction maneuvers (see paragraph 2-10).

(4) After giving two slow breaths, which cause the chest to rise, attempt to locate a pulse on the casualty. Feel for a pulse on the side of the casualty's neck closest to you by placing the first two fingers (index and middle fingers) of your hand on the groove beside the casualty's Adam's apple (carotid pulse) (Figure 2-9). (Your thumb should not be used for pulse taking because you may confuse your pulse beat with that of the casualty.) Maintain the airway by keeping your other hand on the casualty's forehead. Allow 5 to 10 seconds to determine if there is a pulse.

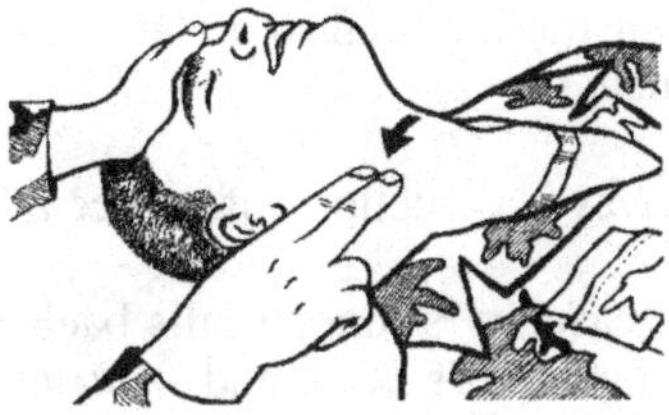

Figure 2-9. Placement of fingers to detect pulse.

(*a*) If signs of circulation are present and a pulse is found and the casualty is breathing—**STOP**; allow the casualty to breathe on his own. If possible, keep him warm and comfortable.

(*b*) If a pulse is found and the casualty is not breathing, continue rescue breathing.

(*c*) If a pulse is not found, seek medically trained personnel for help as soon as possible.

2-8. Mouth-to-Nose Method

Use this method if you cannot perform mouth-to-mouth rescue breathing because the casualty has a severe jaw fracture or mouth wound or his jaws are tightly closed by spasms. The mouth-to-nose method is performed in the same way as the mouth-to-mouth method except that you blow into the nose while you hold the lips closed with one hand at the chin. You then remove your mouth to allow the casualty to exhale passively. It may be necessary to separate the casualty's lips to allow the air to escape during exhalation.

2-9. Heartbeat

If a casualty's heart stops beating, you must immediately seek medical help. **SECONDS COUNT!** Stoppage of the heart is soon followed by cessation of respiration unless it has occurred first. Be calm! Think and act! When a casualty's heart has stopped, there is no pulse at all; the person is unconscious and limp, and the pupils of his eyes are open wide. When evaluating a casualty or when performing the preliminary steps of rescue breathing, feel for a pulse. If you DO NOT detect a pulse, seek medical help.

2-10. Airway Obstructions

In order for oxygen from the air to flow to and from the lungs, the upper airway must be unobstructed.

a. Upper airway obstructions often occur because—

(1) The casualty's tongue falls back into his throat while he is unconscious. The tongue *falls back* and *obstructs* the airway, it is not swallowed by the casualty.

NOTE

Ensure the correct positioning and maintenance of the open airway for an injured or unconscious casualty.

(2) Foreign bodies become lodged in the throat. These obstructions usually occur while eating. Choking on food (usually meat) is associated with—

- Attempting to swallow large pieces of poorly chewed food.
- Drinking alcohol.
- Slipping dentures.

(3) The contents of the stomach are regurgitated and may block the airway.

(4) Blood clots may form as a result of head and facial injuries.

b. Upper airway obstruction may cause either partial or complete airway blockage.

(1) *Partial airway obstruction.* The casualty may still have an air exchange. A good air exchange means that the casualty can cough

forcefully, though he may be wheezing between coughs. You, the rescuer, should not interfere, and should encourage the casualty to cough up the object obstructing his airway on his own. A poor air exchange may be indicated by weak coughing with a high pitched noise between coughs. Further, the casualty may show signs of shock (paragraph 1-6*b*[5]) indicating a need for oxygen. You should assist the casualty and treat him as though he had a complete obstruction.

(2) *Complete airway obstruction.* A complete obstruction (no air exchange) is indicated if the casualty cannot speak, breathe, or cough at all. He may be clutching his neck and moving erratically. In an unconscious casualty, a complete obstruction is also indicated if after opening his airway you cannot ventilate him.

2-11. Opening the Obstructed Airway—Conscious Casualty

Clearing a conscious casualty's airway obstruction can be performed with the casualty either standing or sitting and by following a relatively simple procedure.

WARNING

Once an obstructed airway occurs, the brain will develop an oxygen deficiency resulting in unconsciousness. Death will follow rapidly if breathing is not promptly restored.

a. Ask the casualty if he can speak or if he is choking. Check for the universal choking sign (Figure 2-10).

Figure 2-10. Universal sign of choking.

b. If the casualty can speak, encourage him to attempt to cough; the casualty still has a good air exchange. If he is able to speak or cough effectively, DO NOT interfere with his attempts to expel the obstruction.

c. Listen for high pitched sounds when the casualty breathes or coughs (poor air exchange). If there is poor air exchange or no breathing, CALL FOR HELP and immediately deliver manual thrusts (either an abdominal or chest thrust).

NOTE

The manual thrust with the hands centered between the waist and the rib cage is called an abdominal thrust (or Heimlich maneuver). The chest thrust (the hands are centered in the middle of the breastbone) is used only for an individual in the advanced stages of pregnancy, in the markedly obese casualty, or if there is a significant abdominal wound.

(1) Apply abdominal thrusts. This can be accomplished by using the following procedures:

(*a*) Stand behind the casualty and wrap your arms around his waist.

(*b*) Make a fist with one hand and grasp it with the other. The thumb side of your fist should be against the casualty's abdomen, in the midline and slightly above the casualty's navel, but well below the tip of the breastbone (Figure 2-11).

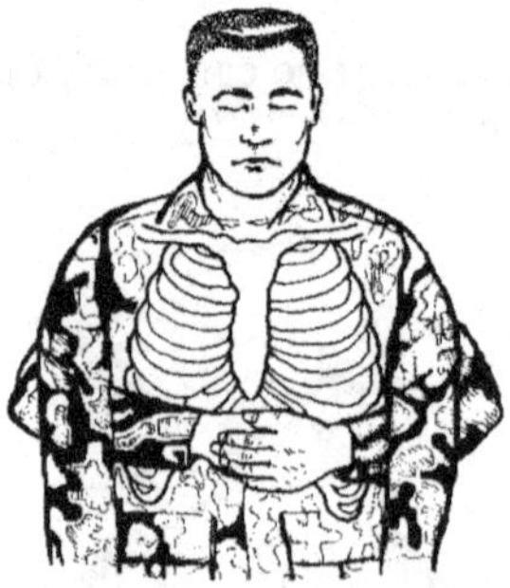

Figure 2-11. Anatomical view of abdominal thrust procedure.

(*c*) Press the fists into the abdomen with a quick backward and upward thrust (Figure 2-12).

Figure 2-12. Profile view of abdominal thrust.

(*d*) Each thrust should be a separate and distinct movement.

NOTE

Continue performing abdominal thrusts until the obstruction is expelled or the casualty becomes unresponsive.

(*e*) If the casualty becomes unresponsive, call for help as you proceed with steps to open the airway, and perform rescue breathing. (Refer to paragraph 2-7 for information on how to perform mouth-to-mouth resuscitation.)

(2) Apply chest thrusts. An alternate technique to the abdominal thrust is the chest thrust. This technique is useful when the casualty has an abdominal wound, when the casualty is pregnant, or when the casualty is so large that you cannot wrap your arms around the abdomen. To apply chest thrusts with casualty sitting or standing:

(*a*) Stand behind the casualty and wrap your arms around his chest with your arms under his armpits.

(*b*) Make a fist with one hand and place the thumb side of the fist in the middle of the breastbone (take care to avoid the tip of the breastbone and the margins of the ribs).

(*c*) Grasp the fist with the other hand and exert thrusts (Figure 2-13).

Figure 2-13. Profile view of chest thrust.

(*d*) Each thrust should be delivered slowly, distinctly, and with the intent of relieving the obstruction.

(*e*) Perform chest thrusts until the obstruction is expelled or the casualty becomes unresponsive.

(*f*) If the casualty becomes unresponsive, call for help as you proceed with steps to open the airway and perform rescue breathing.

2-12. Opening the Obstructed Airway—Casualty Lying Down or Unresponsive

The following procedures are used to expel an airway obstruction in a casualty who is lying down, who becomes unconscious, or who is found unconscious (the cause unknown):

- If a conscious casualty who is choking becomes unresponsive, call for help, open the airway, perform a finger sweep, and attempt rescue breathing (paragraphs 2-4 through 2-8). If you still cannot administer rescue breathing due to an airway blockage, then remove the airway obstruction using the procedures as in *b* below.

- If a casualty is unresponsive when you find him (the cause unknown), assess or evaluate the situation, call for help, position the casualty on his back, open the airway, establish breathlessness, and attempt to perform rescue breathing (paragraphs 2-4 through 2-8).

a. Open the airway and attempt rescue breathing (refer to paragraph 2-7 for information on how to perform mouth-to-mouth resuscitation).

b. If still unable to ventilate the casualty, perform 6 to 10 manual (abdominal or chest) thrusts.

(1) To perform the abdominal thrusts:

(*a*) Kneel astride the casualty's thighs (Figure 2-14).

Figure 2-14. Abdominal thrust on unresponsive casualty.

(*b*) Place the heel of one hand against the casualty's abdomen (in the midline slightly above the navel but well below the tip of the breastbone). Place your other hand on top of the first one. Point your fingers toward the casualty's head.

(*c*) Press into the casualty's abdomen with a quick, forward and upward thrust. You can use your body weight to perform the maneuver. Deliver each thrust quickly and distinctly.

(*d*) Repeat the sequence of abdominal thrusts, finger sweep, and rescue breathing (attempt to ventilate) as long as necessary to remove the object from the obstructed airway.

(*e*) If the casualty's chest rises, proceed to feeling for pulse.

(2) To perform chest thrusts:

(*a*) Place the unresponsive casualty on his back, face up, and open his mouth. Kneel close to the side of the casualty's body.

1. Locate the lower edge of the casualty's ribs with your fingers. Run the fingers up along the rib cage to the notch (Figure 2-15A).

2. Place the middle finger on the notch and the index finger next to the middle finger on the lower edge of the breastbone. Place the heel of the other hand on the lower half of the breastbone next to the two fingers (Figure 2-15B).

3. Remove the fingers from the notch and place that hand on top of the positioned hand on the breastbone, extending or interlocking the fingers (Figure 2-15C).

4. Straighten and lock your elbows with your shoulders directly above your hands without bending the elbows, rocking, or allowing the shoulders to sag. Apply enough pressure to depress the breastbone 1 1/2 to 2 inches, then release the pressure completely (Figure 2-15D). Do this 6 to 10 times. Each thrust should be delivered quickly and distinctly. See Figure 2-16 for another view of the breastbone being depressed.

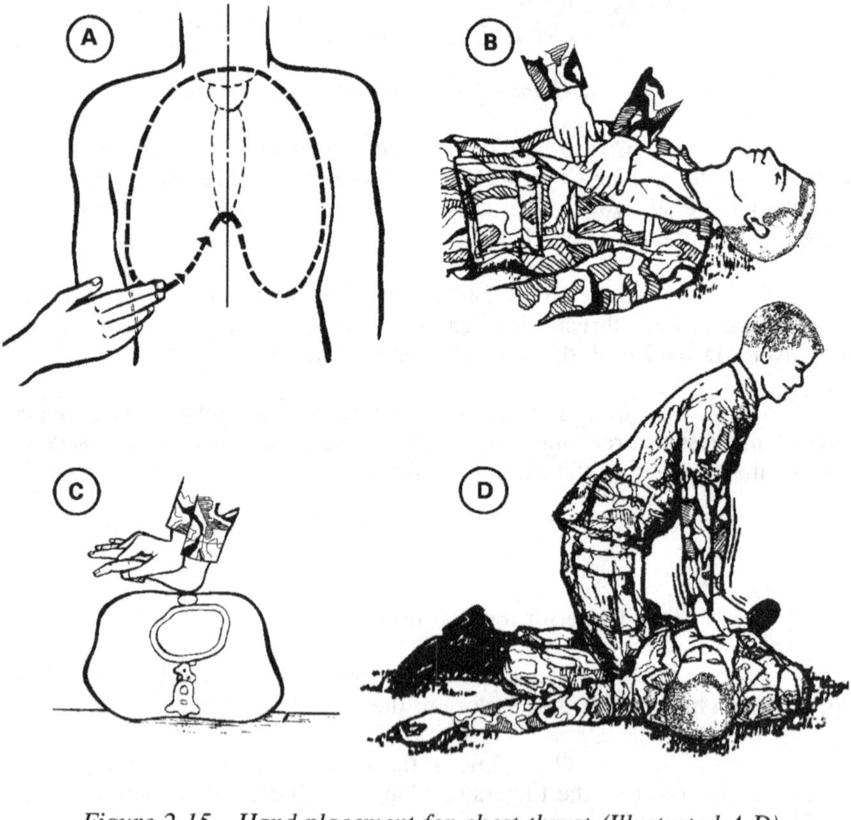

Figure 2-15. Hand placement for chest thrust (Illustrated A-D).

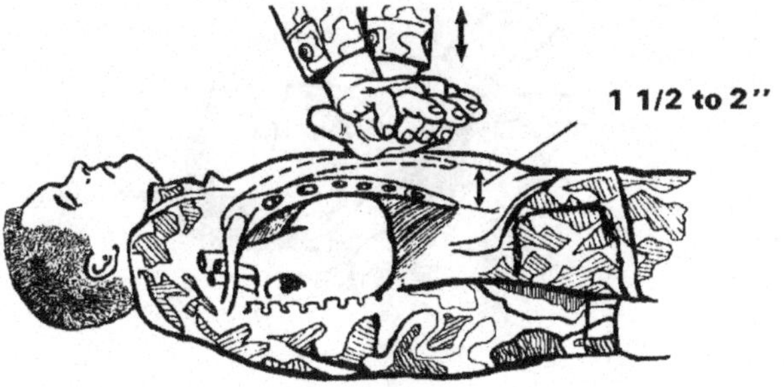

Figure 2-16. Breastbone depressed 1 1/2 to 2 inches.

(*b*) Repeat the sequence of chest thrust, finger sweep, and rescue breathing as long as necessary to clear the object from the obstructed airway. See paragraph (3) below.

(*c*) If the casualty's chest rises, proceed to feeling for his pulse.

(3) If you still cannot administer rescue breathing due to an airway obstruction, then remove the airway obstruction using the procedures in steps (*a*) and (*b*) below.

(*a*) Place the casualty on his back, face up, turn the unresponsive casualty as a unit, and call out for help.

(*b*) Perform finger sweep, keep casualty face up, use tongue-jaw lift to open mouth.

1. Open the casualty's mouth by grasping both his tongue and lower jaw between your thumb and fingers and lifting (tongue-jaw lift) (Figure 2-17). If you are unable to open his mouth, cross your fingers and thumb (crossed-finger method) and push his teeth apart (Figure 2-18) by pressing your thumb against his upper teeth and pressing your finger against his lower teeth.

Figure 2-17. Opening casualty's mouth (tongue-jaw lift).

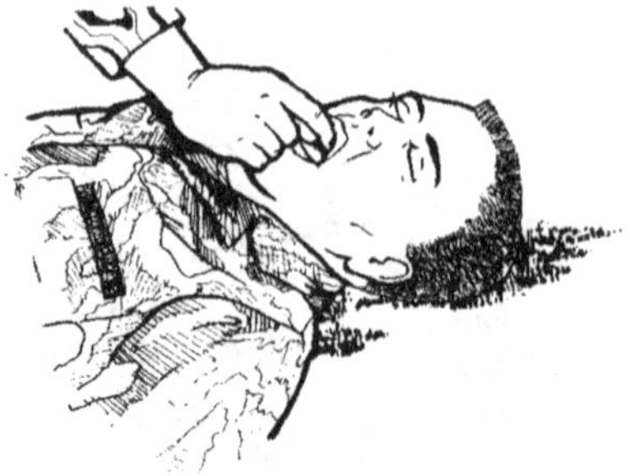

Figure 2-18. Opening casualty's mouth (crossed-finger method).

2. Insert the index finger of the other hand down along the inside of his cheek to the base of the tongue. Use a hooking motion from the side of the mouth toward the center to dislodge the foreign body (Figure 2-19).

Figure 2-19. Using finger to dislodge a foreign body.

WARNING

Take care not to force the object deeper into the airway by pushing it with the finger.

Section II. STOP THE BLEEDING AND PROTECT THE WOUND

2-13. General

The longer a service member bleeds from a major wound, the less likely he will be able to survive his injuries. It is, therefore, important that the first aid provider promptly stop the external bleeding.

2-14. Clothing

In evaluating the casualty for location, type, and size of the wound or injury, cut or tear his clothing and carefully expose the entire area of the wound. This procedure is necessary to properly visualize injury and avoid further contamination. Clothing stuck to the wound should be left in place to avoid further injury. DO NOT touch the wound; keep it as clean as possible.

WARNING

DO NOT REMOVE protective clothing in a chemical environment. Apply dressings over the protective clothing.

2-15. Entrance and Exit Wounds

Before applying the dressing, carefully examine the casualty to determine if there is more than one wound. A missile may have entered at one point and exited at another point. The *EXIT* wound is usually *LARGER* than the entrance wound.

WARNING

The casualty should be continually monitored for development of conditions which may require the performance of necessary basic lifesaving measures, such as clearing the airway and mouth-to-mouth resuscitation. All open (or penetrating) wounds should be checked for a point of entry and exit and first aid measures applied accordingly.

WARNING

If the missile lodges in the body (fails to exit), DO NOT attempt to remove it or probe the wound. Apply a dressing. If there is an object extending from (impaled in) the wound, DO NOT remove the object. Apply a dressing around the object and use additional improvised bulky materials/dressings (use the cleanest material available) to build up the area around the object to stabilize the object and prevent further injury. Apply a supporting bandage over the bulky materials to hold them in place.

2-16. Field Dressing

a. Use the casualty's field dressing; remove it from the wrapper and grasp the tails of the dressing with both hands (Figure 2-20).

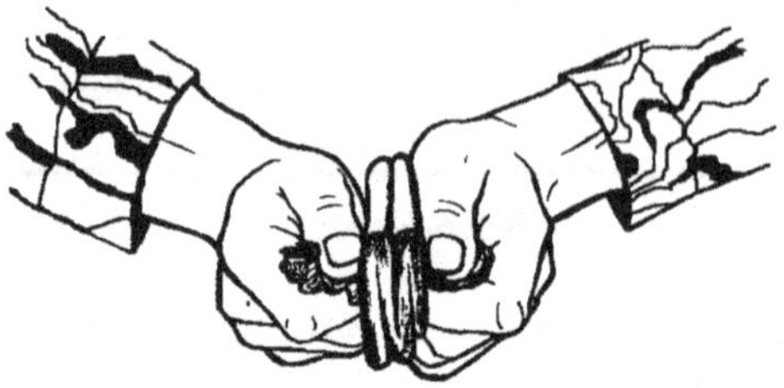

Figure 2-20. Grasping tails of dressing with both hands.

> **WARNING**
>
> **DO NOT touch the white (sterile) side of the dressing, and DO NOT allow it to come in contact with any surface other than the wound.**

b. Hold the dressing directly over the wound with the white side down. Pull the dressing open (Figure 2-21) and place it directly over the wound (Figure 2-22).

Figure 2-21. Pulling dressing open.

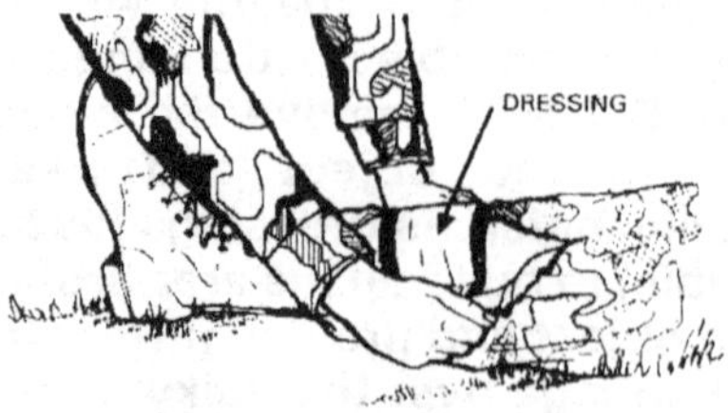

Figure 2-22. Placing dressing directly on wound.

c. Hold the dressing in place with one hand. Use the other hand to wrap one of the tails around the injured part, covering about one-half of the dressing (Figure 2-23). Leave enough of the tail for a knot. If the casualty is able, he may assist by holding the dressing in place.

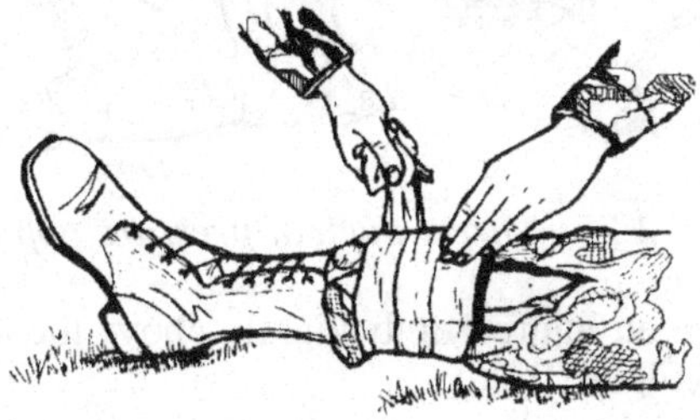

Figure 2-23. Wrapping tail of dressing around injured part.

d. Wrap the other tail in the opposite direction until the remainder of the dressing is covered. The tails should seal the sides of the dressing to keep foreign material from getting under it.

e. Tie the tails into a nonslip knot over the outer edge of the dressing (Figure 2-24). **DO NOT TIE THE KNOT OVER THE WOUND**. In order to allow blood to flow to the rest of an injured limb, tie the dressing firmly enough to prevent it from slipping but without causing a tourniquet-like effect; that is, the skin beyond the injury should not becomes cool, blue, or numb.

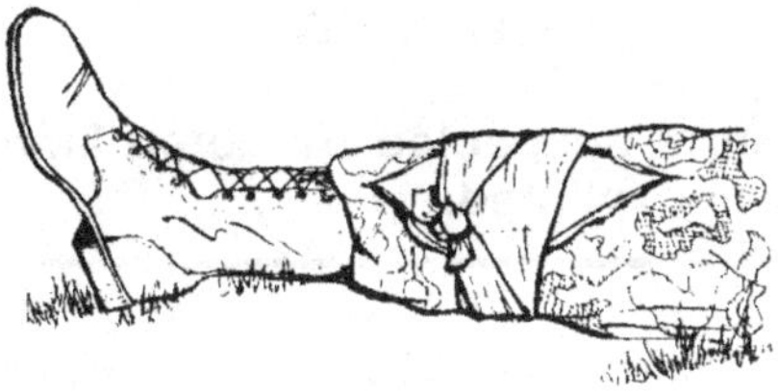

Figure 2-24. Tails tied into nonslip knot.

2-17. Manual Pressure

a. If bleeding continues after applying the sterile field dressing, direct manual pressure may be used to help control bleeding. Apply such pressure by placing a hand on the dressing and exerting firm pressure for 5 to 10 minutes (Figure 2-25). The casualty may be asked to do this himself if he is conscious and can follow instructions.

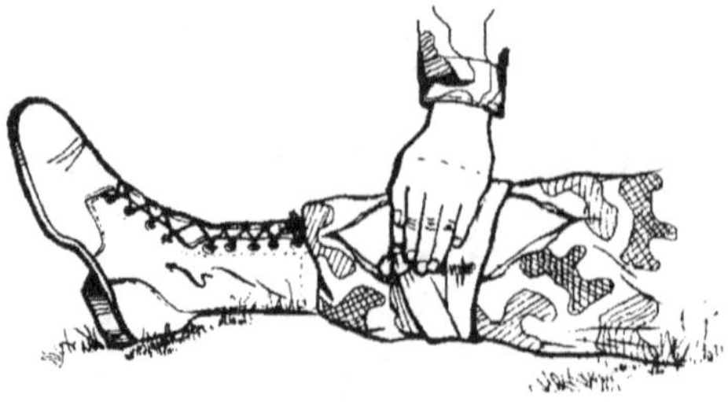

Figure 2-25. Direct manual pressure applied.

b. Elevate an injured limb slightly above the level of the heart to reduce the bleeding (Figure 2-26).

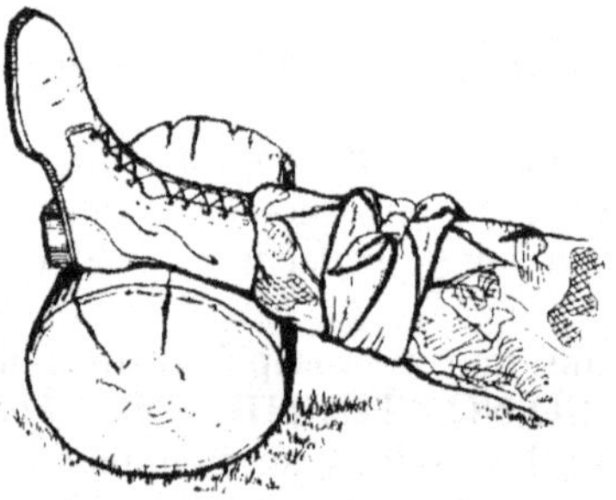

Figure 2-26. Injured limb elevated.

WARNING

DO NOT elevate a suspected fractured limb unless it has been properly splinted.

c. If the bleeding stops, check shock; administer first aid for shock as necessary. If the bleeding continues, apply a pressure dressing.

2-18. Pressure Dressing

Pressure dressings aid in blood clotting and compress the open blood vessel. If bleeding continues after the application of a field dressing, manual pressure, and elevation, then a pressure dressing must be applied as follows:

a. Place a wad of padding on top of the field dressing, directly over the wound (Figure 2-27). Keep the injured extremity elevated.

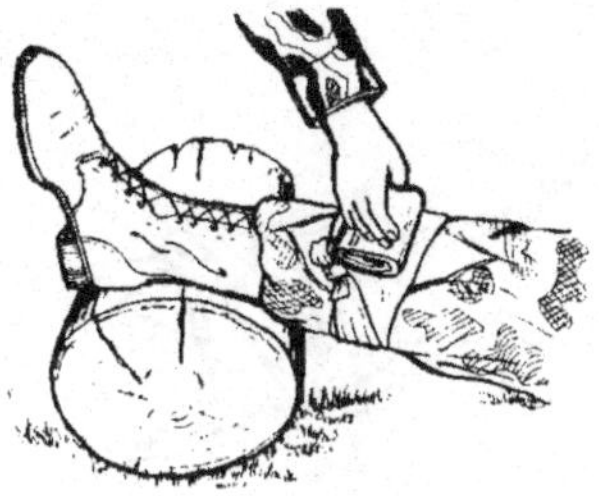

Figure 2-27. Wad of padding on top of field dressing.

NOTE

Improvised bandages may be made from strips of cloth. These strips may be made from T-shirts, socks, or other garments.

b. Place an improvised dressing (or cravat, if available) over the wad of padding (Figure 2-28). Wrap the ends tightly around the injured limb, covering the previously placed field dressing (Figure 2-29).

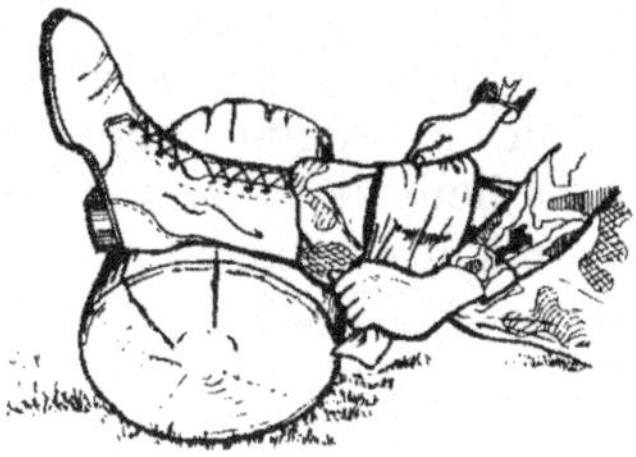

Figure 2-28. Improvised dressing over wad of padding

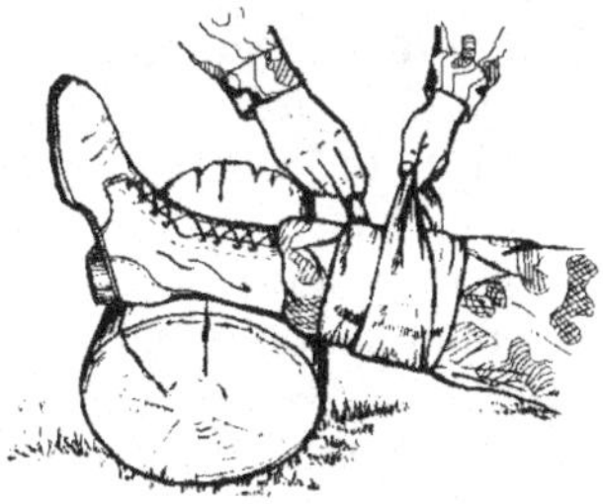

Figure 2-29. Ends of improvised dressing wrapped tightly around limb.

c. Tie the ends together in a nonslip knot, directly over the wound site (Figure 2-30). DO NOT tie so tightly that it has a tourniquet-like effect. If bleeding continues and all other measures have failed, or if the limb is severed, then apply a tourniquet. Use the tourniquet as a **LAST RESORT**. When the bleeding stops, check for shock; administer first aid for shock as necessary.

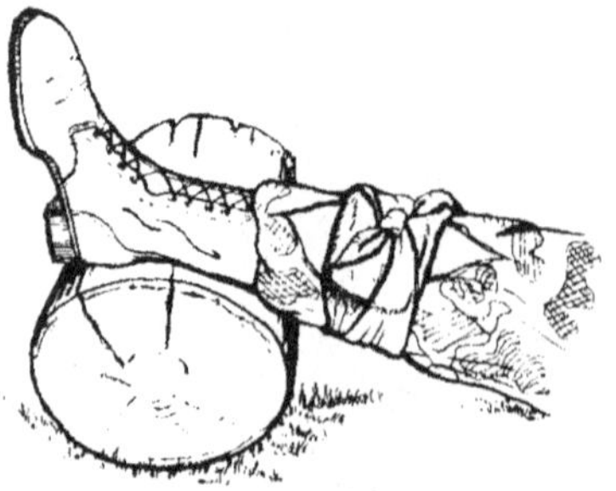

Figure 2-30. Ends of improvised dressing tied together in nonslip knot.

NOTE

Distal end of wounded extremities (fingers and toes) should be checked periodically for adequate circulation. The dressing must be loosened if the extremity becomes cool, blue, or numb.

NOTE

If bleeding continues and all other measures have failed (dressings and covering wound, applying direct manual pressure, elevating the limb above the heart level, and applying a pressure dressing while maintaining limb elevation) *then apply digital pressure* (see paragraph 2-19).

2-19. Digital Pressure

Digital pressure (often called "pressure points") is an alternative method to control bleeding. This method uses pressure from the fingers, thumbs, or hands to press at the site or point where a main artery supplying the wounded area lies near the skin surface or over bone (Figure 2-31). This pressure may help shut off or slow down the flow of blood from the heart to the wound and is used in combination with direct pressure and elevation. It may help in instances where bleeding is not easily controlled, where a pressure dressing has not yet been applied, or where pressure dressings are not readily available.

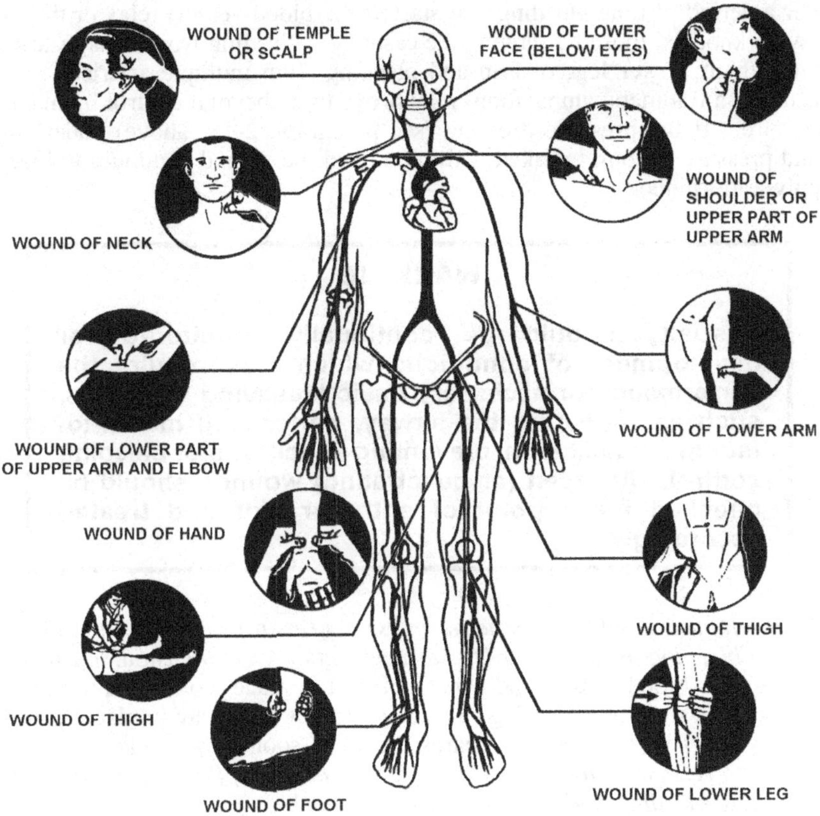

Figure 2-31. Digital pressure (pressure with fingers, thumbs or hands).

2-20. Tourniquet

DANGER

A tourniquet is only used on an arm or leg where there is a danger of the casualty losing his life (bleeding to death).

A tourniquet is a constricting band placed around an arm or leg to control bleeding. A service member whose arm or leg has been completely amputated may not be bleeding when first discovered, but a tourniquet should be applied anyway. This absence of bleeding is due to the body's normal defenses (contraction or clotting of blood vessels) as a result of the amputation, but after a period of time bleeding will start as the blood vessels relax or the clot may be knocked loose by moving the casualty. Bleeding from a major artery of the thigh, lower leg, or arm and bleeding from multiple arteries (which occurs in a traumatic amputation) may prove to be beyond control by manual pressure. If the pressure dressing (see paragraph 2-18, above) under firm hand pressure becomes soaked with blood and the wound continues to bleed, apply a tourniquet.

WARNING

Casualty should be continually monitored for development of conditions which may require the performance of necessary basic lifesaving measures, such as: clearing the airway, performing mouth-to-mouth resuscitation, preventing shock, and/or bleeding control. All open (or penetrating) wounds should be checked for a point of entry or exit and treated accordingly.

The tourniquet should not be used unless a pressure dressing has failed to stop the bleeding or an arm or leg has been cut off. On occasion, tourniquets have injured blood vessels and nerves. If left in place too long, a tourniquet can cause loss of an arm or leg. Once applied, it must stay in place, and the casualty must be taken to the nearest MTF as soon as possible. *DO NOT loosen or release a tourniquet after it has been applied as release could precipitate bleeding and potentially lead to shock.*

a. Improvising a Tourniquet. In the absence of a specially designed tourniquet, a tourniquet may be made from a strong, pliable material, such as gauze or muslin bandages, clothing, or cravats. An improvised tourniquet is used with a rigid stick-like object. To minimize skin damage, ensure that the improvised tourniquet is at least 2 inches wide.

WARNING

The tourniquet must be easily identified or easily seen.

WARNING

DO NOT use wire or shoestring for a tourniquet band.

b. Placing the Improvised Tourniquet.

(1) Place the tourniquet around the limb, between the wound and the body trunk (or between the wound and the heart). Never place it directly over a wound, a fracture, or joint. Tourniquets, for maximum effectiveness, should be placed on the upper arm or above the knee on the thigh (Figure 2-32).

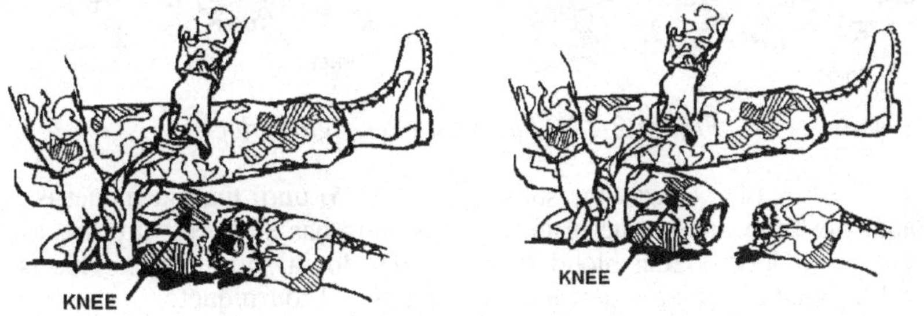

Figure 2-32. Tourniquet above knee.

(2) The tourniquet should be well-padded. If possible, place the tourniquet over the smoothed sleeve or trouser leg to prevent the skin from being pinched or twisted. If the tourniquet is long enough, wrap it around the limb several times, keeping the material as flat as possible. Damaging the skin may deprive the surgeon of skin required to cover an amputation. Protection of the skin also reduces pain.

c. Applying the Tourniquet.

(1) Tie a half-knot. (A half-knot is the same as the first part of tying a shoe lace.)

(2) Place a stick (or similar rigid object) on top of the half-knot (Figure 2-33).

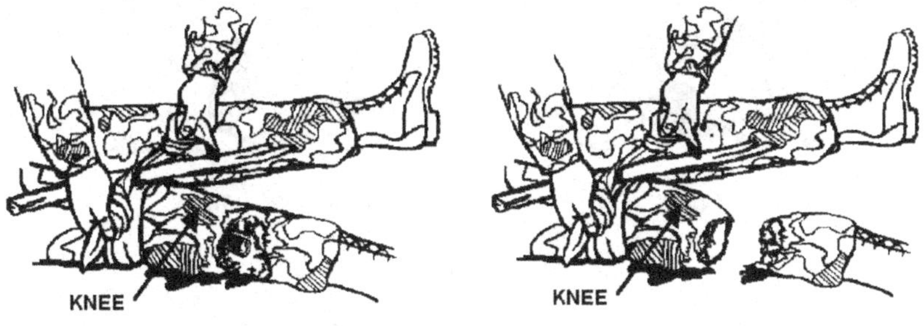

Figure 2-33. Rigid object on top of half-knot.

(3) Tie a full knot over the stick (Figure 2-34).

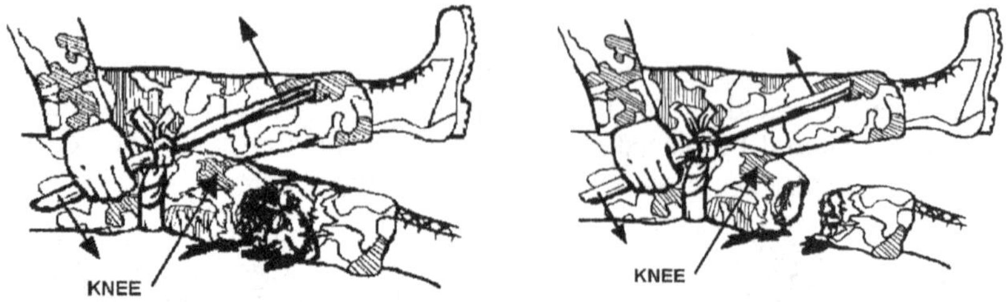

Figure 2-34. Full knot over rigid object.

(4) Twist the stick (Figure 2-35) until the tourniquet is tight around the limb and/or the bright red bleeding has stopped. In the case of amputation, dark oozing blood may continue for a short time. This is the blood trapped in the area between the wound and tourniquet.

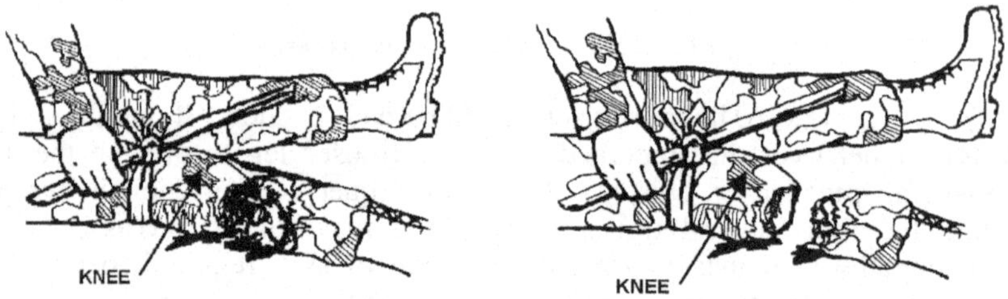

Figure 2-35. Stick twisted.

(5) Fasten the tourniquet to the limb by looping the free ends of the tourniquet over the ends of the stick. Then bring the ends around the limb to prevent the stick from loosening. Tie them together on the side of the limb (Figure 2-36).

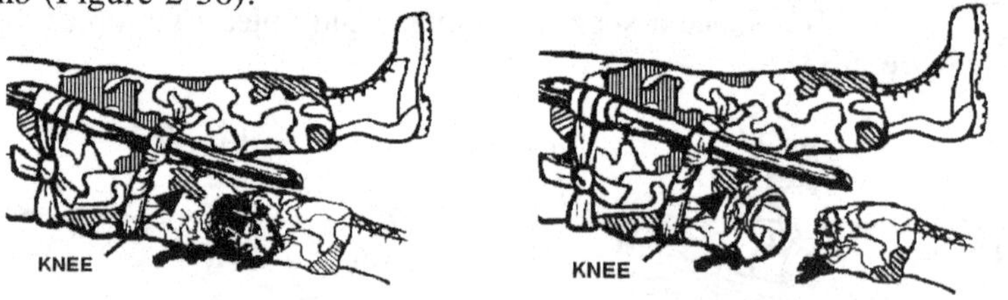

Figure 2-36. Tie free ends on side of limb.

NOTE

Other methods of securing the stick may be used as long as the stick does not unwind and no further injury results.

NOTE

If possible, save and transport any severed (amputated) limbs or body parts with (but out of sight of) the casualty.

(6) DO NOT cover the tourniquet—you should leave it in full view. If the limb is missing (total amputation), apply a dressing to the stump. All wounds should have a dressing to protect the wound from contamination.

(7) Mark the casualty's forehead with a "T" and the time to indicate a tourniquet has been applied. If necessary, use the casualty's blood to make this mark.

(8) Check and treat for shock.

(9) Seek medical aid.

CAUTION

Only appropriately skilled medical personnel may adjust or otherwise remove/release the tourniquet in the appropriate setting.

Section III. CHECK FOR SHOCK AND ADMINISTER FIRST AID MEASURES

2-21. General

The term *shock* has a variety of meanings. In medicine, it refers to a collapse of the body's cardiovascular system which includes an inadequate supply of blood to the body's tissues. Shock stuns and weakens the body. When the normal blood flow in the body is upset, death can result. Early recognition and proper first aid may save the casualty's life.

2-22. Causes and Effects

a. There are three basic mechanisms associated with shock. These are—

- The heart is damaged and fails to work as a pump.

- Blood loss (heavy bleeding) causes the volume of fluid within the vascular system to be insufficient.

- The blood vessels dilate (open wider) so that the blood within the system (even though it is a normal volume [the casualty is not bleeding or dehydrated]) is insufficient to provide adequate circulation within the body.

b. Shock may be the result of a number of conditions. These include—

- Dehydration.

- Allergic reaction to foods, drugs, insect stings, and snakebites.

- Significant loss of blood.

- Reaction to the sight of a wound, blood, or other traumatic scene.

- Traumatic injuries, such as—

  - Burns.

  - Gunshot or shrapnel wounds.

  - Crush injuries.

  - Blows to the body (which can cause broken bones or damage to internal organs).

  - Head injuries.

  - Penetrating wounds (such as from a knife, bayonet, or missile).

2-23. Signs and Symptoms of Shock

Examine the casualty to see if he has any of the following signs and symptoms:

- Sweaty but cool skin (clammy skin).

- Weak and rapid pulse.
- Paleness of skin (in dark-skinned individuals they may have a grayish look to their skin).
- Restlessness, nervousness.
- Thirst.
- Loss of blood (bleeding).
- Confusion (or loss of awareness).
- Faster-than-normal breathing rate.
- Blotchy or bluish skin (especially around the mouth and lips).
- Nausea and/or vomiting.

2-24. First Aid Measures for Shock

In the field, the first aid procedures administered for shock are identical to procedures that would be performed *to prevent shock*. When treating a casualty, assume that shock is present or will occur shortly. By waiting until actual signs and symptoms of shock are noticeable, the rescuer may jeopardize the casualty's life.

a. Position the Casualty. (DO NOT move the casualty or his limbs if suspected fractures have not been splinted. See Chapter 4 for details.)

(1) Move the casualty to cover, if cover is available and the situation permits.

(2) Lay the casualty on his back.

NOTE

A casualty in shock from a chest wound or one who is experiencing breathing difficulty, may breathe easier in a sitting position. If this is the case, allow him to sit upright, but monitor carefully in case his condition worsens.

(3) Elevate the casualty's feet higher than the level of his heart. Use a stable object (field pack or rolled up clothing) so that his feet will not slip off (Figure 2-37).

WARNING

DO NOT elevate legs if the casualty has an unsplinted broken leg, head injury, or abdominal injury.

Figure 2-37. Clothing loosened and feet elevated.

WARNING

Check casualty for leg fracture(s) and splint, if necessary, before elevating his feet. For a casualty with an abdominal wound, place his knees in an upright (flexed) position.

(4) Loosen clothing at the neck, waist, or wherever it may be binding.

CAUTION

DO NOT loosen or remove protective clothing in a chemical environment.

(5) Prevent chilling or overheating. The key is to maintain body temperature. In cold weather, place a blanket or other like item over him to keep him warm and under him to prevent chilling (Figure 2-38). However, if a tourniquet has been applied, leave it exposed (if possible). In hot weather, place the casualty in the shade and protect him from becoming chilled; however, avoid the excessive use of blankets or other coverings.

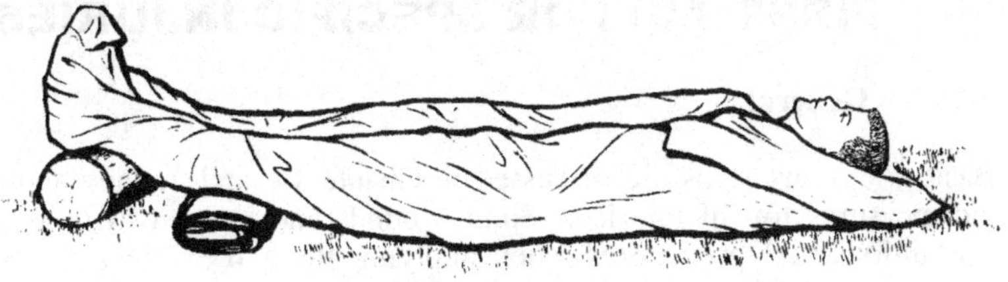

Figure 2-38. Body temperature maintained.

(6) Calm the casualty. Throughout the entire procedure of providing first aid for a casualty, the rescuer should reassure the casualty and keep him calm. This can be done by being authoritative (taking charge) and by showing self-confidence. Assure the casualty that you are there to help him.

(7) Seek medical aid.

b. Food and/or Drink. When providing first aid for shock, DO NOT give the casualty any food or drink. If you must leave the casualty or if he is unconscious, turn his head to the side to prevent him from choking if he vomits (Figure 2-39).

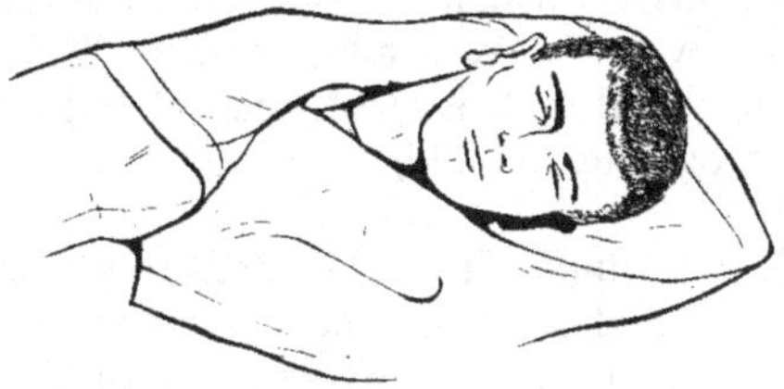

Figure 2-39. Casualty's head turned to side.

c. Evaluate Casualty. Continue to evaluate the casualty until medical personnel arrives or the casualty is transported to an MTF.

CHAPTER 3

FIRST AID FOR SPECIFIC INJURIES

3-1. General

Basic lifesaving steps are discussed in Chapters 1 and 2; they apply to first aid measures for all injuries. Some wounds and burns will require special precautions and procedures when applying these measures. This chapter discusses specific first aid procedures for wounds of the head, face, and neck; chest and stomach wounds; and burns. It also discusses the techniques for applying dressings and bandages to specific parts of the body.

3-2. Head, Neck, and Facial Injuries

a. Head Injuries.

(1) Head injuries range from minor abrasions or cuts on the scalp to severe brain injuries that may result in unconsciousness and sometimes death. Head injuries are classified as open or closed wounds. An open wound is one that is visible, has a break in the skin, and usually has evidence of bleeding. A closed wound may be visible (such as a depression in the skull) or the first aid provider may not be able to see any apparent injury (such as internal bleeding). Some head injuries result in unconsciousness; however, a service member may have a serious head wound and still be conscious. Casualties with head and neck injuries should be treated as though they also have a spinal injury. The casualty should not be moved until the head and neck is stabilized unless he is in immediate danger (such as close to a burning vehicle).

(2) Prompt first aid measures should be initiated for casualties with suspected head and neck injuries. The conscious casualty may be able to provide information on the extent of his injuries. However, as a result of the head injury, he may be confused and unable to provide accurate information. The signs and symptoms a first aid provider might observe are—

- Nausea and vomiting.
- Convulsions or twitches.
- Slurred speech.
- Confusion and loss of memory. (Does he know who he is? Does he know where he is? Does he know what day it is?)
- Recent unconsciousness.

- Dizziness.
- Drowsiness.
- Blurred vision, unequal pupils, or bruising (black eyes).
- Paralysis (partial or full).
- Complaint of headache.
- Bleeding or other fluid discharge from the scalp, nose, or ears.
- Deformity of the head (depression or swelling).
- Staggering while walking.

b. Neck Injuries. Neck injuries may result in heavy bleeding. Apply pressure above and below the injury, *but do not interfere with the breathing process*, and attempt to control the bleeding. Apply a dressing. Always evaluate the casualty for a possible neck fracture/spinal cord injury; if suspected, seek medical treatment immediately.

NOTE

Establish and maintain the airway in cases of facial or neck injuries. If a neck fracture or spinal cord injury is suspected, immobilize the injury and, if necessary, perform basic life support measures.

c. Facial Injuries. Soft tissue injuries of the face and scalp are common. Abrasions (scrapes) of the skin cause no serious problems. Contusions (injury without a break in the skin) usually cause swelling. A contusion of the scalp looks and feels like a lump. Laceration (cut) and avulsion (torn away tissue) injuries are also common. Avulsions are frequently caused when a sharp blow separates the scalp from the skull beneath it. Because the face and scalp are richly supplied with blood vessels (arteries and veins), wounds of these areas usually bleed heavily.

3-3. General First Aid Measures

a. General Considerations. The casualty with a head injury (or suspected head injury) should be continually monitored for the development of conditions that *may require* basic lifesaving measures. After initiating first aid measures, request medical assistance and evacuation. If dedicated medical evacuation assets are not available, transport the casualty to an MTF as soon as the situation permits. The first aid provider should not attempt to remove a protruding object from the head or give the casualty anything to eat or drink. Further, the first aid provider should be prepared to—

- Clear the airway.
- Control bleeding (external).
- Administer first aid measures for shock.
- Keep the casualty warm.
- Protect the wound.

b. *Unconscious Casualty.* An unconscious casualty does not have control of all of his body's functions and may choke on his tongue, blood, vomitus, or other substances. (Refer to Figure 2-39.)

(1) *Breathing.* The brain requires a constant supply of oxygen. A bluish (or in an individual with dark skin—grayish) color of skin around the lips and nail beds indicates that the casualty is not receiving enough oxygen. Immediate action must be taken to clear the airway, to position the casualty on his side, or to initiate rescue breathing.

(2) *Bleeding.* Bleeding from a head injury usually comes from blood vessels within the scalp. Bleeding can also develop inside the skull or within the brain. In most instances visible bleeding from the head can be controlled by application of the field first aid dressing.

CAUTION

DO NOT attempt to put unnecessary pressure on the wound or attempt to push any brain matter back into the head (skull). **DO NOT** apply a pressure dressing.

c. *Concussion.* If an individual receives a heavy blow to the head or face, he may suffer a brain concussion (an injury to the brain that involves a temporary loss of some or all of the brain's ability to function). For example, the casualty may not breathe properly for a short period of time, or he may become confused and stagger when he attempts to walk. Symptoms of a concussion may only last for a short period of time. However, if a casualty is suspected of having suffered a concussion, he should be transported to an MTF as soon as conditions permit.

d. *Convulsions*. Convulsions (seizures/involuntary jerking) may occur even after a mild head injury. When a casualty is convulsing, protect him from hurting himself. Take the following measures:

(1) Ease him to the ground if he is standing or sitting.

(2) Support his head and neck.

(3) Maintain his airway.

(4) Protect him from further injury (such as hitting close-by objects).

NOTE

DO NOT forcefully hold the arms and legs if they are jerking because this can lead to broken bones. **DO NOT** force anything between the casualty's teeth—especially if they are tightly clenched because this may obstruct the casualty's airway. Maintain the casualty's airway if necessary.

e. *Brain Damage*. In *severe* head injuries where brain tissue is protruding, *leave the wound alone*; carefully place a loose moistened dressing (moistened with sterile normal saline if available) and also a first aid dressing over the tissue to protect it from further contamination. ***DO NOT** remove or disturb any foreign matter that may be in the wound*. Position the casualty so that his head is higher than his body. Keep him warm and *seek medical assistance immediately*.

NOTE

If there is an object extending from the wound, **DO NOT** remove the object. Improvise bulky dressings from the cleanest material available and place this material around the protruding object for support, then apply the field dressing.

3-4. Chest Wounds

Blunt trauma, bullet or missile wounds, stab wounds, or falls may cause chest injuries. These injuries can be serious and may cause death quickly if first aid is not administered in a timely manner. A casualty with a chest injury may complain of pain in the chest or shoulder area; he may have difficulty breathing. His chest may not rise normally when he breathes. The injury may cause the casualty to cough up blood and to have a rapid or a weak heartbeat. A casualty with an open chest wound has a punctured chest wall. The sucking sound heard when he breathes is caused by air leaking into his chest cavity. This particular type of wound is dangerous and will collapse the injured lung (Figure 3-1). Breathing becomes difficult for the casualty because the wound is open. The service members life may depend upon how quickly you apply an occlusive dressing over the wound (refer to paragraph 3-5).

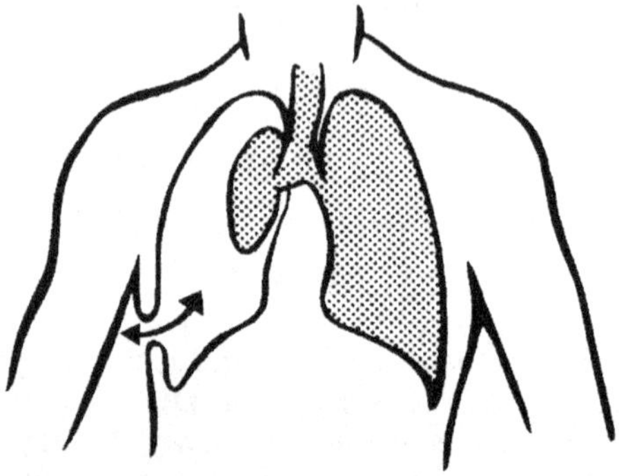

Figure 3-1. Collapsed lung.

3-5. First Aid for Chest Wounds

a. Evaluate the Casualty. Be prepared to perform first aid measures. These measures may include clearing the airway, rescue breathing, treatment for shock, and/or bleeding control.

b. Expose the Wound. If appropriate, cut or remove the casualty's clothing to expose the wound. Remember, **DO NOT** remove clothing that is stuck to the wound because additional injury may result. **DO NOT** attempt to clean the wound.

NOTE

Examine the casualty to see if there is an entry and exit wound. If there are two wounds (entry, exit), perform the same procedure for both wounds. Treat the more serious (heavier bleeding, larger) wound first. It may be necessary to improvise a dressing for the second wound by using strips of cloth, such as a torn T-shirt, or whatever material is available. Also, listen for sucking sounds to determine if the chest wall is punctured.

> **CAUTION**
>
> If there is an object impaled in the wound, **DO NOT** remove it. Apply a dressing around the object and use additional improvised bulky materials/dressings (use the cleanest materials available) to build up the area around the object. Apply a supporting bandage over the bulky materials to hold them in place.

> **CAUTION**
>
> **DO NOT REMOVE** protective clothing in a chemical environment. Apply dressings *over* the protective clothing.

c. Open the Casualty's Field Dressing Plastic Wrapper. In cases where there is a sucking chest wound, the plastic wrapper is used with the field dressing to create an occlusive dressing. If a plastic wrapper is not available, or if an additional wound needs to be treated; cellophane, foil, the casualty's poncho, or similar material may be used. The covering should be wide enough to extend 2 inches or more beyond the edges of the wound in all directions.

(1) Tear open one end of the casualty's plastic wrapper covering the field dressing. Be careful not to destroy the wrapper and **DO NOT** touch the inside of the wrapper.

(2) Remove the inner packet (field dressing).

(3) Complete tearing open the empty plastic wrapper using as much of the wrapper as possible to create a flat surface.

d. Place the Wrapper Over the Wound. Place the inside surface of the plastic wrapper directly over the wound *when the casualty exhales* and hold it in place (Figure 3-2). The casualty may hold the plastic wrapper in place if he is able.

Figure 3-2. Open chest wound sealed with an occlusive dressing.

e. Apply the Dressing to the Wound.

(1) Use your free hand and shake open the field dressing (Figure 3-3).

Figure 3-3. Shaking open the field dressing.

(2) Place the white side of the dressing on the plastic wrapper covering the wound (Figure 3-4).

Figure 3-4. Field dressing placed on plastic wrapper.

NOTE

Use the casualty's field dressing, not your own.

(3) Have the casualty breathe normally.

(4) While maintaining pressure on the dressing, grasp one tail of the field dressing with the other hand and wrap it around the casualty's back. If tape is available, tape three sides of the plastic wrapper to the chest wall to provide occlusive type dressing. Leave one side untapped to provide emergency escape for air that may build up in the chest. If tape is not available, secure wrapper on three sides with field dressing leaving the fourth side as a flap.

(5) Wrap the other tail in the opposite direction, bringing both tails over the dressing (Figure 3-5).

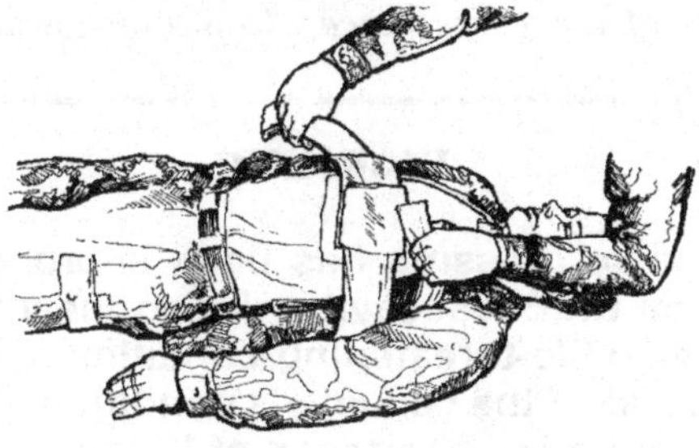

Figure 3-5. Tails of field dressing wrapped around casualty in opposite direction.

(6) Tie the tails into a square knot in the center of the dressing *after* the casualty exhales and *before* he inhales. This will aid in maintaining pressure on the bandage after it has been tied (Figure 3-6). Tie the dressing firmly enough to secure the dressing without interfering with the casualty's breathing.

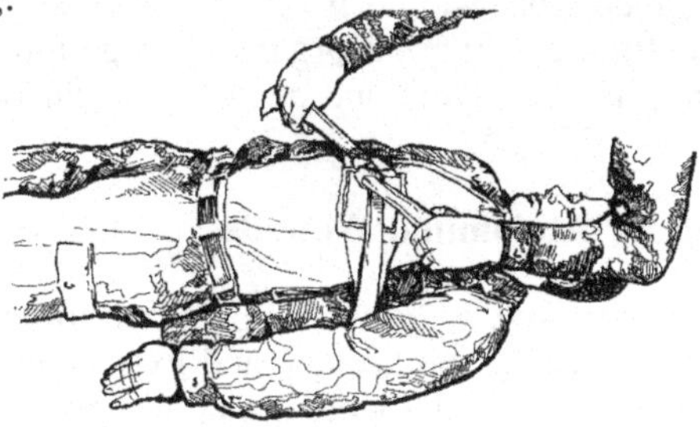

Figure 3-6. Tails of dressing tied into square knot over center of dressing.

NOTE

When practical, apply direct manual pressure over the dressing for 5 to 10 minutes to help control the bleeding.

f. *Position the Casualty*. Position the casualty on his injured side or in a sitting position, whichever makes breathing easier (Figure 3-7).

Figure 3-7. Casualty positioned (lying) on injured side.

g. Seek Medical Assistance. Contact medical personnel.

WARNING

If an occlusive dressing has been improperly placed, air may enter the chest cavity with no means of escape. This causes a life-threatening condition called *tension pneumothorax*. If the casualty's condition (for example, difficulty breathing, shortness of breath, restlessness, or blueness/grayness of the skin) worsens after placing the dressing, quickly lift or remove, and then replace the occlusive dressing.

3-6. Abdominal Wounds

The most serious abdominal wound is one in which an object penetrates the abdominal wall and pierces internal organs or large blood vessels. In these instances, bleeding may be severe and death can occur rapidly.

3-7. First Aid for Abdominal Wounds

a. Evaluate the Casualty. Be prepared to perform basic first aid measures. Always check for both entry and exit wounds. If there are two wounds (entry and exit), treat the wound that appears more serious first (for example, the heavier bleeding, protruding organs, larger wound, and so forth). It may be necessary to improvise dressings for the second wound by using strips of cloth, a T-shirt, or the cleanest material available.

b. Position the Casualty. Place and maintain the casualty on his back with his knees in an upright (flexed) position (Figure 3-8). The knees-up position helps relieve pain, assists in the treatment of shock, prevents further exposure of the bowel (intestines) or abdominal organs, and helps relieve abdominal pressure by allowing the abdominal muscles to relax.

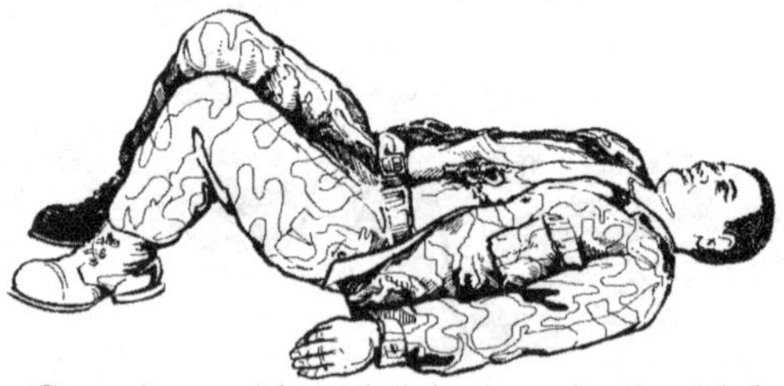

Figure 3-8. Casualty positioned (lying) on back with knees (flexed) up.

c. Expose the Wound.

(1) Remove the casualty's loose clothing to expose the wound. However, **DO NOT** attempt to remove clothing that is stuck to the wound; removing it may cause further injury.

CAUTION

DO NOT REMOVE protective clothing in a chemical environment. Apply dressings *over* the protective clothing.

(2) Gently pick up any organs that may be on the ground. Do this with a clean, dry dressing or with the cleanest available material. Place the organs on top of the casualty's abdomen (Figure 3-9).

Figure 3-9. Protruding organs placed near wound.

NOTE

DO NOT probe, clean, or try to remove any foreign object from the abdomen. **DO NOT** touch with bare hands any exposed organs. **DO NOT** push organs back inside the body.

d. Apply the Field Dressing. Use the casualty's field dressing, not your own. If the field dressing is not large enough to cover the entire wound, the plastic wrapper from the dressing may be used to cover the wound first (placing the field dressing on top). Open the plastic wrapper carefully without touching the inner surface. If necessary, other improvised dressings may be made from clothing, blankets, or the cleanest materials available.

WARNING

If there is an object extending from the wound, DO NOT remove it. Place as much of the wrapper over the wound as possible without dislodging or moving the object. DO NOT place the wrapper over the object.

(1) Grasp the tails in both hands.

(2) Hold the dressing with the white side down directly over the wound. **DO NOT** touch the white (sterile) side of the dressing or allow anything except the wound to come in contact with it.

(3) Pull the dressing open and place it directly over the wound (Figure 3-10). If the casualty is able, he may hold the dressing in place.

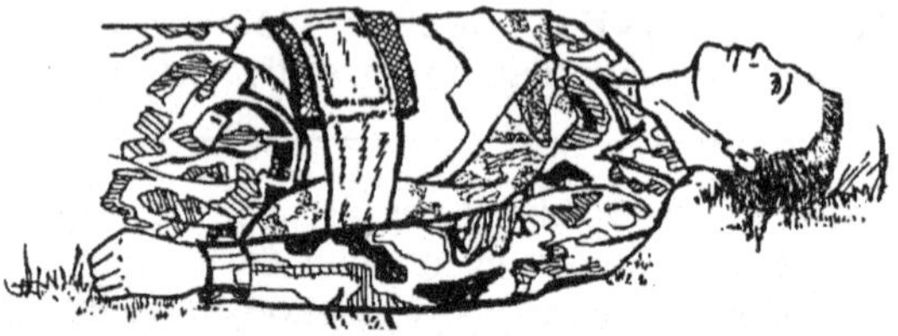

Figure 3-10. Dressing placed directly over the wound.

(4) Hold the dressing in place with one hand and use the other hand to wrap one of the tails around the body.

(5) Wrap the other tail in the opposite direction until the dressing is completely covered. Leave enough of the tail for a knot.

(6) Loosely tie the tails with a square knot at the casualty's side (Figure 3-11).

Figure 3-11. Dressing applied and tails tied with a square knot.

WARNING

When the dressing is applied, DO NOT put pressure on the wound or exposed internal parts, because pressure could cause further injury (vomiting, ruptured intestines, and so forth). Therefore, tie the dressing ties (tails) loosely at casualty's side, not directly over the dressing.

(7) Tie the dressing firmly enough to prevent slipping without applying pressure to the wound site (Figure 3-12).

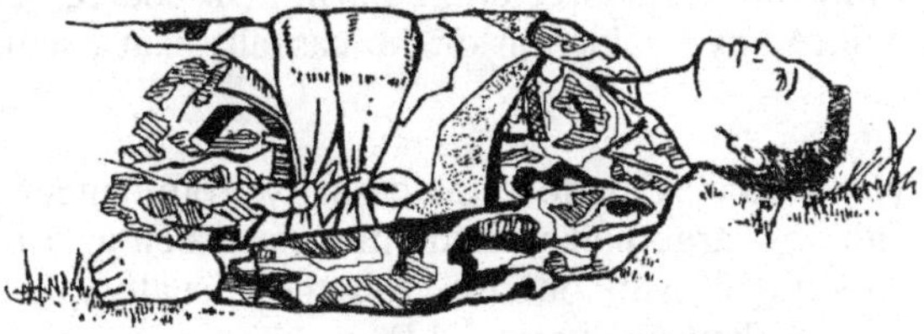

Figure 3-12. Field dressing covered with improvised material and loosely tied.

Field dressings can be covered with improvised reinforcement material (cravats, strips of torn T-shirt, or other cloth) for additional support and protection. Tie improvised bandage on the opposite side of the dressing ties firmly enough to prevent slipping but without applying additional pressure to the wound.

> **CAUTION**
>
> **DO NOT** give casualties with abdominal wounds food or water (moistening the lips is allowed).

e. *Seek Medical Assistance.* Notify medical personnel.

3-8. Burn Injuries

Burns often cause extreme pain, scarring, or even death. Before administering first aid, you must be able to recognize the type of burn. There are four types of burns:

- Thermal burns caused by fire, hot objects, hot liquids, and gases; or by nuclear blast or fireball.
- Electrical burns caused by electrical wires, current, or lightning.
- Chemical burns caused by contact with wet or dry chemicals or white phosphorus (WP)—from marking rounds and grenades.
- Laser burns (eye [ocular] injury).

3-9. First Aid for Burns

a. Eliminate the Source of the Burn. The source of the burn must be eliminated before any evaluation of the casualty can occur and first aid administered.

(1) Quickly remove the casualty from danger and cover the *thermal burn* with any large nonsynthetic material, such as a field jacket. If the casualty's clothing is still on fire, roll the casualty on the ground to smother (put out) the flames (Figure 3-13).

Figure 3-13. Casualty covered and rolled on ground.

> **CAUTION**
>
> Synthetic materials, such as nylon, may melt and cause further injury.

(2) Remove the *electrical burn* casualty from the electrical source by turning off the electrical current. **DO NOT** attempt to turn off the electricity if the source is not close by. Speed is critical, so **DO NOT** waste unnecessary time. If the electricity cannot be turned off, wrap any *nonconductive* material (*dry* rope, clothing, wood, and so forth) around the casualty's back and shoulders and drag the casualty away from the electrical source (Figure 3-14). **DO NOT** make body-to-body contact with the casualty or touch any wires because you could also become an electrical burn casualty.

Figure 3-14. Casualty removed from electrical source (using nonconductive material).

WARNING
High voltage electrical burns may cause temporary unconsciousness, difficulties in breathing, or difficulties with the heart (heartbeat).

(3) Remove the *chemical* from the *burned casualty*. Remove *liquid* chemicals by flushing with as much water as possible. Remove *dry* chemicals by brushing off loose particles (**DO NOT** use the bare surface of your hand because you could become a chemical burn casualty) and then flush with large amounts of water, if available. If large amounts of water are not available, then **NO** water should be applied because small amounts of water applied to a dry chemical burn may cause a chemical reaction. When WP strikes the skin, smother with a wet cloth or mud. Keep WP covered with a wet material to exclude air; this should help prevent the particles from burning.

(4) Remove the *laser burn* casualty from the source. When removing the casualty from the laser beam source, be careful not to enter the

beam or you may become a casualty. Never look directly at the beam source and if possible, wear appropriate eye protection.

NOTE

After the casualty is removed from the source of the burn, he should be evaluated for conditions requiring basic first aid measures.

b. Expose the Burn. Cut and gently lift away any clothing covering the burned area, without pulling clothing over the burns. Leave in place any clothing that is stuck to the burn. If the casualty's hands or wrists have been burned, remove jewelry if possible without causing further injury (rings, watches, and so forth) and place in his pockets. This prevents the necessity to cut off jewelry since swelling usually occurs as a result of a burn.

CAUTION

DO NOT lift or cut away clothing if in a chemical environment. Apply the dressing directly over the casualty's protective clothing. **DO NOT** attempt to decontaminate skin where blisters have formed.

c. Apply a Field Dressing to the Burn.

(1) Grasp the tails of the casualty's dressing in both hands.

(2) Hold the dressing directly over the wound with the white side down, pull the dressing open, and place it directly over the wound. **DO NOT** touch the white (sterile) side of the dressing or allow anything except the wound to come in contact with it. If the casualty is able, he may hold the dressing in place.

(3) Hold the dressing in place with one hand and use the other hand to wrap one of the tails around the limbs or the body.

(4) Wrap the other tail in the opposite direction until the dressing is completely covered.

(5) Tie the tails into a square knot over the outer edge of the dressing. The dressing should be applied lightly over the burn. Ensure that dressing is applied firmly enough to prevent it from slipping.

NOTE

Use the cleanest improvised dressing material available if a field dressing is not available or if it is not large enough for the entire wound.

d. *Take the Following Precautions*:

- **DO NOT** place the dressing over the face or genital area.
- **DO NOT** break the blisters.
- **DO NOT** apply grease or ointments to the burns.
- For electrical burns, check for both an entry and exit burn from the passage of electricity through the body. Exit burns may appear on any area of the body despite location of entry burn.
- For burns caused by wet or dry chemicals, flush the burns with large amounts of water and cover with a dry dressing.
- For burns caused by WP, flush the area with water, then cover with a wet material, dressing, or mud to exclude the air and keep the WP particles from burning.
- For laser burns, apply a field dressing.
- If the casualty is conscious and not nauseated, give him small amounts of water.

e. *Seek Medical Assistance*. Notify medical personnel.

3-10. Dressings and Bandages

a. *Head Wounds*.

(1) *Position the casualty*.

WARNING

DO NOT move the casualty if you suspect he has sustained a neck, spine, or head injury (which produces any signs or symptoms other than minor bleeding).

- If the casualty has a minor (superficial) scalp wound and is conscious:

  - Have the casualty sit up (unless other injuries prohibit or he is unable to).

  - If the casualty is lying down and is not accumulating fluids or drainage in his throat, elevate his head slightly.

  - If the casualty is bleeding from or into his mouth or throat, turn his head to the side or position him on his side so that the airway will be clear. Avoid putting pressure on the wound and place him on his uninjured side (Figure 3-15).

Figure 3-15. Casualty lying on side opposite injury.

- If the casualty is unconscious or has a severe head injury, then suspect and treat him as having a potential neck or spinal injury, *immobilize and **DO NOT** move the casualty.*

NOTE

If the casualty is choking or vomiting or is bleeding from or into his mouth (thus compromising his airway), position him on his uninjured side to allow for drainage and to help keep his airway clear.

> **WARNING**
>
> **If it is necessary to turn a casualty with a suspected neck/spine injury; roll the casualty gently onto his side, keeping the head, neck, and body aligned while providing support for the head and neck. DO NOT roll the casualty by yourself but seek assistance. *Move him only if absolutely necessary*, otherwise keep the casualty immobilized to prevent further damage to the neck/spine.**

(2) *Expose the wound.* Remove the casualty's helmet (if necessary). In a nuclear, biological, and chemical (NBC) environment, the first aid provider must leave the casualty as much protection (such as protective mask, mission-oriented protective posture [MOPP] overgarments) as possible. What items of protective equipment can be removed is dependent upon the casualty's injuries (where on the body and what type), the MOPP level, integrity of protective equipment (such as tears in the garment or mask seal), availability of chemical protective shelters, and the tactical situation.

WARNING

DO NOT attempt to clean the wound or remove a protruding object.

NOTE

Always use the casualty's field dressing, not your own.

(3) *Apply a dressing to a wound of the forehead or back of head.* To apply a dressing to a wound of the forehead or back of the head—

(*a*) Remove the dressing from the wrapper.

(*b*) Grasp the tails of the dressing in both hands.

(*c*) Hold the dressing (white side down) directly over the wound. **DO NOT** touch the white (sterile) side of the dressing or allow anything except the wound to come in contact with it.

(*d*) Place it directly over the wound.

(*e*) Hold it in place with one hand. If the casualty is able, he may assist.

(*f*) Wrap the first tail horizontally around the head; ensure the tail covers the dressing (Figure 3-16).

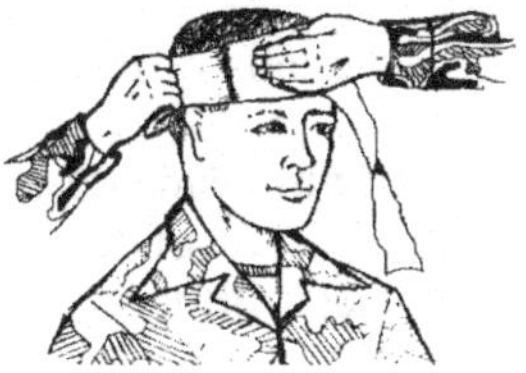

Figure 3-16. First tail of dressing wrapped horizontally around head.

(*g*) Hold the first tail in place and wrap the second tail in the opposite direction, covering the dressing (Figure 3-17).

Figure 3-17. Second tail wrapped in opposite direction.

(*h*) Tie a square knot and secure the tails at the side of the head, making sure they **DO NOT** cover the eyes or ears (Figure 3-18).

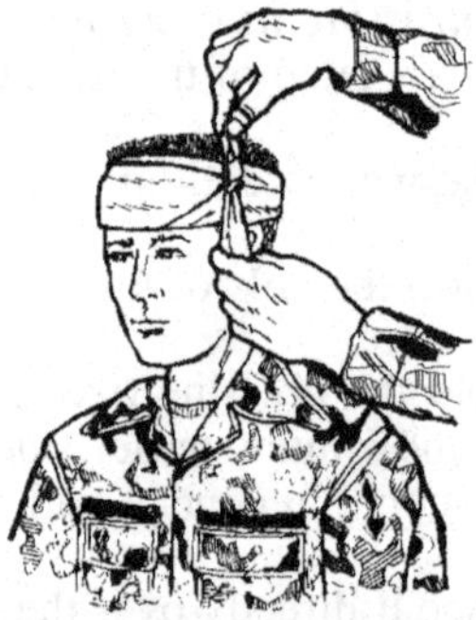

Figure 3-18. Tails tied in square knot at side of head.

(4) *Apply a dressing to a wound on top of the head.* To apply a dressing to a wound on top of the head—

(*a*) Remove the dressing from the wrapper.

(*b*) Grasp the tails of the dressing in both hands.

(*c*) Hold it (white side down) directly over the wound. **DO NOT** touch the white (sterile) side of the dressing or allow anything except the wound to come in contact with it.

(*d*) Place it over the wound (Figure 3-19).

Figure 3-19. Dressing placed over wound.

(*e*) Hold it in place with one hand. If the casualty is able, he may assist.

(*f*) Wrap one tail down under the chin (Figure 3-20), up in front of the ear, over the dressing, and in front of the other ear.

Figure 3-20. One tail of dressing wrapped under chin.

WARNING

Ensure the tails remain wide and close to the front of the chin to avoid choking the casualty.

(*g*) Wrap the remaining tail under the chin in the opposite direction and up the side of the face to meet the first tail (Figure 3-21).

Figure 3-21. Remaining tail wrapped under chin in opposite direction.

(*h*) Cross the tails (Figure 3-22), bringing one around the forehead (above the eyebrows) and the other around the back of the head (at the base of the skull) to a point just above and in front of the opposite ear, and tie them using a square knot (Figure 3-23).

Figure 3-22. Tails of dressing crossed with one around forehead.

Figure 3-23. Tails tied in square knot (in front of and above ear).

(5) *Apply a triangular bandage to the head.* To apply a triangular bandage to the head—

(*a*) Turn the base (longest side) of the bandage up and center its base on the center of the forehead, letting the point (apex) fall on the back of the neck (Figure 3-24A).

(*b*) Take the ends behind the head and cross the ends over the apex.

(*c*) Take them over the forehead and tie them (Figure 3-24B).

(*d*) Tuck the apex behind the crossed part of the bandage or secure it with a safety pin, if available (Figure 3-24C).

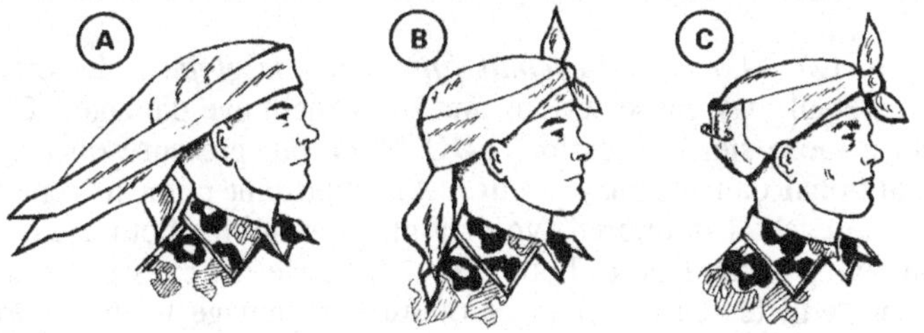

Figure 3-24. Triangular bandage applied to head (Illustrated A—C)

(6) *Apply a cravat bandage to the head.* To apply a cravat bandage to the head—

(*a*) Place the middle of the bandage over the dressing (Figure 3-25A).

(*b*) Cross the two ends of the bandage in opposite directions completely around the head (Figure 3-25B).

(*c*) Tie the ends over the dressing (Figure 3-25C).

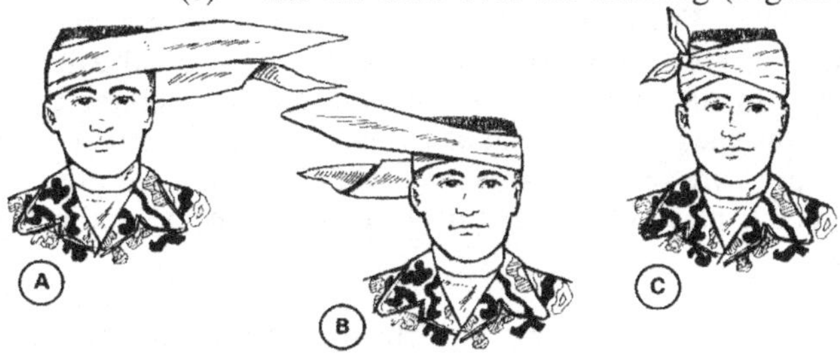

Figure 3-25. Cravat bandage applied to head (Illustrated A—C).

b. *Eye Injuries*. The eye is a vital sensory organ, and blindness is a severe physical handicap. Timely first aid of the eye may relieve pain and may also help to prevent shock, permanent eye injury, and possible loss of vision. Because the eye is very sensitive, any injury can be easily aggravated if it is improperly handled. Injuries of the eye may be quite severe. Cuts of the eyelids can appear to be very serious, but if the eyeball is not involved, a person's vision usually will not be damaged. However, lacerations (cuts) of the eyeball can cause permanent damage or loss of sight.

(1) *Lacerated/torn eyelids*. Lacerated eyelids may bleed heavily, but bleeding usually stops quickly. Cover the injured eye with a sterile dressing. **DO NOT** put pressure on the wound because you may injure the eyeball. Handle torn eyelids very carefully to prevent further injury. Place any detached pieces of the eyelid on a clean bandage or dressing and immediately send them with the casualty to the medical facility.

(2) *Lacerated eyeball (injury to the globe)*. Lacerations or cuts to the eyeball may cause serious and permanent eye damage. Cover the injury with a loose sterile dressing. **DO NOT** put pressure on the eyeball because additional damage may occur. An important point to remember is that when one eyeball is injured, you should immobilize both eyes. This is done by applying a bandage to both eyes. Because the eyes move together, covering both will lessen the chances of further damage to the injured eye. (However, in hazardous surroundings, leave uninjured eye uncovered to enable casualty to see.)

CAUTION

DO NOT apply pressure when there is a possible laceration of the eyeball. The eyeball contains fluid. Pressure applied over the eye will force the fluid out, resulting in permanent injury. **APPLY PROTECTIVE DRESSING WITHOUT ADDED PRESSURE**.

(3) *Extruded eyeballs*. Service members may encounter casualties with severe eye injuries that include an extruded eyeball (eyeball out-of-socket). In such instances you should gently cover the extruded eye with a loose moistened dressing and also cover the unaffected eye. **DO NOT** bind or exert pressure on the injured eye while applying the dressing. Keep the casualty quiet, place him on his back, treat for shock, and evacuate him immediately.

(4) *Burns of the eyes*. Chemical burns, thermal (heat) burns, and light burns can affect the eyes.

(*a*) *Chemical burns*. Injuries from chemical burns require immediate first aid. Mainly acids or alkalies cause chemical burns. The first aid measures consist of flushing the eyes immediately with large amounts of water for at least 5 to 20 minutes, or as long as necessary to flush out the chemical and, once flushed, bandaging the eyes. If the burn is an acid burn, you should flush the eye for at least 5 to 10 minutes. If the burn is an alkali burn, you should flush the eye for at least 20 minutes. After the eye has been flushed evacuate the casualty immediately.

(*b*) *Thermal burns*. When an individual suffers burns of the face from a fire, the eyes will close quickly due to extreme heat. This reaction is a natural reflex to protect the eyeballs; however, the eyelids remain exposed and are frequently burned. If a casualty receives burns of the eyelids or face—

- **DO NOT** apply a dressing.
- **DO NOT** touch.
- **SEEK** medical assistance immediately.

(*c*) *Light burns*. Exposure to intense light can burn an individual. Infrared rays, eclipse light (if the casualty has looked directly at the sun), or laser burns cause injuries of the exposed eyeball. Ultraviolet rays from arc welding can cause a superficial burn to the surface of the eye. These injuries are generally not painful but may cause permanent damage to the eyes. Immediate first aid is usually not required. Loosely bandaging the eyes may make the casualty more comfortable and protect his eyes from further injury caused by exposure to other bright lights or sunlight.

CAUTION

With impaled objects or significant sized foreign bodies, both eyes are usually bandaged to help secure the foreign body in the injured eye. In a battlefield environment, leave the uninjured eye uncovered so that the casualty can see.

c. *Side-of-Head or Cheek Wound*. Facial injuries to the side of the head or the cheek may bleed profusely (Figure 3-26). Prompt action is necessary to ensure that the airway remains open and also to control the bleeding. It may be necessary to apply a dressing. To apply a dressing—

(1) Remove the dressing from its wrapper.

(2) Grasp the tails in both hands.

(3) Hold the dressing directly over the wound with the white side down and place it directly on the wound (Figure 3-27A). **DO NOT** touch the white (sterile) side of the dressing or allow anything except the wound to come in contact with it.

(4) Hold the dressing in place with one hand (the casualty may assist if able). Wrap the top tail over the top of the head and bring it down in front of the ear (on the side opposite the wound), under the chin (Figure 3-27B) and up over the dressing to a point just above the ear (on the wound side).

Figure 3-26. Side of head or cheek wound.

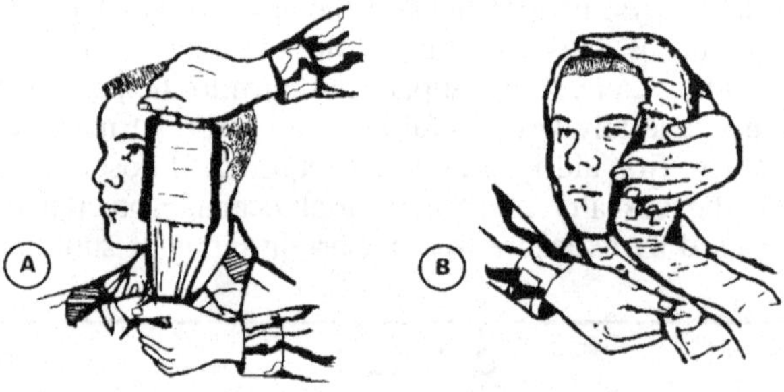

Figure 3-27. Dressing placed directly on wound. Top tail wrapped over top of head, down in front of ear, and under chin (Illustrated A—B).

NOTE

When possible, avoid covering the casualty's ear with the dressing, as this will decrease his ability to hear.

(5) Bring the second tail under the chin, up in front of the ear (on the side opposite the wound), and over the head to meet the other tail (on the wounded side) (Figure 3-28).

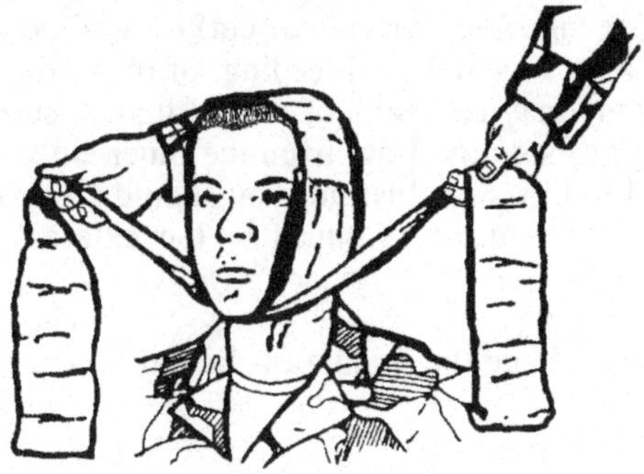

Figure 3-28. Bringing second tail under the chin.

(6) Cross the two tails (on the wound side) (Figure 3-29) and bring one end across the forehead (above the eyebrows) to a point just in front of the opposite ear (on the uninjured side).

Figure 3-29. Crossing the tails on the side of the wound.

(7) Wrap the other tail around the back of the head (at the base of the skull), and tie the two ends just in front of the ear on the uninjured side with a square knot (Figure 3-30).

Figure 3-30. Tying the tails of the dressing in a square knot.

d. Ear Injuries. Lacerated (cut) or avulsed (torn) ear tissue may not, in itself, be a serious injury. Bleeding, or the drainage of fluids from the ear canal, however, may be a sign of a head injury, such as a skull fracture. **DO NOT** attempt to stop the flow from the inner ear canal nor put anything into the ear canal to block it. Instead, you should cover the ear lightly with a dressing. For minor cuts or wounds to the external ear, apply a cravat bandage as follows:

(1) Place the middle of the bandage over the ear (Figure 3-31A).

(2) Cross the ends, wrap them in opposite directions around the head, and tie them (Figures 3-31B and 3-31C).

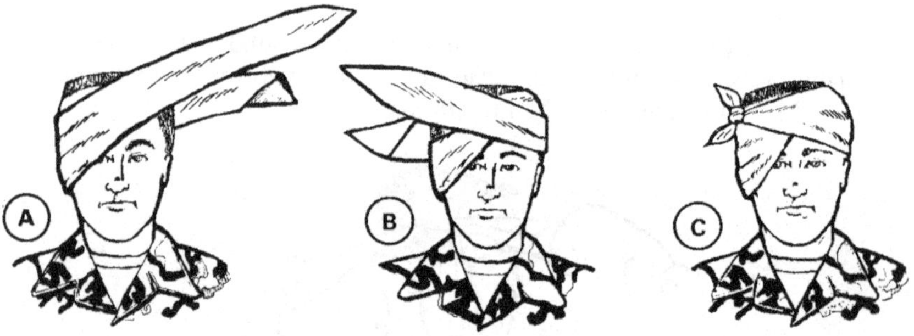

Figure 3-31. Applying cravat bandage to ear (Illustrated A—C).

(3) If possible, place some dressing material between the back of the ear and the side of the head to avoid crushing the ear against the head with the bandage.

e. Nose Injuries. Nose injuries generally produce bleeding. The bleeding may be controlled by placing an ice pack (if available) over the nose, or pinching the nostrils together. The bleeding may also be controlled by placing torn gauze (rolled) between the upper teeth and the lip.

CAUTION

DO NOT attempt to remove objects inhaled into the nose. An untrained person who removes such an object could worsen the casualty's condition and cause permanent injury.

f. Jaw Injuries. Before applying a bandage to a casualty's jaw, remove all loose or free-floating foreign material from the casualty's mouth.

If the casualty is unconscious, check for obstructions in the airway and remove if possible. If there is profuse bleeding in the oral cavity, the cavity may require loose packing with soft bandaging material (for example: Kerlix™ gauze) prior to applying a bandage. Care should be taken to avoid occluding the airway. When applying the bandage, allow the jaw enough freedom to permit passage of air and drainage from the mouth.

(1) *Apply bandages attached to field first aid dressing to the jaw.* After dressing the wound, apply the bandages using the same technique illustrated in Figure 3-32A—C.

NOTE

The dressing and bandaging procedure outlined for the jaw serves a twofold purpose. In addition to stopping the bleeding and protecting the wound, it also immobilizes a fractured jaw.

(2) *Apply a cravat bandage to the jaw.*

(*a*) Place the bandage under the chin and pull its ends upward. Adjust the bandage to make one end longer than the other (Figure 3-32A).

(*b*) Take the longer end over the top of the head to meet the short end at the temple and cross the ends over (Figure 3-32B).

(*c*) Take the ends in opposite directions to the other side of the head and tie them over the part of the bandage that was applied first (Figure 3-32C).

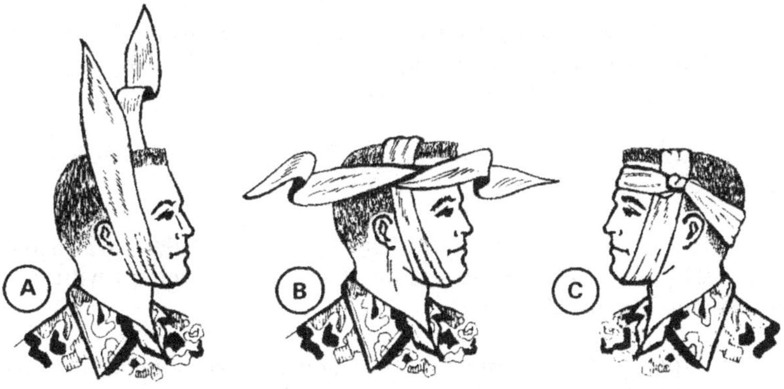

Figure 3-32. Applying a cravat bandage to jaw (Illustrated A—C).

NOTE

The cravat bandage technique is used to immobilize a fractured jaw or to maintain a sterile dressing that does not have tail bandages attached.

3-11. Shoulder Bandage

a. To apply bandages attached to the field first aid dressing—

(1) Take one bandage across the chest and the other across the back and under the arm opposite the injured shoulder.

(2) Tie the ends with a square knot (Figure 3-33).

Figure 3-33. Shoulder bandage.

b. To apply a cravat bandage to the shoulder or armpit—

(1) Make an extended cravat bandage by using two triangular bandages (Figure 3-34A); place the end of the first triangular bandage along the base of the second one (Figure 3-34B).

(2) Fold the two bandages into a single extended bandage (Figure 3-34C).

(3) Fold the extended bandage into a single cravat bandage (Figure 3-34D). After folding, secure the thicker part (overlap) with two or more safety pins (Figure 3-34E).

(4) Place the middle of the cravat bandage under the armpit so that the front end is longer than the back end and safety pins are on the outside (Figure 3-34F).

(5) Cross the ends on top of the shoulder (Figure 3-34G).

(6) Take one of the bandage ends across the back and under the arm on the opposite side and the other end across the chest. Tie the ends (Figure 3-34H).

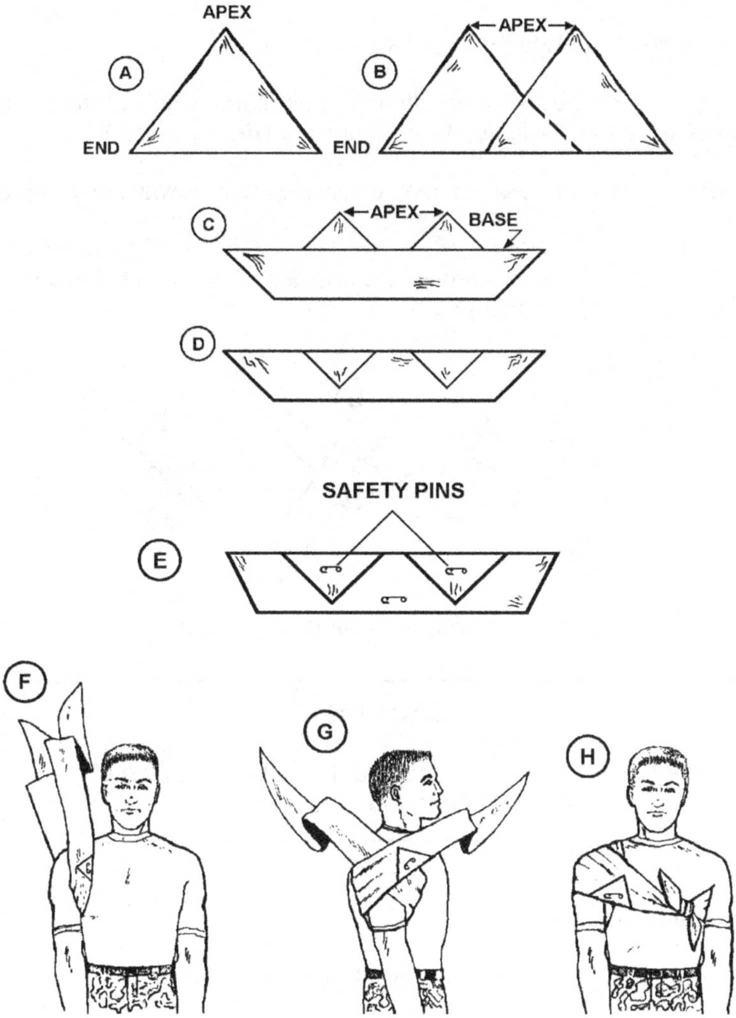

Figure 3-34. Extended cravat bandage applied to shoulder or armpit (Illustrated A—H).

Be sure to place sufficient wadding in the armpit. **DO NOT** tie the cravat bandage too tightly. Avoid compressing the major blood vessels in the armpit.

3-12. Elbow Bandage

To apply a cravat bandage to the elbow—

a. Bend the arm at the elbow and place the middle of the cravat at the point of the elbow bringing the ends upward (Figure 3-35A).

b. Bring the ends across, extending both downward (Figure 3-35B).

c. Take both ends around the arm and tie them with a square knot at the front of the elbow (Figure 3-35C).

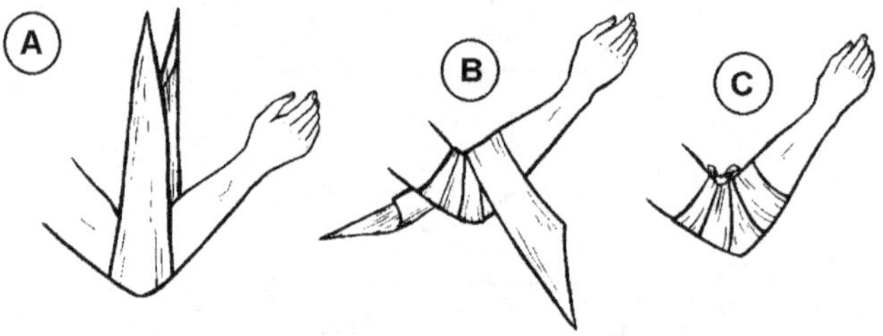

Figure 3-35. Elbow bandage (Illustrated A—C).

CAUTION
If an elbow fracture is suspected, **DO NOT** bend the elbow; bandage it in the position found.

3-13. Hand Bandage

a. To apply a triangular bandage to the hand—

(1) Place the hand in the middle of the triangular bandage with the wrist at the base of the bandage (Figure 3-36A). Ensure that the fingers are separated with absorbent material to prevent chafing and irritation of the skin.

(2) Place the apex over the fingers and tuck any excess material into the pleats on each side of the hand (Figure 3-36B).

(3) Cross the ends on top of the hand, take them around the wrist, and tie them (Figures 3-36C—E) with a square knot.

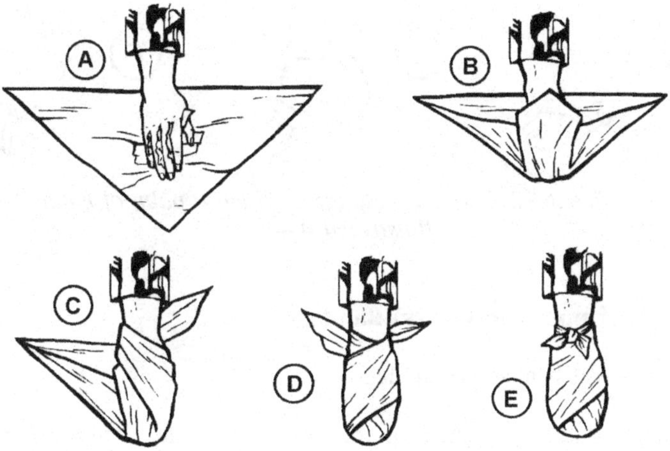

Figure 3-36. Triangular bandage applied to hand (Illustrated A—E).

b. To apply a cravat bandage to the palm of the hand—

(1) Lay the middle of the cravat over the palm of the hand with the ends hanging down on each side (Figure 3-37A).

(2) Take the end of the cravat at the little finger across the back of the hand, extending it upward over the base of the thumb; then bring it downward across the palm (Figure 3-37B).

(3) Take the thumb end across the back of the hand, over the palm, and through the hollow between the thumb and palm (Figure 3-37C).

(4) Take the ends to the back of the hand and cross them; then bring them up over the wrist and cross them again (Figure 3-37D).

(5) Bring both ends down and tie them with a square knot on top of the wrist (Figure 3-37E—F).

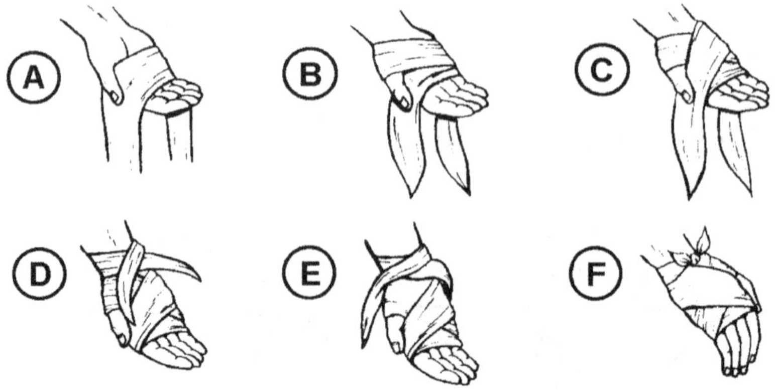

Figure 3-37. Cravat bandage applied to palm of hand (Illustrated A—F).

3-14. Leg (Upper and Lower) Bandage

To apply a cravat bandage to the leg—

a. Place the center of the cravat over the dressing (Figure 3-38A).

b. Take one end around and up the leg in a spiral motion and the other end around and down the leg in a spiral motion, overlapping part of each preceding turn (Figure 3-38B).

c. Bring both ends together and tie them (Figure 3-38C) with a square knot.

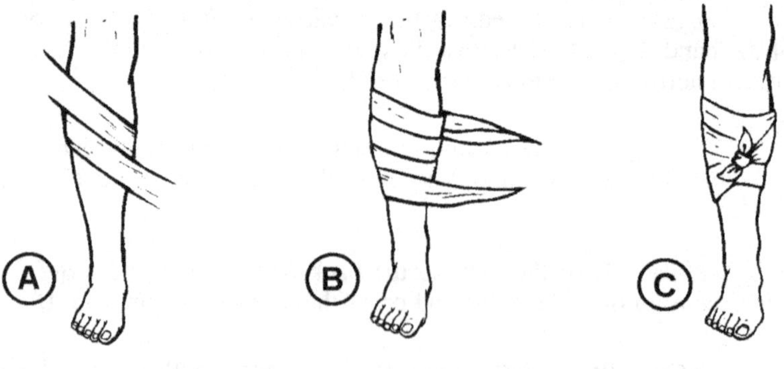

Figure 3-38. Cravat bandage applied to leg (Illustrated A—C).

3-15. Knee Bandage

To apply a cravat bandage to the knee as illustrated in Figure 3-39, use the same technique applied in bandaging the elbow.

> **CAUTION**
>
> If a fracture of the kneecap is suspected, **DO NOT** bend the knee; bandage it in the position found.

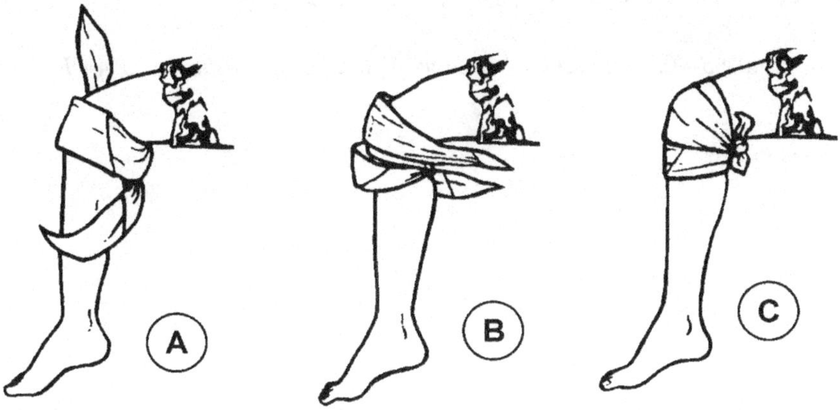

Figure 3-39. Cravat bandage applied to knee (Illustrated A—C).

3-16. Foot Bandage

To apply a triangular bandage to the foot—

a. Place the foot in the middle of the triangular bandage with the heel well forward of the base (Figure 3-40A). Ensure that the toes are separated by absorbent material to prevent chafing and irritation of the skin.

b. Place the apex over the top of the foot and tuck any excess material into the pleats on each side of the foot (Figure 3-40B).

c. Cross the ends on top of the foot, take them around the ankle, and tie them at the front of the ankle (Figure 3-40C—E).

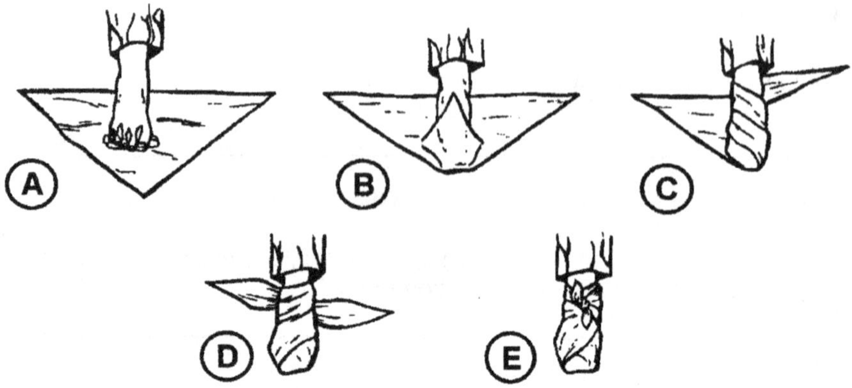

Figure 3-40. Triangular bandage applied to foot (Illustrated A—E).

CHAPTER 4

FIRST AID FOR FRACTURES

4-1. General

A fracture is any break in the continuity of a bone. Fractures can cause total disability or in some cases death by severing vital organs and/or arteries. On the other hand, they can most often be treated so there is a complete recovery. The potential for recovery depends greatly upon the first aid the individual receives before he is moved. First aid includes immobilizing the fractured part in addition to applying lifesaving measures when necessary. The basic splinting principle is to immobilize the joints above and below the fracture.

4-2. Kinds of Fractures

Figure 4-1 depicts types of fractures.

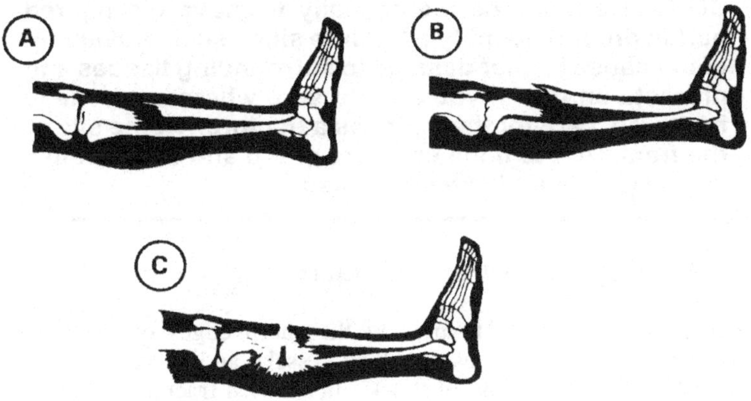

Figure 4-1. Types of fractures (Illustrated A—C).

a. *Closed Fracture* (Figure 4-1A). A closed fracture is a broken bone that does not break the overlying skin. The tissue beneath the skin may be damaged. A *dislocation* is when a joint, such as a knee, ankle, or shoulder, is not in the proper position. A *sprain* is when the connecting tissues of the joints have been torn. Dislocations and sprains (swelling, possible deformity, and discoloration) should be treated as closed fractures.

b. *Open Fracture* (Figure 4-1B and 4-1C). An open fracture is a broken bone that breaks (pierces) the overlying skin. The broken bone may come through the skin or a missile such as a bullet or shell fragment may go through the flesh and break the bone.

NOTE

An open fracture is contaminated and subject to infection.

4-3. Signs and Symptoms of Fractures

Indications of a fracture are deformity, tenderness, swelling, pain, inability to move the injured part, protruding bone, bleeding, or discolored skin at the injury site. A sharp pain when the service member attempts to move the part is also a sign of a fracture.

WARNING

DO NOT encourage the casualty to move the injured part in order to identify a fracture since such movement could cause further damage to surrounding tissues and promote shock. If you are not sure whether a bone is fractured, care for the injury as a fracture. At the site of the fracture, the bone ends are sharp and could cause vessel (artery and/or vein) damage.

4-4. Purposes of Immobilizing Fractures

A fracture is immobilized to prevent the sharp edges of the bone from moving and cutting tissue, muscle, blood vessels, and nerves. This reduces pain and helps prevent or control shock. In a closed fracture, immobilization keeps bone fragments from causing an open wound, which can become contaminated and subject to infection.

4-5. Splints, Padding, Bandages, Slings, and Swathes

a. Splints. Splints may be improvised from such items as boards, poles, sticks, tree limbs, or cardboard. If nothing is available for a splint, the chest wall can be used to immobilize a fractured arm and the uninjured leg can be used to immobilize (to some extent) the fractured leg.

b. Padding. Padding may be improvised from such items as a jacket, blanket, poncho, shelter half, or leafy vegetation.

c. Bandages. Bandages may be improvised from belts, rifle slings, kerchiefs, or strips torn from clothing or blankets. Narrow materials such as wire or cord should not be used to secure a splint in place. The application of wire and/or narrow material to an extremity could cause tissue damage and a tourniquet effect.

d. Slings. A sling is a bandage suspended from the neck to support an upper extremity. If a bandage is not available, a sling can be improvised by using the tail of a coat or shirt or pieces of cloth torn from such items as clothing and blankets. The triangular bandage is ideal for this purpose. Remember that the casualty's hand should be higher than his elbow, and the fingers should be showing at all times. The sling should be applied so that the supporting pressure is on the uninjured side.

e. Swathes. Swathes are any bands (pieces of cloth or load bearing equipment [LBE]) that are used to further immobilize a splinted fracture. Triangular and cravat bandages are often used and are called *swathe bandages*. The purpose of the swathe is to immobilize; therefore, the swathe bandage is placed above and/or below the fracture—not over it.

4-6. Procedures for Splinting Suspected Fractures

Before beginning first aid procedures for a fracture, gather whatever splinting materials are available. Ensure that splints are long enough to immobilize the joint above and below the suspected fracture. If possible, use at least four ties (two above and two below the fracture) to secure the splints. The ties should be square knots and should be tied away from the body on the splint. Distal pulses of the affected extremity should be checked before and after the application of the splint.

a. Evaluate the Casualty. Be prepared to perform any necessary lifesaving measures. Monitor the casualty for development of conditions that may require you to perform necessary lifesaving measures.

WARNING

Unless there is immediate life-threatening danger, such as a fire or an explosion, DO NOT move the casualty with a suspected back or neck injury. Improper movement may cause permanent paralysis or death.

WARNING

In a chemical environment, DO NOT remove any protective clothing. Apply the dressings and splints over the garments.

b. Locate the Site of the Suspected Fracture.

(1) Ask the casualty for the location of the injury.

- Does he have any pain?
- Where is it tender?
- Can he move the extremity?

NOTE

With the presence of an obvious deformity, do not make the casualty move extremity.

(2) Look for an unnatural position of the extremity.

(3) Look for a bone sticking out (protruding).

c. Prepare the Casualty for Splinting the Suspected Fracture.

(1) Reassure the casualty. Tell him that you will be providing first aid for him and that medical help is on the way.

(2) Loosen any tight or binding clothing.

(3) Remove all jewelry from the injured part and place it in the casualty's pocket. Tell the casualty you are doing this because if the jewelry is not removed and swelling occurs later, he may not be able to get it off and further bodily injury could result.

(4) Boots should not be removed from the casualty unless they are needed to stabilize a neck injury or there is actual bleeding from the foot.

d. Gather Splinting Materials. If standard splinting materials (splints, padding, and cravats) are not available, gather improvised materials. If splinting material is not available and the suspected fracture **CANNOT** be splinted, then swathes, or a combination of swathes and slings can be used to immobilize the extremity.

e. Pad the Splints. Pad the splints where they touch any bony part of the body, such as the elbow, wrist, knee, ankle, crotch, or armpit areas. Padding prevents excessive pressure on the area, which could lead to circulation problems.

f. Check the Circulation Below the Site of the Injury.

(1) Note any pale, white, or bluish-gray color of the skin, which may indicate impaired circulation. Circulation can also be checked by depressing the toe or fingernail beds and observing how quickly the color returns. A slower return of color to the injured side when compared with the uninjured side indicates a problem with circulation. The fingernail bed is the method to use to check the circulation in a dark-skinned casualty.

(2) Check the temperature of the injured extremity. Use your hand to compare the temperature of the injured side with the uninjured side. The body area below the injury may be colder to the touch indicating poor circulation.

(3) Question the casualty about the presence of numbness, tightness, cold, or tingling sensations.

WARNING

Casualties with fractures of the extremities may show impaired circulation, such as numbness, tingling, cold or pale to bluish skin tone. These casualties should be evacuated by medical personnel and treated as soon as possible. Prompt medical treatment may prevent possible loss of the limb.

WARNING

If it is an open fracture and the bone is protruding from the skin, *DO NOT ATTEMPT TO PUSH THE BONE BACK UNDER THE SKIN.* Apply a field dressing over the wound to protect the area.

g. Apply the Splint in Place.

(1) Splint the fracture in the position found. **DO NOT** attempt to reposition or straighten the injury. If it is an open fracture, stop the bleeding and protect the wound. Cover all wounds with field dressings before applying a splint. Remember to use the casualty's field dressing, not your own.

(2) Place one splint on each side of the fracture. Make sure that the splints reach, if possible, beyond the joints above and below the fracture.

(3) Tie the splints. Secure each splint in place above and below the fracture site with improvised (or actual) cravats. Improvised cravats, such as strips of cloth, belts, or whatever else you have, may be used. With minimal motion to the injured areas, place and tie the splints with the bandages. Push cravats through and under the natural body curvatures, and then gently position improvised cravats and tie in place. Use square knots. Tie all knots on the splint away from the casualty (Figure 4-2). **DO NOT** tie cravats directly over the suspected fracture site.

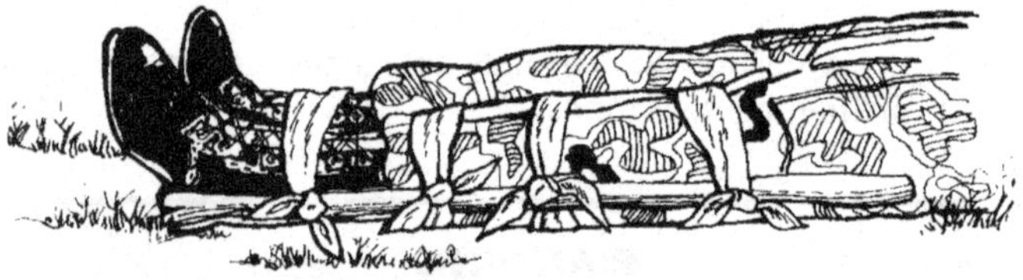

Figure 4-2. Square knots tied away from casualty.

h. Check the Splint for Tightness.

(1) **CHECK** to be sure that bandages are tight enough to securely hold splinting materials in place, but not so tight that circulation is impaired.

(2) **RECHECK** the circulation after application of the splint. Check the skin color and temperature. This is to ensure that the bandages holding the splint in place have not been tied too tightly. A fingertip check can be made by inserting the tip of the finger between the bandaged knot and the skin.

(3) **MAKE** any necessary adjustment without allowing the splint to become ineffective.

i. Apply a Sling. An improvised sling may be made from any available nonstretching piece of cloth, such as a battle dress uniform (BDU) shirt or trousers, poncho, or shelter half. Slings may also be improvised using the tail of a coat, belt, or a piece of cloth. Figure 4-3 depicts a shirttail used for support. A trousers belt or LBE may also be used for support (Figure 4-4). A sling should place the supporting pressure on the casualty's uninjured side. The supported arm should have the hand positioned slightly higher than the elbow showing the fingers.

Figure 4-3. Shirttail used for support.

Figure 4-4. Belt used for support.

(1) Insert the splinted arm in the center of the sling (Figure 4-5).

Figure 4-5. Arm inserted in center of improvised sling.

(2) Bring the ends of the sling up and tie them at the side (or hollow) of the neck on the uninjured side (Figure 4-6).

Figure 4-6. Ends of improvised sling tied to side of neck.

(3) Twist and tuck the corner of the sling at the elbow (Figure 4-7).

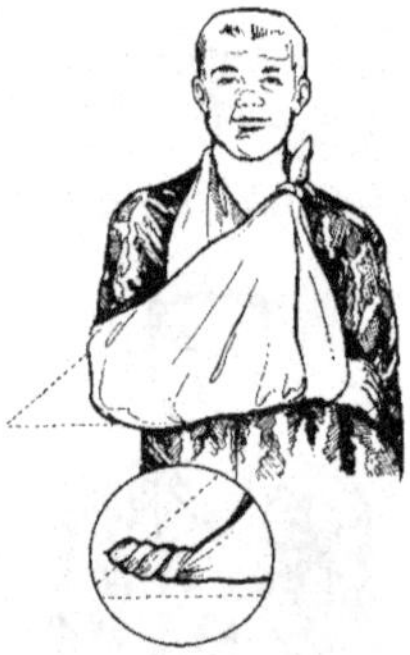

Figure 4-7. Corner of sling twisted and tucked at elbow.

j. Apply a Swathe. You may use any large piece of cloth, service member's belt, or pistol belt, to improvise a swathe.

WARNING

The swathe should not be placed directly on top of the injury, but positioned either above or below the fracture site.

(1) Apply swathes to the injured arm by wrapping the swathe over the injured arm, around the casualty's back, and under the arm on the uninjured side. Tie the ends on the uninjured side (Figure 4-8).

Figure 4-8. Arm immobilized with strip of clothing.

(2) A swathe is applied to an injured leg by wrapping the swathe around both legs and securing it on the uninjured side.

k. Seek Medical Assistance. Notify medical personnel, watch closely for development of life-threatening conditions and/or impaired circulation to the injured extremity. (Refer to Chapter 1 for additional information on life-threatening conditions.)

4-7. Upper Extremity Fractures

Figures 4-9 through 4-17 show how to apply slings, splints, and cravats (swathes) to immobilize and support fractures of the upper extremities. *Although the padding is not visible in some of the illustrations, it is always preferable to apply padding along the injured part for the length of the splint and especially where it touches any bony parts of the body.*

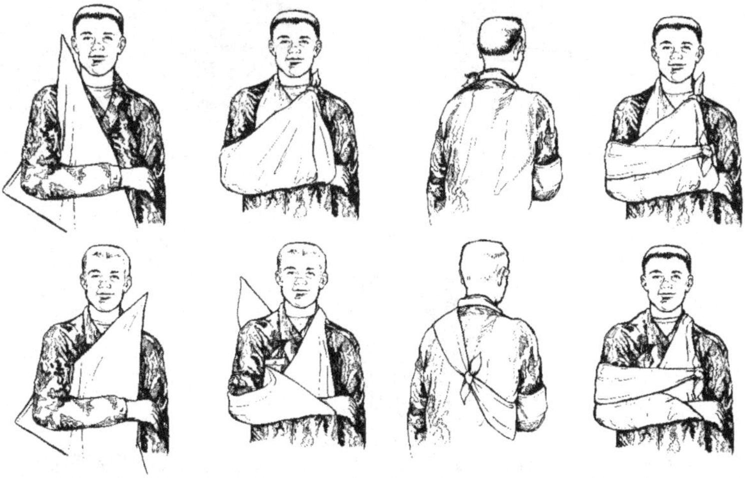

Figure 4-9. Application of triangular bandage to form sling (two methods).

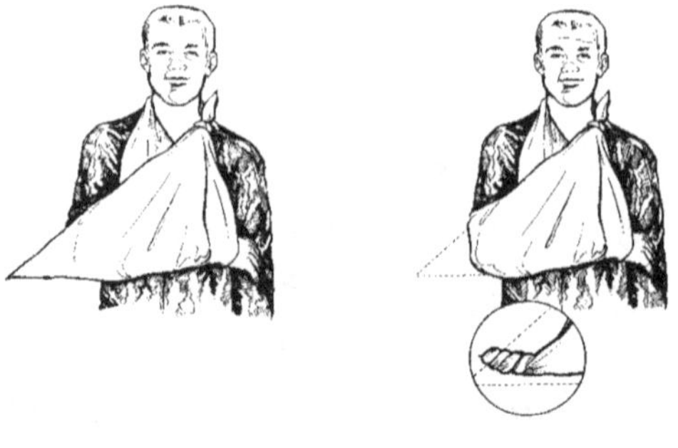

Figure 4-10. Completing sling sequence by twisting and tucking the corner of the sling at the elbow.

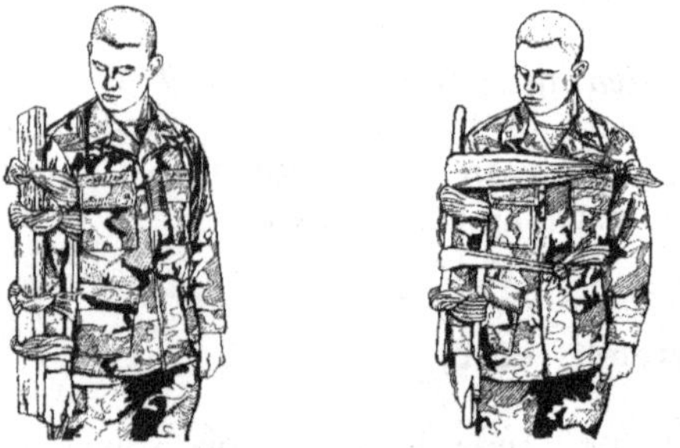

Figure 4-11. Board splints applied to fractured elbow when elbow is not bent (two methods).

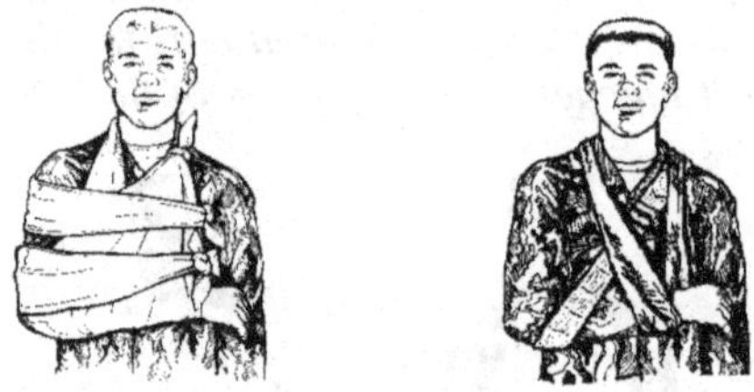

Figure 4-12. Chest wall used as splint for upper arm fracture when no splint is available.

Figure 4-13. Chest wall, sling, and cravat used to immobilize fractured elbow when elbow is bent.

Figure 4-14. Board splint applied to fractured forearm.

Figure 4-15. Fractured forearm or wrist splinted with sticks and supported with tail of shirt and strips of material.

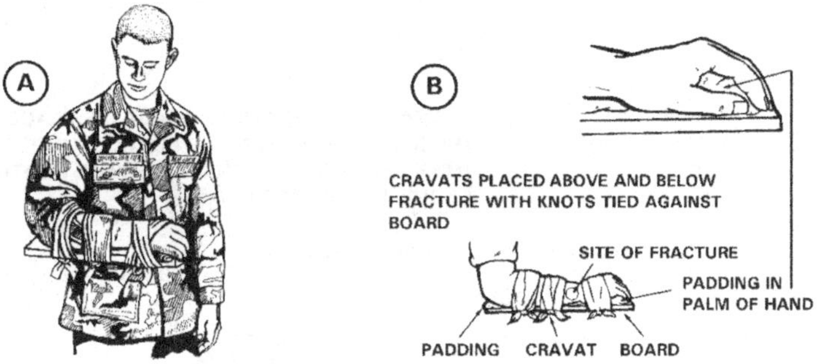

Figure 4-16. Board splint applied to fractured wrist and hand (Illustrated A—B).

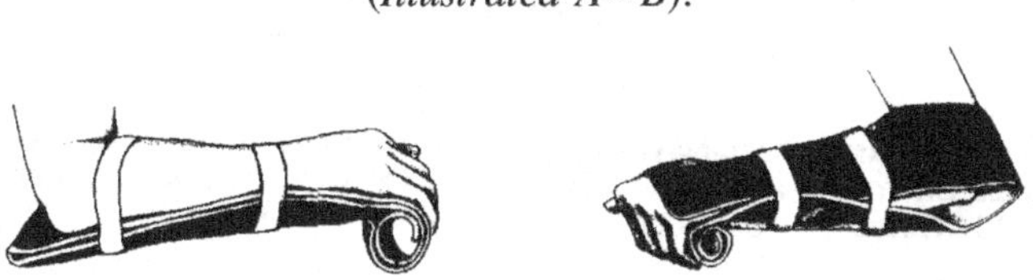

Figure 4-17. SAM® splint applied to fractured wrist or forearm.

4-8. Lower Extremity Fractures

Figures 4-18 through 4-24 show how to apply splints to immobilize fractures of the lower extremities. *Although padding is not visible in some of the illustrations, it is always preferable to apply padding along the injured part for the length of the splint and especially where it touches any bony parts of the body.*

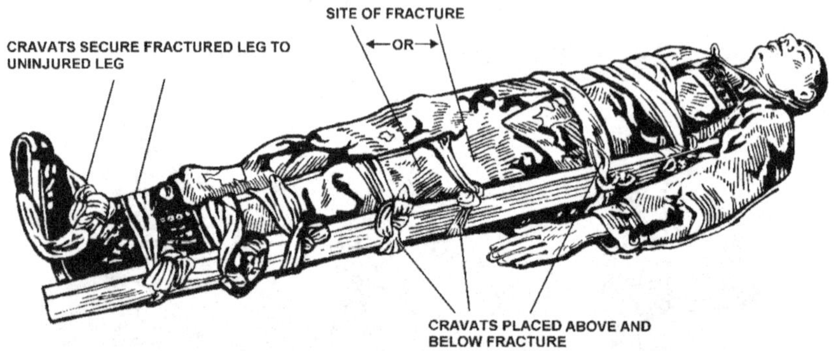

Figure 4-18. Board splints applied to fractured hip or thigh.

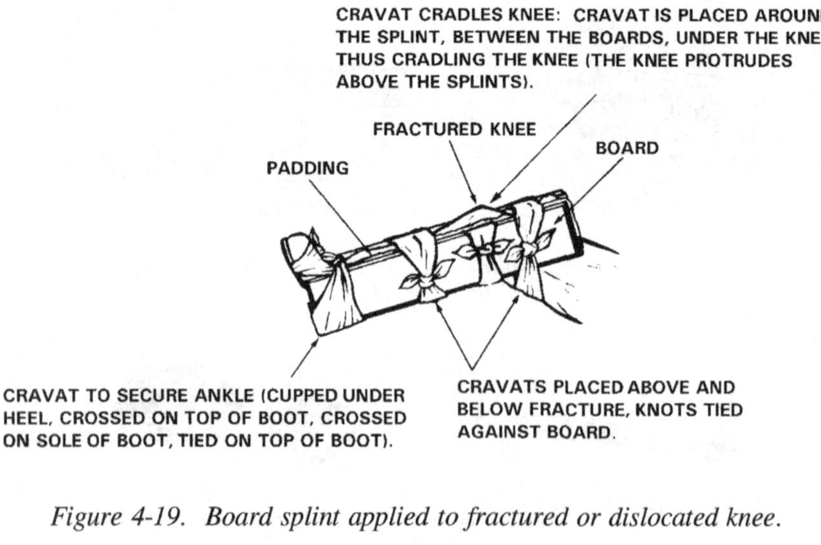

Figure 4-19. Board splint applied to fractured or dislocated knee.

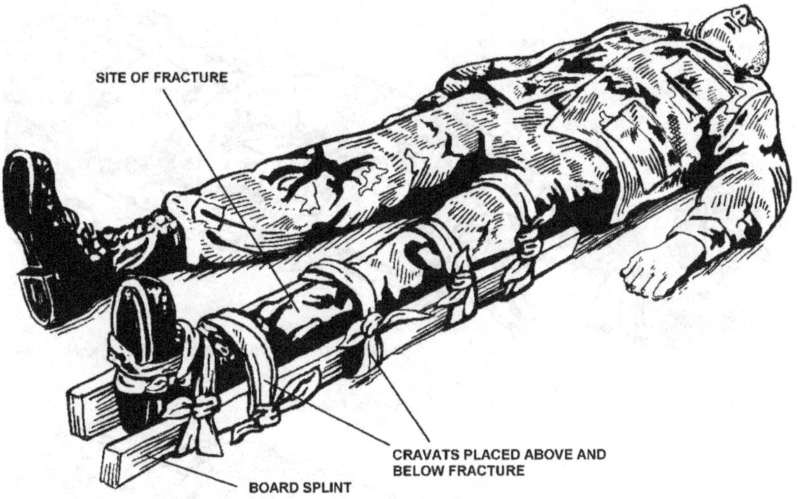

Figure 4-20. Board splints applied to fractured lower leg or ankle.

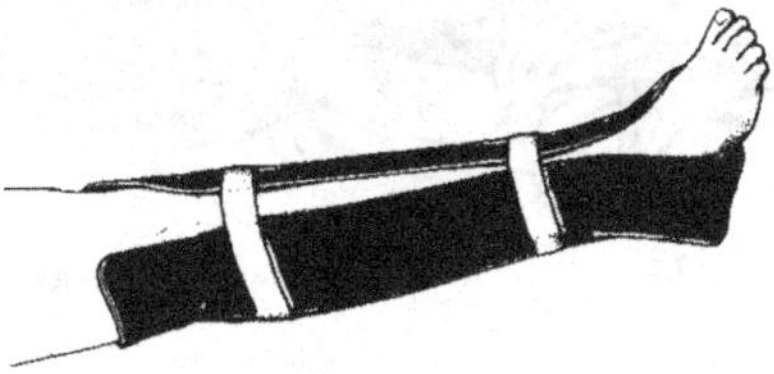

Figure 4-21. SAM® splint applied to fractured lower leg or ankle.

Figure 4-22. Improvised splints applied to fractured lower leg or ankle.

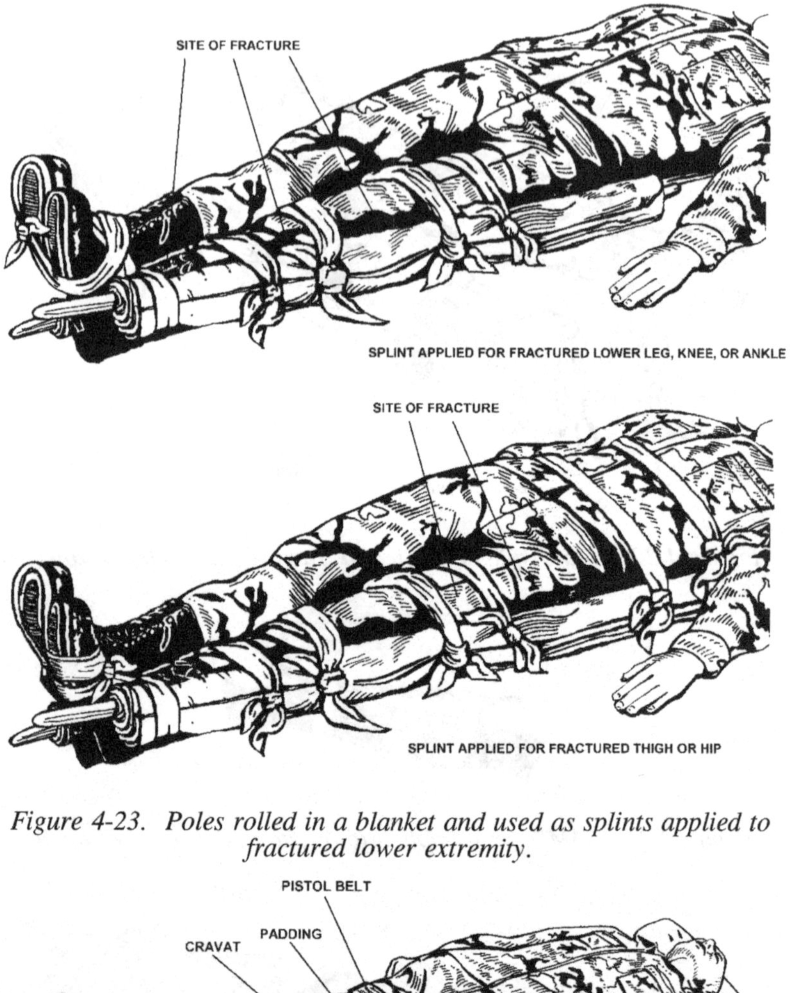

Figure 4-23. Poles rolled in a blanket and used as splints applied to fractured lower extremity.

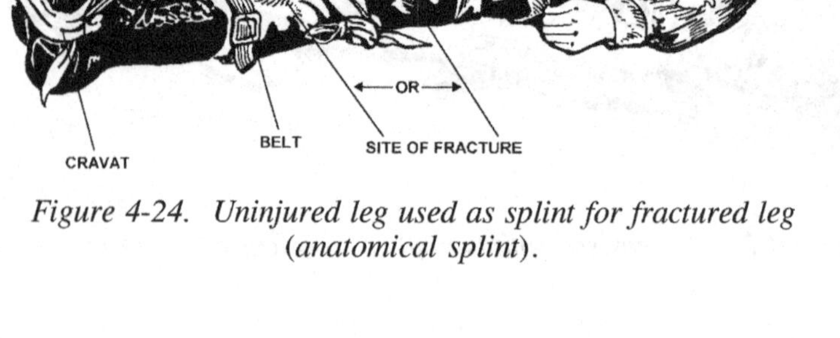

Figure 4-24. Uninjured leg used as splint for fractured leg (anatomical splint).

4-9. Jaw, Collarbone, and Shoulder Fractures

a. Apply a cravat to immobilize a fractured jaw as illustrated in Figure 4-25. Direct all bandaging support to the top of the casualty's head, not to the back of his neck. If incorrectly placed, the bandage will pull the casualty's jaw back and interfere with his breathing.

Figure 4-25. Fractured jaw immobilized.

WARNING
Casualties with lower jaw (mandible) fractures cannot be laid flat on their backs because facial muscles will relax and may cause an airway obstruction.

b. Apply two belts, a sling, and a cravat to immobilize a fractured collarbone, as illustrated in Figure 4-26.

Figure 4-26. Application of belts, sling, and cravat to immobilize a fractured collarbone.

c. Apply a sling and a cravat to immobilize a fractured or dislocated shoulder, using the technique illustrated in Figure 4-27.

Figure 4-27. Application of sling and cravat to immobilize a fractured or dislocated shoulder.

4-10. Spinal Column Fractures

It is often impossible to be sure a casualty has a fractured spinal column. Be suspicious of any back injury, especially if the casualty has fallen or if his back has been sharply struck or bent. If a casualty has received such an injury and does not have feeling in his legs or cannot move them, you can be reasonably sure that he has a severe back injury, which should be managed as a fracture. Remember, that the possibility of a neck fracture or injury to the back should always be suspected, and it is often impossible to be sure if a casualty has a fractured spinal column. If the spine is fractured, bending it can cause the sharp bone fragments to bruise or cut the spinal cord and result in permanent paralysis or death (Figure 4-28A). The spinal column must maintain normal spinal position at the lower back (lumbar region) to help remove pressure from the spinal cord.

a. If the casualty is not to be transported until medical personnel arrive—

- Caution him not to move. Ask him if he is in pain or if he is unable to move any part of his body.

- Leave him in the position in which he is found. **DO NOT** move any part of his body, unless he is in imminent danger.

- If the casualty is lying face up, slip a blanket or other supporting material under the arch of his lower back to help support the spine in a normal position (Figure 4-28B). Take care not to place so much bulky padding as to cause potential damage by causing undo pressure on the spine. If he is lying face down, **DO NOT** put anything under any part of his body.

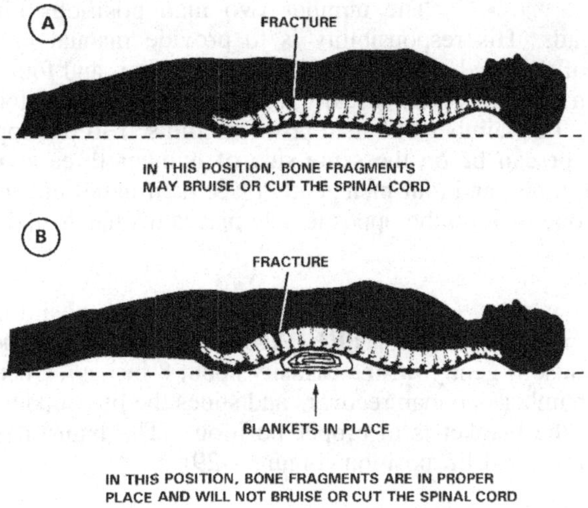

Figure 4-28. Spinal column must maintain a normal spine position.

b. If the casualty must be transported to a safe location before medical personnel arrive and if the casualty is in a—

- Face-up position, transport him by litter or use a firm substitute, such as a wide board or a door longer than his height. Loosely tie the casualty's wrists together over his waistline, using a cravat or a strip of cloth. Tie his feet together to prevent the accidental dropping or shifting of his legs. Lay a folded blanket across the litter where the arch of his back is to be placed. Using a four-man team (Figure 4-29), place the casualty on the litter without bending his spinal column or his neck.

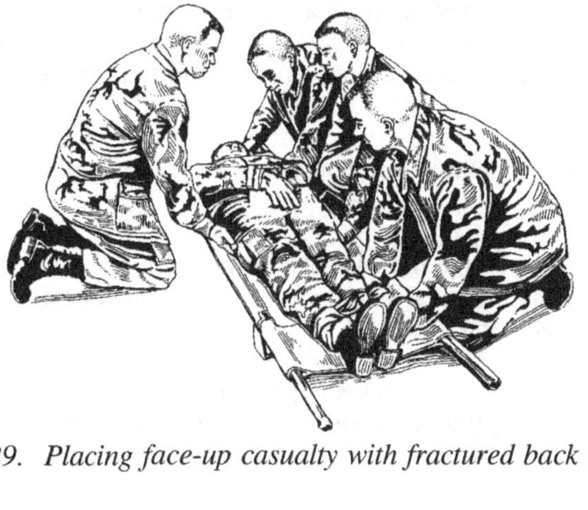

Figure 4-29. Placing face-up casualty with fractured back onto litter.

• The number two man positions himself at the casualty's head. His responsibility is to provide manual in-line (neutral) stabilization of the head and neck. The number three, and four men position themselves on one side of the casualty; all kneel on one knee along the side of the casualty. The number one man positions himself to the opposite side of the casualty (or can be on the same side of number three and four). The numbers two, three, and four men gently place their hands under the casualty. The number one man on the opposite side places his hands under the injured part to assist.

• When all four men are in position to lift, the number two man commands, "**PREPARE TO LIFT**" and then, "**LIFT**." All men, in unison, gently lift the casualty about 8 inches. Once the casualty is lifted, the number one man recovers and slides the litter under the casualty, ensuring that the blanket is in proper position. The number one man then returns to his original lift position (Figure 4-29).

• When the number two man commands, "**LOWER CASUALTY**," all men, in unison, gently lower the casualty onto the litter.

• Facedown position, he must be transported in this same position. The four-man team lifts him onto a regular or improvised litter, keeping the spinal column in a normal spinal position. If a regular litter is used, first place a folded blanket on the litter at the point where the chest will be placed.

4-11. Neck Fractures

A fractured neck is extremely dangerous. Bone fragments may bruise or cut the spinal cord just as they might in a fractured back.

a. If the casualty is not to be transported until medical personnel arrive—

• Caution him not to move. Moving may cause permanent injury or death.

• Leave the casualty in the position in which he is found. If his neck and head (cervical spine) are in an abnormal position, immediately immobilize the neck and head.

• Keep his head still, if the casualty is lying face up, raise his shoulders slightly, and slip a roll of cloth that has the bulk of a bath towel under his neck (Figure 4-31). The roll should be thick enough to arch his neck only slightly, leaving the back of his head on the ground. **DO NOT** bend his neck or head forward. **DO NOT** raise or twist his head. Immobilize the casualty's head (Figure 4-32). Do this by padding heavy objects (such as rocks or his boots filled with dirt, sand, gravel, or rock) and placing them on each side of his head. If it is necessary to use boots, after filling them, tie the top tightly or stuff with pieces of cloth to secure the contents.)

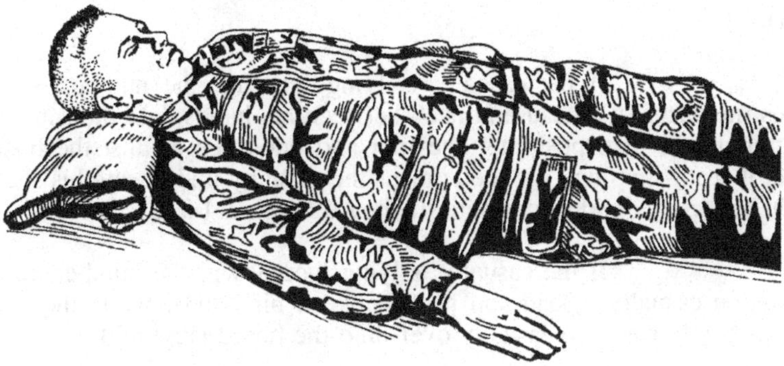

Figure 4-30. Casualty with roll of cloth (bulk) under neck.

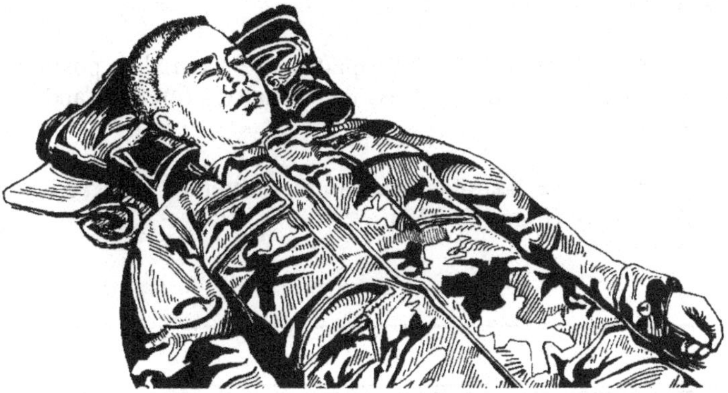

Figure 4-31. Immobilization of fractured neck.

• **DO NOT** move him if the casualty is lying face down. Immobilize the head and neck by padding heavy objects and placing them on each side of his head. DO NOT put a roll of cloth under the neck. **DO NOT** bend the neck or head, nor roll the casualty onto his back.

b. If the casualty must be prepared for transportation before medical personnel arrive—

- If the casualty has a fractured neck, at least two persons are needed because the casualty's head and trunk must be moved in unison. The two persons must work in close coordination (Figure 4-32) to avoid bending of the neck.

- A wide board is placed lengthwise beside the casualty. It should extend at least 4 inches beyond the casualty's head and feet (Figure 4-32A).

- If the casualty is lying face up, the number one man steadies the casualty's head and neck between his hands. At the same time, the number two man positions one foot and one knee against the board to prevent it from slipping. He then grasps the casualty underneath his shoulder and hip and gently slides him onto the board (Figure 4-32B).

- If the casualty is lying face down, the number one man steadies the casualty's head and neck between his hands, while the number two man gently rolls the casualty over onto the board (Figure 4-32C).

- The number one man continues to steady the casualty's head and neck. The number two man simultaneously raises the casualty's shoulders slightly, places padding under his neck, and immobilizes the casualty's head (Figures 4-32D—E).

- Any improvised supports are secured in position with a cravat or strip of cloth extended across the casualty's forehead and under the board (Figure 4-32D).

- The board is lifted onto a litter or blanket in order to transport the casualty (Figure 4-32E).

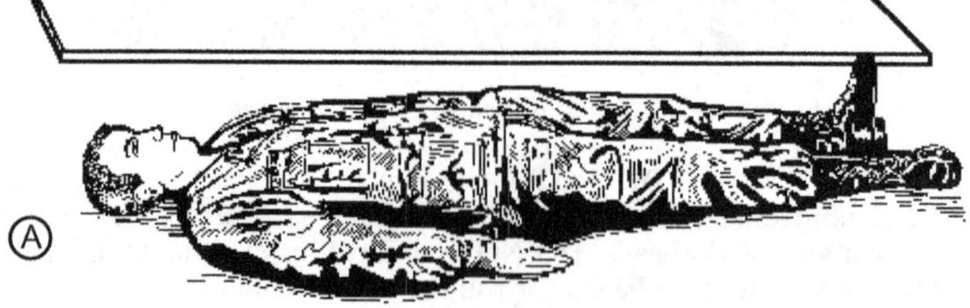

Figure 4-32. Preparing casualty with fractured neck for transportation (Illustrated A—E).

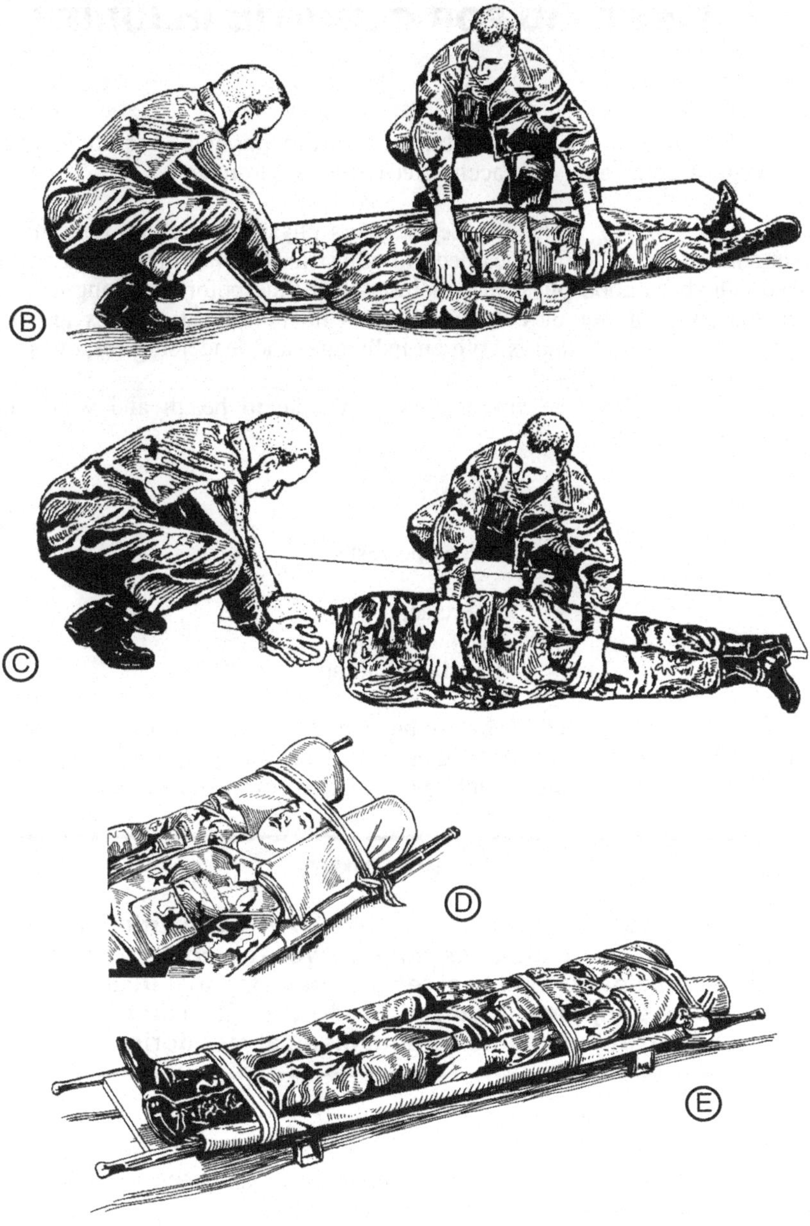

Figure 4-32. Preparing casualty with fractured neck for transportation (Illustrated A—E) (Continued).

CHAPTER 5

FIRST AID FOR CLIMATIC INJURIES

5-1. General

a. It is desirable, but not always possible, for an individual's body to become adjusted (acclimated) to an environment.

(1) The service members physical condition determines the amount of time their bodies need to adjust to the environment. Even those individuals in good physical condition need time before working or training in extremes of hot or cold weather. Climate-related injuries are usually preventable; prevention is both an individual and leadership responsibility.

(2) Several factors contribute to health and well-being in any environment—

- Diet.
- Sleep and rest.
- Exercise.
- Suitable clothing.

(3) Diet should be suited to an individual's needs in a particular climate. A special diet started for any purpose (such as weight reduction) should be done with appropriate medical supervision.

WARNING

Service members should use extreme caution when starting fad diets or taking over-the-counter herbal supplements. Medical records revealed that deaths and severe injuries occurred in individuals using dietary/ herbal supplements without medical monitoring.

NOTE

Weight loss and the use of weight loss supplements should be supervised by a trained health care provider.

(4) Specialized clothing and equipment (such as cold weather gear) for a specific environment should be obtained and used properly.

b. For information on the prevention of heat and cold injuries, refer to FM 21-10/Marine Corps Reference Publication (MCRP) 4-11.1D.

5-2. Heat Injuries

a. Heat injuries are environmental injuries. They may result when a service member—

- Is exposed to extreme heat, such as from the sun or from high temperatures.
- Does not wear proper clothing.
- Is in MOPP gear.
- Is inside closed spaces, such as inside an armored vehicle.
- Wears body armor.

b. Heat injury can be divided into three categories: heat cramps, heat exhaustion, and heatstroke.

c. Each service member must be able to recognize and give first aid for heat injuries.

WARNING

The heat casualty should be continually monitored for development of conditions which may require the performance of necessary basic lifesaving measures.

CAUTION

Do not use salt solutions in first aid procedures for heat injuries.

(1) Check the casualty for signs and symptoms of *cramping*.

- *Signs and symptoms.* Cramping is caused by an imbalance of chemicals (called electrolytes) in the body as a result of excessive sweating. This condition causes the casualty to exhibit:

- Cramping in the extremities (arms and legs).
- Abdominal (stomach) cramps.
- Excessive sweating.

NOTE

Thirst may or may not occur. Cramping can occur without the service member being thirsty.

- *First aid measures.*

- Move the casualty to a cool, shady area or improvise shade if none is available.

- Loosen his clothing (if not in a chemical environment).

NOTE

In a chemical environment, transport the heat casualty to a noncontaminated area as soon as the mission permits.

- Have him slowly drink at least one canteen full of water. (The body absorbs cool water faster than warm or cold water; therefore, cool water is preferred if it is available.)

- Seek medical assistance should cramps continue.

(2) Check the casualty for signs and symptoms of *heat exhaustion*.

- *Signs and symptoms.* Heat exhaustion is caused by loss of body fluids (dehydration) through sweating without adequate fluid replacement. It can occur in an otherwise fit individual who is involved in physical exertion in any hot environment especially if the service member is not acclimatized to that environment. These signs and symptoms are—

- Excessive sweating with pale, moist, cool skin.
- Headache.
- Weakness.

- Dizziness.
- Loss of appetite.
- Cramping.
- Nausea (with or without vomiting).
- Urge to defecate.
- Chills (gooseflesh).
- Rapid breathing.
- Tingling of hands and/or feet.
- Confusion.

- *First aid measures.*

    - Move the casualty to a cool, shady area or improvise shade if none is available.
    - Loosen or remove his clothing and boots (unless in a chemical environment); pour water on him and fan him.
    - Have him slowly drink at least one canteen of water.
    - Elevate his legs.
    - If possible, the casualty should not participate in strenuous activity for the remainder of the day.
    - Monitor the casualty until the symptoms are gone, or medical assistance arrives.

(3) Check the casualty for signs and symptoms of *heatstroke*.

WARNING

Heatstroke is a medical emergency which may result in death if care is delayed.

- *Signs and symptoms.* A service member suffering from heatstroke has been exposed to high temperatures (such as direct sunlight) or been dressed in protective overgarments, which causes the body temperature to rise. Heatstroke occurs more rapidly in service members who are engaged in work or other physical activity in a high heat environment. Heatstroke is caused by a failure of the body's cooling mechanism which includes a decrease in the body's ability to produce sweat. The casualty's skin is red (flushed), hot, and dry. He may experience weakness, dizziness, confusion, headaches, seizures, nausea, stomach pains or cramps, and his respiration and pulse may be rapid and weak. Unconsciousness and collapse may occur suddenly.

- *First aid measures.* Cool casualty immediately by—

  - Moving him to a cool, shady area or improvising shade if none is available.

  - Loosening or removing his clothing (except in a chemical environment).

  - Spraying or pouring water on him; fanning him to permit the coolant effect of evaporation.

  - Massaging his extremities and skin, which increases the blood flow to those body areas, thus aiding the cooling process.

  - Elevating his legs.

  - Having him slowly drink at least one canteen full of water if he is conscious.

NOTE

Start cooling casualty immediately. Continue cooling while awaiting transportation and during transport to an MTF.

- *Medical assistance.* Seek medical assistance because the casualty should be transported to an MTF as soon as possible. Do not interrupt the cooling process or lifesaving measures to seek help; if someone else is present send them for help. The casualty should be continually monitored for development of conditions that may require the performance of necessary basic lifesaving measures.

d. *Table.* See Table 5-1 for further information.

Table 5-1. Heat Injuries

INJURIES	SIGNS AND SYMPTOMS	FIRST AID[1]
HEAT CRAMPS	THE CASUALTY EXPERIENCES MUSCLE CRAMPS OF THE ARMS, LEGS, AND/OR STOMACH. THE CASUALTY MAY ALSO HAVE EXCESSIVE SWEATING.	1. MOVE THE CASUALTY TO A COOL SHADY AREA OR IMPROVISE SHADE AND LOOSEN CLOTHING.[2] 2. HAVE HIM SLOWLY DRINK AT LEAST ONE CANTEEN FULL OF COOL WATER SLOWLY. 3. MONITOR THE CASUALTY AND GIVE HIM MORE WATER AS TOLERATED.
HEAT EXHAUSTION	THE CASUALTY EXPERIENCES HEAVY SWEATING WITH PALE, MOIST, COOL SKIN; HEADACHE, WEAKNESS, DIZZINESS, AND/OR LOSS OF APPETITE, HEAT CRAMPS, NAUSEA (WITH OR WITHOUT VOMITING), URGE TO DEFECATE, CHILLS (GOOSE-FLESH), RAPID BREATHING, CONFUSION, AND TINGLING OF THE HANDS AND/OR FEET.	1. MOVE THE CASUALTY TO A COOL, SHADY AREA OR IMPROVISE SHADE AND LOOSEN OR REMOVE HIS CLOTHING.[2] 2. POUR WATER ON HIM AND FAN HIM TO PERMIT THE COOLANT EFFECT OF EVAPORATION. 3. HAVE HIM SLOWLY DRINK AT LEAST ONE CANTEEN FULL OF COOL WATER. 4. ELEVATE THE CASUALTY'S LEGS. 5. SEEK MEDICAL ASSISTANCE IF SYMPTOMS CONTINUE; MONITOR UNTIL SYMPTOMS ARE GONE OR MEDICAL ASSISTANCE ARRIVES.
HEATSTROKE[3] (SUNSTROKE)	THE CASUALTY STOPS SWEATING (RED [FLUSHED] HOT, DRY SKIN). HE FIRST MAY EXPERIENCE HEADACHE, DIZZINESS, NAUSEA, FAST PULSE AND RESPIRATION, SEIZURES, AND MENTAL CONFUSION. HE MAY COLLAPSE	1. MOVE THE CASUALTY TO A COOL, SHADY AREA OR IMPROVISE SHADE AND LOOSEN OR REMOVE HIS CLOTHING, REMOVE THE OUTER GARMENTS AND PROTECTIVE

Table 5-1. Heat Injuries (Continued)

INJURIES	SIGNS AND SYMPTOMS	FIRST AID[1]
	AND SUDDENLY BECOME UNCONSCIOUS. **THIS IS A MEDICAL EMERGENCY.**	CLOTHING IF THE SITUATION PERMITS.[2] 2. START COOLING THE CASUALTY IMMEDIATELY. SPRAY OR POUR WATER ON HIM. FAN HIM. MASSAGE HIS EXTREMITIES AND SKIN. 3. ELEVATE HIS LEGS. 4. IF CONSCIOUS, HAVE HIM SLOWLY DRINK AT LEAST ONE CANTEEN FULL OF COOL WATER. 5. SEEK MEDICAL AID. CONTINUE COOLING WHILE AWAITING TRANSPORT AND CONTINUE FIRST AID WHILE EN ROUTE.

LEGEND:

1 THE FIRST AID PROCEDURE FOR HEAT RELATED INJURIES CAUSED BY WEARING INDIVIDUAL PROTECTIVE EQUIPMENT (IPE) IS TO MOVE THE CASUALTY TO A CLEAN AREA AND GIVE HIM WATER TO DRINK.
2 WHEN IN A CHEMICAL ENVIRONMENT, DO NOT LOOSEN OR REMOVE THE CASUALTY'S CLOTHING.
3 CAN BE FATAL IF NOT PROVIDED FIRST AID AND MEDICAL TREATMENT PROMPTLY.

5-3. Cold Injuries

Cold injuries are most likely to occur when conditions are moderately cold, but accompanied by wet or windy conditions. Cold injuries can usually be prevented. Well-disciplined and well-trained service members can be protected even in the most adverse circumstances. They and their leaders must know the hazards of exposure to the cold. They must know the importance of personal hygiene, exercise, care of the feet and hands, and the use of protective clothing.

a. Contributing Factors.

(1) Temperature, humidity, precipitation, and wind greatly increase likelihood of cold injuries, and the service members with wet clothing are at great risk of cold injuries. Riverine operations (river, swamp, and stream crossings) increase likelihood of cold injuries. Low temperatures and low relative humidity (dry cold) promote frostbite. Higher temperatures, together with moisture, promote immersion syndrome. Windchill accelerates the loss of body heat and may aggravate cold injuries.

(2) Relatively stationary activities such as being in an observation post or on guard duty increase the service member's vulnerability to cold injury. Also, a service member is more likely to receive a cold injury if he is—

- In contact with the ground (such as marching, performing guard duty, or engaging in other outside activities).
- Immobile for long periods (such as while riding in an unheated or open vehicle).
- Standing in water, such as in a foxhole.
- Out in the cold for days without being warmed.
- Deprived of an adequate diet and rest.
- Not able to take care of his personal hygiene.

(3) Physical fatigue contributes to apathy, which leads to inactivity, personal neglect, carelessness, and reduced heat production. In turn, these increase the risk of cold injury. Service members with prior cold injuries have a higher-than-normal risk of subsequent cold injury; not necessarily involving the body part previously injured.

(4) Depressed or unresponsive service members are also vulnerable because they are less active. These service members tend to be careless about precautionary measures, especially warming activities, when cold injury is a threat.

(5) Excessive use of alcohol or drugs leading to faulty judgment or unconsciousness in a cold environment increases the risk of becoming a cold injury casualty.

b. *Signs and Symptoms*. Once a service member becomes familiar with the factors that contribute to cold injury, he must learn to recognize cold injury signs and symptoms.

(1) Many service members suffer cold injury without realizing what is happening to them. They may be cold and generally uncomfortable. These service members often do not notice the injured part because it is already numb from the cold.

(2) Superficial cold injury usually can be detected by numbness or tingling sensations. These signs and symptoms often can be relieved simply by loosening boots or other clothing and by exercising to improve circulation. In more advanced cases involving deep cold injury, the service member often is not aware that there is a problem until the affected part feels like a stump or block of wood.

(3) Outward signs of cold injury include discoloration of the skin at the site of injury. In light-skinned persons, the skin first reddens and then becomes pale or waxy white. In dark-skinned persons, grayness in the skin is usually evident. An injured foot or hand feels cold to the touch. Swelling may be an indication of deep injury. Also note that blisters may occur after rewarming the affected parts. Service members should work in pairs (buddy teams) to check each other for signs of discoloration and other symptoms.

c. First Aid Measures. First aid for cold injuries depends on whether they are superficial or deep. Rewarming the affected part using body heat can adequately treat cases of superficial cold injury. (For example, this can be done by covering cheeks with hands, putting fingertips in armpits, or placing the casualty's feet under the clothing of a buddy [next to his belly].) The injured part should **NOT** be massaged, exposed to a fire or stove, rubbed with snow, slapped, chafed, or soaked in cold water. Walking on injured feet should be avoided. Deep cold injury (frostbite) is very serious and requires prompt first aid to avoid or to minimize the loss of parts or all of the fingers, toes, hands, or feet. The sequence for treating cold injuries depends on whether the condition is life-threatening. The first priority in managing cold injuries is to remove the casualty from the cold environment (such as building an improvised shelter). Other injuries the casualty may have are provided first aid simultaneously while waiting for transportation or evacuation. If the casualty is to be transported in a nonmedical vehicle, first aid measures should be continued en route to the MTF.

d. Conditions Caused by Cold. Conditions caused by cold include chilblain, immersion syndrome (immersion foot and trench foot), frostbite, snow blindness, dehydration, and hypothermia.

(1) *Chilblain.*

- *Signs and symptoms.* Chilblain is caused by repeated prolonged exposure of bare skin at temperatures from 60° Fahrenheit (F) to

32°F, or 20°F for acclimated, dry, unwashed skin. The area may be acutely swollen, red, tender, and hot with itchy skin. There may be no loss of skin tissue in untreated cases but continued exposure may lead to infected, ulcerated, or bleeding lesions.

- *First aid measures.* Within minutes, the area usually responds to locally applied body heat. Rewarm the affected part by applying firm steady pressure with your hands, or placing the affected part under your arms or against the stomach of a buddy. **DO NOT** rub or massage affected areas.

NOTE

Medical personnel should evaluate the injury, because signs and symptoms of tissue damage may be slow to appear.

(2) *Immersion syndrome (immersion foot and trench foot).* Immersion foot and trench foot are injuries that result from fairly long exposure of the feet to wet conditions at temperatures from approximately 32°F to 50°F. Inactive feet in damp or wet socks and boots, or tightly laced boots which impair circulation, are even more susceptible to injury. This injury can be very serious; it can lead to loss of toes or parts of the feet. If exposure of the feet has been prolonged and severe, the feet may swell so much that pressure closes the blood vessels and cuts off circulation. Should an immersion injury occur, dry the feet thoroughly and transport the casualty to an MTF by the fastest means possible.

- *Signs and symptoms.* At first, the parts of the affected foot are cold and painless, the pulse is weak, and numbness may be present. Second, the parts may feel hot, and burning and shooting pains may begin. In later stages, the skin is pale with a bluish cast and the pulse decreases. Other signs and symptoms that may follow are blistering, swelling, redness, heat, hemorrhaging (bleeding), and gangrene.

- *First aid measures.* First aid measures are required for all stages of immersion syndrome injury. Rewarm the injured part gradually by exposing it to warm air. Protect it from trauma and secondary infections. Dry, loose clothing or several layers of warm coverings are preferable to extreme heat. Under no circumstances should the injured part be exposed to an open fire. Elevate the injured part to relieve the swelling. Transport the casualty to an MTF as soon as possible. When the part is rewarmed, the casualty often feels a burning sensation and pain. Symptoms may persist for days or weeks even after rewarming.

NOTE

When providing first aid for immersion foot and trench foot—**DO NOT** massage the injured part. **DO NOT** moisten the skin. **DO NOT** apply heat or ice.

(3) *Frostbite*. Frostbite is the injury of tissue caused from exposure to cold, usually below 32°F depending on the windchill factor, duration of exposure, and adequacy of protection. Individuals with a history of cold injury are likely to suffer an additional cold injury. The body parts most easily frostbitten are the cheeks, nose, ears, chin, forehead, wrists, hands, and feet. Frostbite may involve only the skin (superficial), or it may extend to a depth below the skin (deep). Deep frostbite is very serious and requires prompt first aid to avoid or to minimize the loss of parts or all of the fingers, toes, hands, or feet.

- *Signs and symptoms*.
  - Loss of sensation (numb feeling) in any part of the body.
  - Sudden blanching (whitening) of the skin of the affected part, followed by a momentary tingling sensation.
  - Redness of skin in light-skinned service members; grayish coloring in dark-skinned service members.
  - Blisters.
  - Swelling or tender areas.
  - Loss of previous sensation of pain in affected area.
  - Pale, yellowish, waxy-looking skin.
  - Frozen tissue that feels solid (or wooden) to the touch.

CAUTION

Deep frostbite is a very serious injury and requires immediate first aid and subsequent medical treatment to avoid or minimize loss of body parts.

- *First aid measures.*

  - *Face, ears, and nose.* Cover the casualty's affected area with his and/or your bare hands until sensation and color return.

  - *Hands.* Open the casualty's field jacket and shirt. (In a chemical environment, do not loosen or remove the clothing and protective overgarments.) Place the affected hands under the casualty's armpits. Close the field jacket and shirt to prevent additional exposure.

  - *Feet.* Remove the casualty's boots and socks if he does not need to walk any further to receive additional treatment. (Thawing the casualty's feet and forcing him to walk on them will cause additional pain and injury.) Place the affected feet under clothing and against the body of another service member.

WARNING

DO NOT attempt to thaw the casualty's feet or other frozen areas if he will be required to walk or travel to an MTF for additional medical treatment. The possibility of additional injury from walking is less when the feet are frozen than when they are thawed. (However, if possible avoid walking.) Thawing in the field increases the possibilities of infection, gangrene, or other injury.

NOTE

Thawing may occur spontaneously during transportation to the MTF; this cannot be avoided since the body in general must be kept warm.

In all of the above areas, ensure that the casualty is kept warm and that he is covered (to avoid further injury). Seek medical treatment as soon as possible. Reassure the casualty, protect the affected area from further injury by covering it lightly with a blanket or any dry clothing, and seek shelter out of the wind. Remove or loosen constricting clothing (except in a contaminated environment) and increase insulation. Ensure the casualty exercises as much as possible, avoiding trauma to the injured part, and is prepared for pain when thawing occurs. Protect the frostbitten part from additional injury. **DO NOT**—

- Rub the injured part with snow or apply cold water soaks.

- Warm the part by massage or exposure to open fire because the frozen part may be burned due to the lack of feeling.

- Use ointments or other salves.

- Manipulate the part in any way to increase circulation.

- Use alcohol or tobacco because this reduces the body's resistance to cold.

NOTE

Remember, when freezing extends to a depth below the skin, it is a much more serious injury. Extra care is required to reduce or avoid the chances of losing all or part of the toes or feet. This also applies to the fingers and hands.

(4) *Snow blindness.* Snow blindness is the effect that glare from an ice field or snowfield has on the eyes. It is more likely to occur in hazy, cloudy weather than when the sun is shining. Glare from the sun will cause an individual to instinctively protect his eyes. However, in cloudy weather, he may be overconfident and expose his eyes longer than when the threat is more obvious. He may also neglect precautions such as the use of protective eyewear. Waiting until discomfort (pain) is felt before using protective eyewear is dangerous because a deep burn of the eyes may already have occurred.

- *Signs and symptoms.* Symptoms of snow blindness are a sensation of grit in the eyes with pain in and over the eyes, made worse by moving the eyeball. Other signs and symptoms are watering, redness, headache, and increased pain on exposure to light.

- *First aid measures.* First aid measures consist of blindfolding or covering the eyes with a dark cloth which stops painful eye movement. Complete rest is desirable. If further exposure to light is not preventable, the eyes should be protected with dark bandages or the darkest glasses available. Once unprotected exposure to sunlight stops, the condition usually heals in a few days without permanent damage. The casualty should be evacuated to the nearest MTF.

(5) *Dehydration.* Dehydration occurs when the body loses too much fluid. A certain amount of body fluid is lost through normal body processes. A normal daily intake of liquids replaces these losses. When individuals are engaged in any strenuous exercises or activities, fluid is lost

through sweating and this loss creates an imbalance of fluids in the body, and if not matched by rehydration it can contribute to dehydration. The danger of dehydration is as prevalent in cold regions as it is in hot regions. In hot weather, the individual is aware of his body losing fluids through sweat. In cold weather, however, it is extremely difficult to realize that this condition exists since sweating is not as apparent as in a hot environment. The danger of dehydration in cold weather operations is a serious problem. In cold climates, sweat evaporates so rapidly or is absorbed so thoroughly by layers of heavy clothing that it is rarely visible on the skin. Dehydration also occurs during cold weather operations because drinking is inconvenient. Dehydration will weaken or incapacitate a casualty for a few hours, or sometimes, several days. Because rest is an important part of the recovery process, casualties must take care that limited movement during their recuperative period does not enhance the risk of becoming a cold injury casualty.

- *Signs and symptoms*. The symptoms of cold weather dehydration are similar to those encountered in heat exhaustion. The mouth, tongue, and throat become parched and dry, and swallowing becomes difficult. The casualty may have nausea (with or without vomiting) along with extreme dizziness and fainting. The casualty may also feel generally tired and weak and may experience muscle cramps. Focusing the eyes may also become difficult.

- *First aid measures*. The casualty should be kept warm and his clothes should be loosened (if not in a chemical environment) to allow proper circulation. Shelter from wind and cold must be provided. Fluid replacement should begin immediately and the service member transported to an MTF as soon as possible.

(6) *Hypothermia (general cooling)*. When exposed to prolonged cold weather a service member may become both mentally and physically numb, thus neglecting essential tasks or requiring more time and effort to achieve them. Under some conditions (particularly cold water immersion), even a service member in excellent physical condition may die in a matter of minutes. The destructive influence of cold on the body is called *hypothermia*. This means bodies lose heat faster than they can produce it. Hypothermia can occur from exposure to temperatures either above or below freezing, especially from immersion in cold water, wet-cold conditions, or from the effect of wind. Physical exhaustion and insufficient food intake may also increase the risk of hypothermia. General cooling of the entire body to a temperature below 95°F is caused by continued exposure to low or rapidly dropping temperatures, cold moisture, snow, or ice. Fatigue, poor physical condition, dehydration, faulty blood circulation, alcohol or other drug use, trauma, and immersion can cause hypothermia. Remember, cold

may affect the body systems slowly and almost without notice. Service members exposed to low temperatures for extended periods may suffer ill effects even if they are well protected by clothing.

- *Signs and symptoms*. As the body cools, there are several stages of progressive discomfort and impairment. A sign that is noticed immediately is shivering. Shivering is an attempt by the body to generate heat. The pulse is faint or very difficult to detect. People with temperatures around 90°F may be drowsy and mentally slow. Their ability to move may be hampered, stiff, and uncoordinated, but they may be able to function minimally. Their speech may be slurred. As the body temperature drops further, shock becomes evident as the person's eyes assume a glassy state, breathing becomes slow and shallow, and the pulse becomes weaker or absent. The person becomes very stiff and uncoordinated. Unconsciousness may follow quickly. As the body temperature drops even lower, the extremities freeze, and a deep (or core) body temperature (below 85°F) increases the risk of irregular heart action. This irregular heart action or heart standstill can result in sudden death.

- *First aid measures*. Except in cases of the most severe hypothermia (marked by coma or unconsciousness and a weak pulse), first aid measures for hypothermia are directed towards protecting the casualty from further loss of body heat. For the casualty who is conscious, first aid measures are directed at rewarming the body evenly and without delay. Provide heat by using a hot water bottle or field expedient or another service member's body heat.

CAUTION

DO NOT expose the casualty to an open fire, as he may become burned.

NOTE

When using a hot water bottle or field expedient (canteen filled with warm water), the bottle or canteen must be wrapped in cloth prior to placing it next to the casualty. This will reduce the chance of burning the casualty's skin.

Always call or send for help as soon as possible and protect the casualty immediately with dry clothing or a sleeping bag. Then, move him to a warm place. Evaluate other injuries and provide first aid as required. First aid measures can be performed while the casualty is waiting transportation or while he is en route. In the case of an accidental breakthrough into ice water, or other hypothermic accident, strip the casualty of wet clothing immediately and bundle him into a sleeping bag. Rescue breathing should be started at once if the casualty's breathing has stopped or is irregular or shallow. Warm liquids (**NOT HOT**) may be given gradually if the casualty is conscious. **DO NOT** force liquids on an unconscious or semiconscious casualty because he may choke. The casualty should be transported on a litter because the exertion of walking may aggravate circulation problems. Medical personnel should immediately treat any hypothermia casualty. Hypothermia is life threatening until normal body temperature has been restored. The first aid measures for a casualty with severe hypothermia are based upon the following principles: attempt to avoid further heat loss, handle the casualty gently, and transport the casualty as soon as possible to the nearest MTF. If at all possible, the casualty should be evacuated by medical personnel.

WARNING

Rewarming a severely hypothermic casualty is extremely dangerous in the field due to the possibility of such complications as rewarming, shock and disturbances in the rhythm of the heartbeat. These conditions require treatment by medical personnel.

NOTE

Resuscitation of casualties with hypothermic complications is difficult if not impossible to do outside of an MTF setting.

CAUTION

The casualty is unable to generate his own body heat. Therefore, merely placing him in a blanket or sleeping bag is not sufficient.

e. *Table.* See Table 5-2 for further information.

Table 5-2. Injuries Caused by Cold and Wet Conditions

INJURIES	SIGNS/SYMPTOMS	FIRST AID
CHILBLAIN	RED SWOLLEN, HOT, TENDER, ITCHING SKIN. CONTINUED EXPOSURE MAY LEAD TO INFECTED (ULCERATED OR BLEEDING) SKIN LESIONS.	1. AREA USUALLY RESPONDS TO LOCALLY APPLIED REWARMING (BODY HEAT). 2. DO NOT RUB OR MASSAGE AREA. 3. SEEK MEDICAL AID.
IMMERSION SYNDROME (IMMERSION FOOT/TRENCH FOOT)	AFFECTED PARTS ARE COLD, NUMB, AND PAINLESS. PARTS MAY THEN BE HOT, WITH BURNING AND SHOOTING PAINS. ADVANCED STAGE: SKIN PALE WITH BLUISH CAST; PULSE DECREASES; BLISTERING, SWELLING, HEAT, HEMORRHAGING, AND GANGRENE MAY FOLLOW.	1. GRADUAL REWARMING BY EXPOSURE TO WARM AIR. 2. DO NOT MASSAGE OR MOISTEN SKIN. 3. PROTECT AFFECTED PARTS FROM TRAUMA. 4. DRY FEET THOROUGHLY, AVOID WALKING. 5. SEEK MEDICAL AID.
FROSTBITE	LOSS OF SENSATION (NUMB FEELING) IN ANY PART OF THE BODY. SUDDEN BLANCHING (WHITENING) OF THE SKIN OF THE AFFECTED PART, FOLLOWED BY A MOMENTARY TINGLING SENSATION. REDNESS OF SKIN IN LIGHT-SKINNED SERVICE MEMBERS; GRAYISH COLORING IN DARK-SKINNED SERVICE MEMBERS. BLISTERS. SWELLING OR TENDER AREAS. LOSS OF PREVIOUS SENSATION OF PAIN IN THE AFFECTED AREA. PALE YELLOWISH, WAXY-LOOKING SKIN. FROZEN TISSUE THAT FEELS SOLID (WOODEN) TO THE TOUCH.	1. WARM THE AREA AT THE FIRST SIGN OF FROSTBITE, USING FIRM, STEADY PRESSURE OF THE HAND, UNDERARM, OR ABDOMEN. 2. FACE, EARS, NOSE: COVER AREA WITH HANDS (CASUALTY'S OWN OR BUDDY'S). 3. HANDS: OPEN FIELD JACKET AND PLACE CASUALTY'S HANDS AGAINST HIS BODY, THEN CLOSE THE JACKET TO PREVENT HEAT LOSS. 4. FEET: REMOVE THE CASUALTY'S BOOTS AND SOCKS AND PLACE HIS FEET AGAINST THE BODY OF ANOTHER SERVICE MEMBER.

Table 5-2. Injuries Caused by Cold and Wet Conditions (Continued)

INJURIES	SIGNS/SYMPTOMS	FIRST AID
		5. WARNING: DO NOT ATTEMPT TO THAW THE CASUALTY'S FEET OR OTHER FROZEN AREAS IF HE WILL BE REQUIRED TO WALK OR TRAVEL TO AN MTF FOR ADDITIONAL TREATMENT. THE POSSIBILITY OF INJURY FROM WALKING IS LESS WHEN THE FEET ARE FROZEN THAN WHEN THEY HAVE BEEN THAWED. (HOWEVER, IF POSSIBLE AVOID WALKING.) THAWING IN THE FIELD INCREASES THE POSSIBILITY OF INFECTION, GANGRENE, OR INJURY. 6. LOOSEN OR REMOVE CONSTRICTING CLOTHING AND REMOVE ANY JEWELRY. 7. INCREASE INSULATION (COVER WITH BLANKET OR OTHER DRY MATERIAL). ENSURE CASUALTY EXERCISES AS MUCH AS POSSIBLE, AVOIDING TRAUMA TO INJURED PART.
SNOW BLINDNESS	EYES MAY FEEL SCRATCHY. WATERING, REDNESS, HEADACHE, AND INCREASED PAIN WITH EXPOSURE TO LIGHT CAN OCCUR.	1. COVER THE EYES WITH A DARK CLOTH. 2. SEEK MEDICAL AID.
DEHYDRATION	SIMILAR TO HEAT EXHAUSTION (REFER TO TABLE 5-1).	1. KEEP WARM. 2. CASUALTY NEEDS FLUID REPLACEMENT, REST, AND PROMPT MEDICAL AID.
HYPOTHERMIA	CASUALTY IS COLD. SHIVERING. CORE TEMPERATURE IS LOW. CONSCIOUSNESS MAY BE	**MILD HYPOTHERMIA** 1. REWARM BODY

Table 5-2. Injuries Caused by Cold and Wet Conditions (Continued)

INJURIES	SIGNS/SYMPTOMS	FIRST AID
	ALTERED. UNCOORDINATED MOVEMENTS MAY OCCUR. SHOCK AND COMA MAY RESULT AS BODY TEMPERATURE DROPS.	EVENLY AND WITHOUT DELAY. (NEED TO PROVIDE HEAT SOURCE; CASUALTY'S BODY UNABLE TO GENERATE HEAT.) 2. KEEP DRY, PROTECT FROM THE ELEMENTS. 3. WARM (NOT HOT) LIQUIDS MAY BE GIVEN GRADUALLY (TO CONSCIOUS CASUALTIES ONLY). 4. BE PREPARED TO START BASIC LIFE SUPPORT MEASURES FOR THE CASUALTY. 5. SEEK MEDICAL TREATMENT IMMEDIATELY. **SEVERE HYPOTHERMIA** 1. STABILIZE THE TEMPERATURE. 2. ATTEMPT TO AVOID FURTHER HEAT LOSS. 3. HANDLE THE CASUALTY GENTLY. 4. EVACUATE TO THE NEAREST MTF AS SOON AS POSSIBLE. **5. WARNING: HYPOTHERMIA IS A MEDICAL EMERGENCY. PROMPT MEDICAL ATTENTION IS NECESSARY.**

CHAPTER 6

FIRST AID FOR BITES AND STINGS

6-1. General

Snakebites, insect bites, or stings can cause intense pain and/or swelling. If not treated promptly and correctly, they can cause serious illness or death. The severity of a snakebite depends upon: whether the snake is poisonous or nonpoisonous, the type of snake, the location of the bite, and the amount of venom injected. Bites from humans and other animals, such as dogs, cats, bats, raccoons, and rats, can cause severe bruises and infection and tears or lacerations of tissue. Awareness of the potential sources of injuries can reduce or prevent them from occurring. Knowledge and prompt application of first-aid measures can lessen the severity of injuries from bites and stings and keep the service member from becoming a serious casualty.

6-2. Types of Snakes

a. Nonpoisonous Snakes. There are approximately 130 different varieties of nonpoisonous snakes in the United States. They have oval-shaped heads and round eyes. Unlike poisonous snakes, discussed below, nonpoisonous snakes do not have fangs with which to inject venom. Figure 6-1 depicts the characteristics of a nonpoisonous snake.

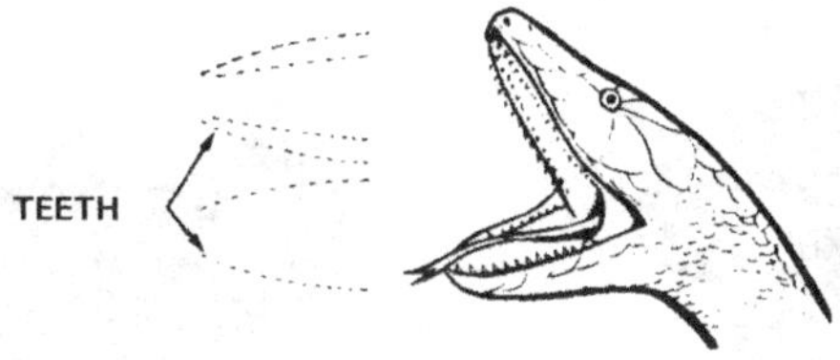

Figure 6-1. Characteristics of nonpoisonous snake.

b. Poisonous Snakes. Poisonous snakes are found throughout the world, primarily in tropical to moderate climates. Within the United States, there are four kinds: rattlesnakes, copperheads, water moccasins (cottonmouth), and coral snakes. Poisonous snakes in other parts of the world include sea snakes, the fer-de-lance, the bushmaster, and the tropical rattlesnake in tropical Central America; the Malayan pit viper in the tropical Far East; the cobra in Africa and Asia; the mamba (or black mamba) in central and southern Africa; and the krait in India and Southeast Asia. Refer to Figure 6-2 for characteristics of a poisonous pit viper.

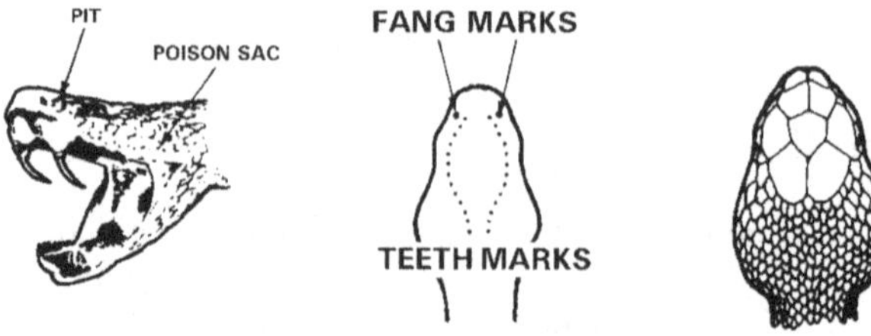

Figure 6-2. Characteristics of poisonous pit viper.

c. Pit Vipers (Poisonous). Figure 6-3 depicts a variety of poisonous snakes.

Figure 6-3. Poisonous snakes.

(1) Rattlesnakes, bushmasters, copperheads, fer-de-lance, Malayan pit vipers, and water moccasins (cottonmouth) are called pit vipers because of the small, deep pits between the nostrils and eyes on each side of the head (Figure 6-2). In addition to their long, hollow fangs, these snakes have other identifying features: thick bodies, slit-like pupils of the eyes, and flat, almost triangular-shaped heads. Color markings and other identifying characteristics, such as rattles or a noticeable white interior of the mouth (cottonmouth), also help distinguish these poisonous snakes. Further

identification is provided by examining the bite pattern of the wound for signs of fang entry. Occasionally there will be only one fang mark, as in the case of a bite on a finger or toe where there is no room for both fangs, or when the snake has broken off a fang.

(2) The casualty's condition provides the best information about the seriousness of the situation, or how much time has passed since the bite occurred. Pit viper bites are characterized by severe burning pain. Discoloration and swelling around the fang marks usually begins within 5 to 10 minutes after the bite. If only minimal swelling occurs within 30 minutes, the bite will almost certainly have been from a nonpoisonous snake or possibly from a poisonous snake which did not inject venom. The venom destroys blood cells, causing a general discoloration of the skin. Blisters and numbness in the affected area follow this reaction. Other signs, which can occur, are weakness, rapid pulse, nausea, shortness of breath, vomiting, and shock.

d. Corals, Cobras, Kraits, and Mambas. Corals (Figure 6-4), cobras (Figure 6-5), kraits, and mambas all belong to the same group even though they are found in different parts of the world. All four inject their venom through short, grooved fangs, leaving a characteristic bite pattern.

Figure 6-4. Coral snake.

(1) The small coral snake, found in the Southeastern US, is brightly colored with bands of red, yellow (or almost white), and black completely encircling the body. Other nonpoisonous snakes have the same coloring, but on the coral snake found in the US, the red ring always touches the yellow ring. To know the difference between a harmless snake and the coral snake found in the United States, remember the following:

"Red on yellow will kill a fellow,
Red on black, venom will lack."

Figure 6-5. Cobra snake.

(2) The venom of corals, cobras, kraits, and mambas produces symptoms different from those of pit vipers. Because there is only minimal pain and swelling, many people believe that the bite is not serious. Delayed reactions in the nervous system normally occur between 1 to 7 hours after the bite. Symptoms include blurred vision, drooping eyelids, slurred speech, drowsiness, and increased salivation and sweating. Nausea, vomiting, shock, respiratory difficulty, paralysis, convulsions, and coma will usually develop if the bite is not treated promptly.

e. Sea Snakes. Sea snakes (Figure 6-6) are found in the warm water areas of the Pacific and Indian oceans, along the coasts, and at the mouths of some larger rivers. Their venom is **VERY** poisonous, but their fangs are only 1/4 inch long. The first aid outlined for land snakes also applies to sea snakes.

Figure 6-6. Sea snake.

6-3. Snakebites

a. Poisonous snakes **DO NOT** always inject venom when they bite or strike a person. However, all snakes may carry tetanus (lockjaw); anyone bitten by a snake, whether poisonous or nonpoisonous, should immediately seek medical attention.

- Poison is injected from the venom sacs through grooved or hollow fangs. Depending on the species, these fangs are either long or short. Pit vipers have long hollow fangs. These fangs are folded against the roof of the mouth and extend when the snake strikes. This allows them to strike quickly and then withdraw. Cobras, coral snakes, kraits, mambas, and sea snakes have short, grooved fangs. These snakes are less effective in their attempts to bite, since they must chew after striking to inject enough venom (poison) to be effective. Figure 6-7 depicts the characteristics of a poisonous snakebite.

- In the event you are bitten, attempt to identify and/or kill the snake. Take it to medical personnel for inspection/identification. This provides valuable information to medical personnel who deal with snakebites. **TREAT ALL SNAKEBITES AS POISONOUS.**

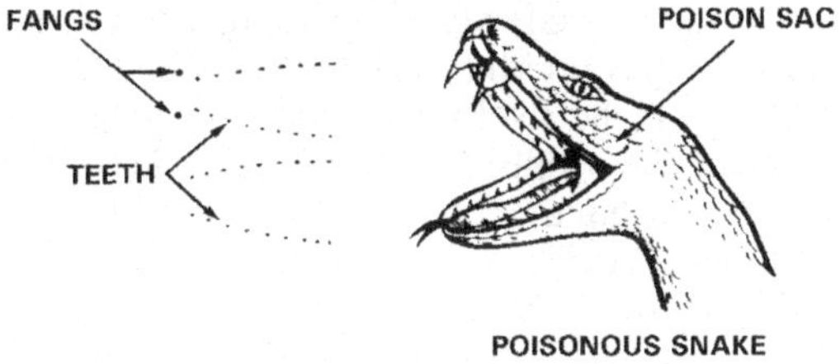

Figure 6-7. Characteristics of poisonous snakebite.

b. The venoms of different snakes cause different effects. Pit viper venom (hemotoxin [blood toxin]) destroys tissue and blood cells. Cobras, adders, and coral snakes inject powerful venom (neurotoxin [nerve toxin]) which affect the central nervous system, causing respiratory paralysis. Water moccasins and sea snakes have venom that is both hemotoxic and neurotoxic.

c. The identification of poisonous snakes is very important since medical treatment will be different for each type of venom. *Unless it can be positively identified, the snake should be killed and saved.* When this is not possible or when doing so is a serious threat to others, identification may sometimes be difficult since many venomous snakes resemble harmless varieties. When dealing with snakebite problems in foreign countries, seek advice, professional or otherwise, which may help identify species in the particular area of operations.

d. Get the casualty to an MTF as soon as possible and with minimum movement. Until evacuation or treatment is possible, have the casualty lie quietly and not move any more than necessary. If the casualty has been bitten on an extremity, **DO NOT** elevate the limb; keep the extremity level with the body. Keep the casualty comfortable and reassure him. If the casualty is alone when bitten, he should go to the medical facility himself rather than wait for someone to find him. Unless the snake has been positively identified, attempt to kill it and send it with the casualty. Be sure that retrieving the snake does not endanger anyone or delay transporting the casualty.

(1) If the bite is on an arm or leg, place a constricting band (narrow cravat [swathe], or narrow gauze bandage) one to two fingerbreadths above and below the bite (Figure 6-8). If the bite is on the hand or foot, place a single band above the wrist or ankle. The band should be tight enough to stop the flow of blood near the skin, but not tight enough to interfere with circulation. In other words, it should not have a tourniquet-like affect. If no swelling is seen, place the bands about 1 inch from either side of the bite. If swelling is present, put the bands on the unswollen part at the edge of the swelling. If the swelling extends beyond the band, move the band to the new edge of the swelling. (If possible, leave the old band on, place a new one at the new edge of the swelling, and then remove and save the old one in case the process has to be repeated.)

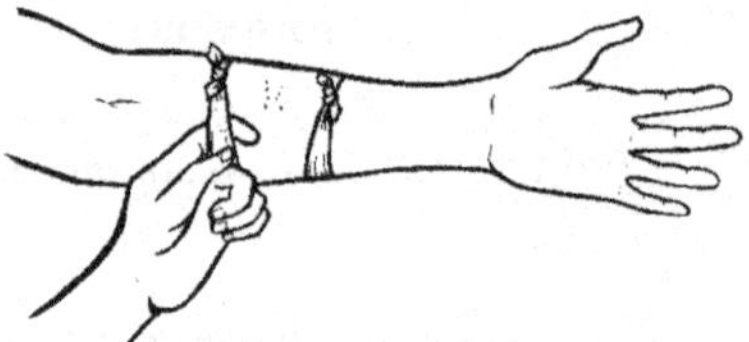

Figure 6-8. Constricting band.

CAUTION

DO NOT attempt to cut open the bite nor suck out the venom. If the venom should seep through any damaged or lacerated tissues in your mouth, you could immediately lose consciousness or even die.

(2) If the bite is located on an arm or leg, immobilize it at a level below the heart. **DO NOT** elevate an arm or leg even with or above the level of the heart.

> **CAUTION**
>
> When a splint is used to immobilize the arm or leg, take **EXTREME** care to ensure the splinting is done properly and does not bind. Watch it closely and adjust it if any changes in swelling occur.

(3) When possible, clean the area of the bite with soap and water. **DO NOT** use ointments of any kind.

(4) **NEVER** give the casualty food, alcohol, stimulants (coffee or tea), drugs, or tobacco.

(5) Remove rings, watches, or other jewelry from the affected limb.

6-4. Human or Animal Bites

Human or other land animal bites may cause lacerations or bruises. In addition to damaging tissue, bites always present the possibility of infection.

a. *Human Bites.* Human bites that break the skin may become seriously infected since the mouth is heavily contaminated with bacteria. Medical personnel **MUST** treat all human bites.

b. *Animal Bites.* Land animal bites can result in both infection and disease. Tetanus, rabies, and various types of fevers can follow an untreated animal bite. Because of these possible complications, the animal causing the bite should, if possible, be captured or killed (without damaging its head) so that it can be tested for disease.

c. *First Aid.*

(1) Cleanse the wound thoroughly with soap.

(2) Flush it well with water.

(3) Cover it with a sterile dressing.

(4) Immobilize the injured arm or leg, if appropriate.

(5) Transport the casualty immediately to an MTF.

NOTE

If unable to capture or kill the animal, provide medical personnel with any information that will help identify it.

6-5. Marine (Sea) Animals

With the exception of sharks and barracuda, most marine animals will not deliberately attack. The most frequent injuries from marine animals are wounds by biting, stinging, or puncturing. Wounds inflicted by marine animals can be very painful, but are rarely fatal.

a. Sharks, Barracuda, and Alligators. Wounds from these marine animals can involve major trauma as a result of bites and lacerations. Bites from large marine animals are potentially the most life threatening of all injuries from marine animals. Major wounds from these animals can be treated by controlling the bleeding, preventing shock, giving basic life support, splinting the injury, and by securing prompt medical aid.

b. Turtles, Moray Eels, and Corals. These animals normally inflict minor wounds. Treat by cleansing the wound(s) thoroughly and by splinting if necessary.

c. Jellyfish, Portuguese Man-of-War, Anemones, and Others. This group of marine animals inflict injury by means of stinging cells in their tentacles. Contact with the tentacles produces burning pain with a rash and small hemorrhages on the skin. Shock, muscular cramping, nausea, vomiting, and respiratory distress may also occur. Gently remove the clinging tentacles with a towel and wash or treat the area. Use diluted ammonia or alcohol, meat tenderizer, and talcum powder. If symptoms become severe or persist, seek medical assistance.

d. Spiny Fish, Urchins, Stingrays, and Cone Shells. These animals inject their venom by puncturing the skin with their spines. General signs and symptoms include swelling, nausea, vomiting, generalized cramps, diarrhea, muscular paralysis, and shock. Deaths are rare. Treatment consists of soaking the wounds in hot water (when available) for 30 to 60 minutes. This inactivates the heat sensitive toxin. In addition, further first aid measures (controlling bleeding, applying a dressing, and so forth) should be carried out as necessary.

CAUTION

Be careful not to scald the casualty with water that is too hot because the pain of the wound will mask the normal reaction to heat.

6-6. Insect (Arthropod) Bites and Stings

An insect bite or sting can cause great pain, allergic reaction, inflammation, and infection. If not treated correctly, some bites/stings may cause serious illness or even death. When an allergic reaction is not involved, first aid is a simple process. In any case, medical personnel should examine the casualty at the earliest possible time. It is important to properly identify the spider, bee, or creature that caused the bite/sting, especially in cases of allergic reaction.

a. Types of Insects. The insects found throughout the world that can produce a bite or sting are too numerous to mention in detail. Commonly encountered stinging or biting insects include brown recluse spiders (Figure 6-9), black widow spiders (Figure 6-10), tarantulas (Figure 6-11), scorpions (Figure 6-12), urticating caterpillars, bees, wasps, centipedes, conenose beetles (kissing bugs), ants, and wheel bugs. Upon being reassigned, especially to overseas areas, take the time to become acquainted with the types of insects to avoid.

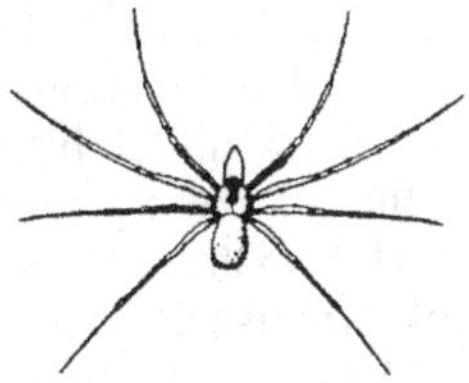

Figure 6-9. Brown recluse spider.

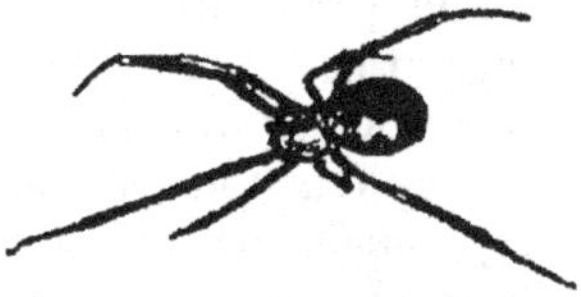

Figure 6-10. Black widow spider.

Figure 6-11. Tarantula.

Figure 6-12. Scorpion.

b. Signs and Symptoms. Discussed in paragraphs (1) and (2) below are the most common effects of insect bites/stings. They can occur alone or in combination with the others.

(1) *Less serious.* Commonly seen signs/symptoms are pain, irritation, swelling, heat, redness, and itching. Hives or wheals (raised areas of the skin that itch) may occur. These are the least severe of the allergic reactions that commonly occur from insect bites/stings. They are usually dangerous only if they affect the air passages (mouth, throat, nose, and so forth), which could interfere with breathing. The bites/stings of bees, wasps, ants, mosquitoes, fleas, and ticks are usually not serious and normally produce mild and localized symptoms. A tarantula's bite is usually no worse than that of a bee sting. Scorpions are rare and their stings (except for a specific species found only in the Southwest desert) are painful but usually not dangerous.

(2) *Serious.* Emergency allergic or hypersensitive reactions sometimes result from the stings of bees, wasps, and ants. Many people are allergic to the venom of these particular insects. Bites or stings from these insects may produce more serious reactions, to include generalized itching and hives, weakness, anxiety, headache, breathing difficulties, nausea, vomiting, and diarrhea. Very serious allergic reactions (called *anaphylactic shock*) can lead to complete collapse, shock, and even death. Spider bites (particularly from the black widow and brown recluse spiders) can also be

serious. Venom from the black widow spider affects the nervous system. This venom can cause muscle cramps, a rigid, nontender abdomen, breathing difficulties, sweating, nausea, and vomiting. The brown recluse spider generally produces local rather than system-wide problems; however, local tissue damage around the bite can be severe and can lead to an ulcer and even gangrene.

c. *First Aid.* There are certain principles that apply regardless of what caused the bite/sting. Some of these are—

- If there is a stinger present (for example, from a bee), remove the stinger by scraping the skin's surface with a fingernail or knife. **DO NOT** squeeze the sac attached to the stinger because it may inject more venom.

- Wash the area of the bite/sting with soap and water (alcohol or an antiseptic may also be used) to help reduce the chances of an infection and remove traces of venom.

- Remove jewelry from bitten extremities because swelling may occur.

- In most cases of insect bites the reaction will be mild and localized; use ice or cold compresses (if available) on the site of the bite/sting. This will help reduce swelling, ease the pain, and slow the absorption of venom. Meat tenderizer (to neutralize the venom) or calamine lotion (to reduce itching) may be applied locally. If necessary, seek medical assistance.

- In more serious reactions (severe and rapid swelling, allergic symptoms, and so forth) treat the bite/sting like you would treat a snakebite; that is, apply constricting bands above and below the site.

- Be prepared to perform basic life-support measures, such as rescue breathing.

- Reassure the casualty and keep him calm.

- In serious reactions, attempt to capture the insect for positive identification; however, be careful not to become a casualty yourself.

- If the reaction to the bite/sting appears serious, seek medical assistance.

WARNING

Insect bites/stings may cause *anaphylactic shock* (a shock caused by a severe allergic reaction). This is a life-threatening event and a TRUE MEDICAL EMERGENCY. Be prepared to perform the basic life-support measures and to immediately transport the casualty to an MTF.

NOTE

Be aware that some allergic or hypersensitive individuals may carry identification or emergency insect bite treatment kits. If the casualty is having an allergic reaction and has such a kit, administer the medication in the kit according to the instructions which accompany the kit.

d. Supplemental Information. For additional information concerning biting insects, see FM 21-10.

6-7. First Aid for Bites and Stings

See the table below for information on bites and stings.

Table 6-1. First Aid Measures for Bites and Stings

TYPES	FIRST AID MEASURES
SNAKEBITE	1. MOVE CASUALTY AWAY FROM THE SNAKE. 2. REMOVE JEWELRY FROM THE AFFECTED AREA, IF APPLICABLE. 3. REASSURE CASUALTY AND KEEP HIM QUIET. 4. APPLY CONSTRICTING BAND, 1-2 FINGERBREADTHS FROM THE BITE. YOU SHOULD BE ABLE TO INSERT A FINGER BETWEEN THE BAND AND THE SKIN. a. *ARM OR LEG BITE.* PLACE ONE BAND ABOVE AND ONE BAND BELOW THE BITE SITE. b. *HAND OR FOOT BITE.* PLACE ONE BAND ABOVE THE WRIST OR ANKLE.

Table 6-1. First Aid Measures for Bites and Stings

TYPES	FIRST AID MEASURES
	5. IMMOBILIZE THE AFFECTED PART IN A POSITION BELOW THE LEVEL OF THE HEART. 6. KILL THE SNAKE (IF POSSIBLE, WITHOUT DAMAGING ITS HEAD OR ENDANGERING YOURSELF) AND SEND IT WITH THE CASUALTY. 7. SEEK MEDICAL ASSISTANCE IMMEDIATELY.
BROWN RECLUSE SPIDER OR BLACK WIDOW SPIDER BITE	1. KEEP CASUALTY QUIET. 2. REMOVE ALL JEWELRY FROM AFFECTED PART, IF APPLICABLE. 3. WASH THE AREA. 4. APPLY ICE OR FREEZE PACK, IF AVAILABLE. 5. SEEK MEDICAL ASSISTANCE.
TARANTULA BITE OR SCORPION STING OR ANT BITE	1. WASH THE AREA. 2. REMOVE ALL JEWELRY FROM AFFECTED PART, IF APPLICABLE. 3. APPLY ICE OR FREEZE PACK, IF AVAILABLE. 4. APPLY BAKING SODA, CALAMINE LOTION, OR MEAT TENDERIZER (IF AVAILABLE) TO BITE SITE TO RELIEVE PAIN AND ITCHING. 5. IF THE SITE OF THE BITE IS ON THE FACE, NECK (POSSIBLE AIRWAY PROBLEMS), OR GENITAL AREA, OR IF LOCAL REACTION SEEMS SEVERE, OR IF THE STING IS BY THE DANGEROUS TYPE OF SCORPION FOUND IN THE SOUTHWEST UNITED STATES DESERT, KEEP THE CASUALTY AS QUIET AS POSSIBLE. SEEK MEDICAL ASSISTANCE.
BEE STING	1. IF THE STINGER IS PRESENT, REMOVE BY SCRAPING WITH A KNIFE OR FINGERNAIL. DO NOT SQUEEZE VENOM SAC ON STINGER; MORE VENOM MAY BE INJECTED. 2. REMOVE ALL JEWELRY FROM AFFECTED PART, IF APPLICABLE.

Table 6-1. First Aid Measures for Bites and Stings

TYPES	FIRST AID MEASURES
	3. WASH THE AREA. 4. APPLY ICE OR FREEZE PACK, IF AVAILABLE. 5. IF ALLERGIC SIGNS OR SYMPTOMS APPEAR, BE PREPARED TO PERFORM BASIC LIFE SUPPORT MEASURES. SEEK IMMEDIATE MEDICAL ASSISTANCE.

CHAPTER 7

FIRST AID IN A NUCLEAR, BIOLOGICAL, AND CHEMICAL ENVIRONMENT

7-1. General

American forces have not been exposed to NBC weapons/agents on the battlefield since World War I. In future conflicts and wars we can expect the use of such agents. Nuclear, biological, and chemical weapons will rapidly degrade unit effectiveness by forcing troops to wear protective clothing and by creating confusion and fear. Through training in protective procedures and first aid, units can maintain their effectiveness on the integrated battlefield.

7-2. First Aid Materials

You may be issued the following materials to protect, decontaminate, and use as first aid for NBC exposure. You must know how to use the items; some items are described in *a* through *d* below. It is equally important that you know when to use them.

a. Nerve Agent Pyridostigmine Pretreatment (*NAPP*). You may be issued a blister pack of pretreatment tablets when your commander directs. The NAPP is a pretreatment; it is not an antidote. It improves the effectiveness of the nerve agent antidote. When ordered to take the pretreatment you must take one tablet every 8 hours, mission permitting. This must be taken prior to exposure to nerve agents, since it may take several hours to develop adequate blood levels.

NOTE

Commanders must follow investigational new drug protocols for use of the NAPP.

b. M291 Skin Decontaminating Kit. The M291 Skin Decontaminating Kit (Figure 7-1) contains six packets of XE-555 decontaminant resin.

WARNING

For external use only. May be slightly irritating to the eyes. Keep decontaminating powder out of eyes. Use water to wash toxic agent out of eyes.

c. Nerve Agent Antidote Kit, MARK I. Each service member is issued three MARK Is for use in first aid for nerve agent poisoning (Figure 7-2 and paragraph 7-6).

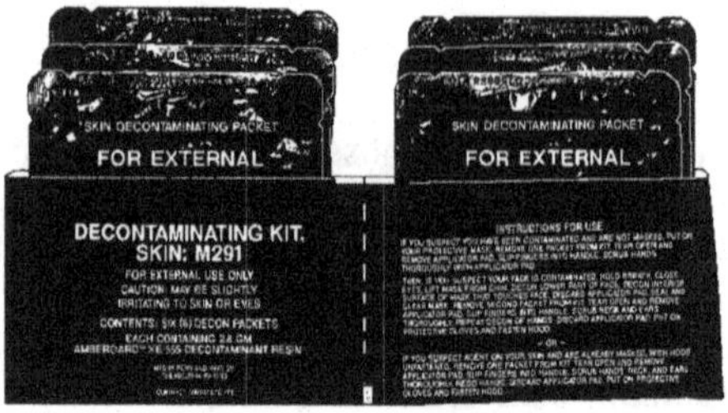

Figure 7-1. M291 Skin Decontamination Kit.

d. Antidote Treatment, Nerve Agent, Autoinjector. A new nerve agent antidote injection device, Antidote Treatment, Nerve Agent, Autoinjector (ATNAA) is currently under development that will replace the MARK I. The ATNAA is a multichambered device with the atropine and pralidoxime chloride in separate chambers. Both antidotes will be administered through a single needle.

7-3. Classification of Chemical and Biological Agents

a. Chemical agents are classified according to the primary physiological effects they produce, such as blistering, choking, vomiting, and incapacitating agents.

b. Biological warfare agents are classified according to the effect they have on man. The effects include their ability to incapacitate and cause death. Most biological warfare agents are delivered as aerosols that effect the respiratory tract; some can be delivered by releasing infected insects, by contaminating food and water, and by injection (injecting material in individuals by terrorist, not mass exposure). These agents are found in living organisms such as fungi, bacteria, and viruses.

WARNING

Swallowing water or food contaminated with nerve, blister, and other chemical agents and with some biological agents can be fatal. NEVER consume water or food that is suspected of being contaminated until it has been tested and found safe for consumption by medical personnel.

7-4. Conditions for Masking Without Order or Alarm

a. Once an attack with a chemical or biological agent is detected or suspected, or information is available that such an agent is about to be used, you must **STOP BREATHING** and mask immediately. **DO NOT WAIT** to receive an order or alarm under the following circumstances:

- Your position is hit by artillery missiles, rockets that produce vapors, smoke, and mists, and aerial sprays.
- Smoke or vapor cloud from an unknown source is present or approaching.
- A suspicious odor, liquid, or solid is present.
- A chemical or biological warfare agent attack is occurring.
- You are entering an area known or suspected of being contaminated.
- When casualties are being received from an area where chemical or biological agents have reportedly been used.
- You have one or more of the following symptoms:
  - An unexplained runny nose.
  - A sudden unexplained headache.
  - A feeling of choking or tightness in the chest or throat.
  - Dimness of vision.
  - Irritation of the eyes.
  - Difficulty in or increased rate of breathing without obvious reasons.
  - Sudden feeling of depression.
  - Dread, anxiety, or restlessness.
  - Dizziness or light-headedness.

- Slurred speech.
- Unexplained laughter or unusual behavior is noted in others.
- Numerous unexplained ill personnel.
- Service members suddenly collapsing without evident cause.
- Animals or birds exhibiting unusual behavior or suddenly dying.

b. For further information on protection and masking procedures, refer to FM 3-4, FM 4-02.7, FM 8-284, and FM 8-285.

7-5. First Aid for a Chemical Attack

Your field protective mask gives protection against biological and chemical warfare agents as well as radiological fallout. With practice you can mask in 9 seconds or less, or put on your mask with hood within 15 seconds.

a. Stop breathing. Don your mask, seal it properly, and clear and check it; then resume breathing. Give the alarm, and continue the mission. Keep your mask on until the "all clear" signal has been given.

NOTE

Keep your mask on until the area is no longer hazardous and you are told to unmask.

b. If symptoms of nerve agent poisoning (paragraph 7-7) appear, immediately give yourself one MARK I or ATNAA.

CAUTION

Do not inject a nerve agent antidote until you are sure you need it.

c. If your eyes and face become contaminated, you must immediately try to get under cover. You need shelter to prevent further contamination while performing decontamination procedures on your face. If no overhead cover is available, put your poncho over your head before beginning the decontamination process. Then you put on the remaining protective clothing. If vomiting occurs, the mask should be lifted momentarily and drained—with your eyes closed and while holding your breath—then replaced, cleared, and sealed.

d. If nerve agents are used, mission permitting, watch for persons needing nerve agent antidotes and immediately follow procedures outlined in paragraph 7-8*b* or *c*.

e. Decontaminate your skin immediately and clothing and equipment as soon as the mission permits.

7-6. Background Information on Nerve Agents

a. Nerve agents are among the deadliest of chemical agents. Nerve agents enter the body by inhalation, by ingestion, and through the skin. Depending on the route of entry and the amount, nerve agents can produce injury or death within minutes. Nerve agents can achieve their effects with small amounts. Nerve agents are absorbed rapidly, and the effects are felt immediately upon entry into the body. You will be issued three MARK Is or three ATNAAs and one Convulsant Antidote for Nerve Agent (CANA). Each MARK I consists of one atropine autoinjector and one pralidoxime chloride (2 PAM Cl) autoinjector (Figure 7-2A). Each ATNAA consist of a multichambered autoinjector with the atropine and pralidoxime chloride in separate chambers (Figure 7-2C). The CANA is a single autoinjector with flanges (Figure 7-2B). Procedures for use of both the MARK I and ATNAA are described below. You will use either the MARK I or the ATNAA in self-aid and buddy aid as issued.

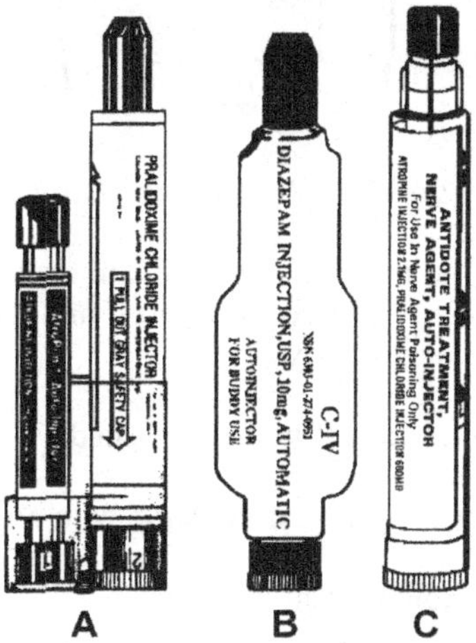

Figure 7-2. Nerve Agent Antidote Kit, MARK I, CANA, and ATNAA.

b. When you have the signs and symptoms of nerve agent poisoning, you should immediately put on the protective mask and then inject yourself with one set of the MARK I or ATNAA. Do not administer the CANA. You should inject yourself in the outer (lateral) thigh muscle (Figure 7-3) or if you are thin, in the upper outer (lateral) part of the buttocks (Figure 7-4).

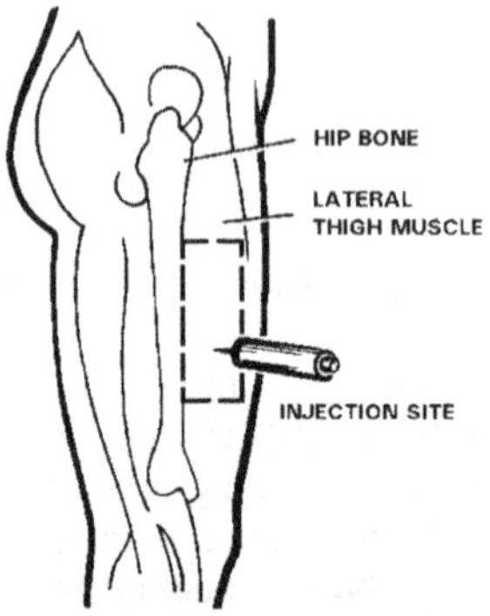

Figure 7-3. Thigh injection site.

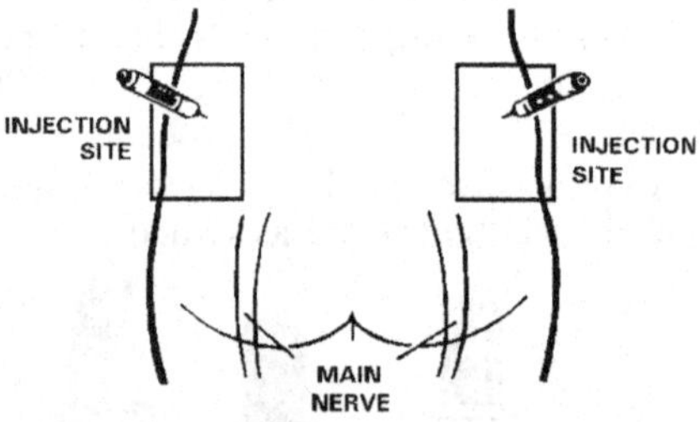

Figure 7-4. Buttocks injection site.

c. Also, you may come upon an unconscious chemical agent casualty who will be unable to care for himself and who will require first aid. You should be able to successfully—

(1) Mask him if he is unmasked.

(2) Inject him, if necessary, with all of **HIS** autoinjectors.

(3) Decontaminate his skin.

(4) Seek medical assistance.

7-7. Signs and Symptoms of Nerve Agent Poisoning

The symptoms of nerve agent poisoning are grouped as **MILD**—those that you recognize and for which you can perform self-aid, and **SEVERE**—those which require buddy aid.

a. MILD Signs and Symptoms.

- Unexplained runny nose.
- Unexplained sudden headache.
- Sudden drooling.
- Difficulty seeing (dimness of vision and miosis).
- Tightness in the chest or difficulty in breathing.
- Localized sweating and muscular twitching in the area of contaminated skin.
- Stomach cramps.
- Nausea.
- Tachycardia followed by bradycardia. (*Tachycardia* is an abnormally rapid heartbeat with a heart rate of over 100 beats per minute. *Bradycardia* is a slow heart rate of less than 60 beats per minute.)

b. SEVERE Signs and Symptoms.

- Strange or confused behavior.
- Wheezing, dyspnea (difficulty in breathing), and coughing.
- Severely pinpointed pupils.
- Red eyes with tearing.
- Vomiting.
- Severe muscular twitching and general weakness.
- Involuntary urination and defecation.

- Convulsions.
- Unconsciousness.
- Respiratory failure.
- Bradycardia.

7-8. First Aid for Nerve Agent Poisoning

First aid for nerve agent poisoning consists of administering the MARK I or ATNAA and CANA.

a. Injection Site. The injection site for administering the antidotes is normally in the outer thigh muscle. The thigh injection site is the area about a hand's width above the knee to a hand's width below the hip joint (Figure 7-3). It is important that the injection be given into a large muscle area. If the individual is thinly built, then the injections should be administered into the upper outer quarter (quadrant) of the buttock (Figure 7-4). Injecting in the buttocks of a thinly built individual avoids injury to the thighbone.

b. Self-Administer MARK I. If you experience any or all of the nerve agent **MILD** symptoms (paragraph 7-7*a*), you must **IMMEDIATELY** put on your protective mask and self-administer one MARK I (Figure 7-2A). Follow the procedure given in Table 7-1. The MARK I is carried in your protective mask carrier, pocket of the MOPP overgarment, or other location as specified in your unit tactical standing operating procedure (TSOP). (In cold weather, the MARK I should be stored in an inside pocket of your clothing to protect the antidote from freezing. A frozen MARK I cannot be immediately used to provide you with antidote, when needed. (However, the MARK I can still be used after complete thawing.)

Table 7-1. Self Aid for Nerve Agent Poisoning

MARK I*	ATNAA*
STEP 1. OBTAIN ONE MARK I.**	STEP 1. OBTAIN ONE ATNAA.**
STEP 2. CHECK INJECTION SITE.	STEP 2. CHECK INJECTION SITE.
STEP 3. HOLD MARK I AT EYE LEVEL WITH NONDOMINANT HAND WITH THE LARGE INJECTOR ON TOP (FIGURE 7-5A).	STEP 3. HOLD ATNAA WITH DOMINANT HAND (FIGURE 7-12A).

Table 7-1. Self Aid for Nerve Agent Poisoning (Continued)

MARK I*	ATNAA*
STEP 4. GRASP SMALL INJECTOR (ATROPINE) (FIGURE 7-5B) AND REMOVE FROM CLIP (FIGURE 7-5C).	STEP 4. GRASP SAFETY CAP WITH NONDOMINANT HAND AND REMOVE FROM INJECTOR (FIGURE 7-12B).
STEP 5. CLEAR HARD OBJECTS FROM INJECTION SITE.	STEP 5. CLEAR HARD OBJECTS FROM INJECTION SITE.
STEP 6. INJECT ATROPINE AT INJECTION SITE APPLYING EVEN PRESSURE TO THE INJECTOR (FIGURE 7-6 OR 7-7). HOLD IN PLACE FOR 10 SECONDS.	STEP 6. INJECT ATNAA AT INJECTION SITE APPLYING EVEN PRESSURE TO THE INJECTOR (FIGURE 7-14 OR 7-15). HOLD IN PLACE FOR 10 SECONDS.
STEP 7. HOLD USED INJECTOR WITH NONDOMINANT HAND.	STEP 7. BEND NEEDLE OF USED INJECTOR BY PRESSING ON A HARD SURFACE TO FORM A HOOK.
STEP 8. GRASP THE LARGE (2 PAM CI) INJECTOR (FIGURE 7-8B) AND PULL IT FROM CLIP (FIGURE 7-8C). DROP CLIP TO GROUND.	STEP 8. ATTACH USED INJECTOR TO BLOUSE POCKET FLAP OF BDO/JSLIST (FIGURE 7-16).
STEP 9. INJECT 2 PAM CI AT INJECTION SITE APPLYING EVEN PRESSURE TO THE INJECTOR (FIGURE 7-6 OR 7-7). HOLD IN PLACE FOR 10 SECONDS.	STEP 9. MASSAGE INJECTION SITE, MISSION PERMITTING.
STEP 10. BEND THE NEEDLES OF ALL USED INJECTORS BY PRESSING ON A HARD SURFACE TO FORM A HOOK.	
STEP 11. ATTACH ALL USED INJECTORS TO BLOUSE POCKET FLAP OF BDO/JSLIST (FIGURE 7-9).	
STEP 12. MASSAGE INJECTION SITE, MISSION PERMITTING.	

* USE STEPS LISTED FOR TYPE OF ANTIDOTE DEVICE ISSUED.

** ONLY ADMINISTER ONE MARK I OR ATNAA AS SELF-AID. DO NOT SELF-ADMINISTER CANA.

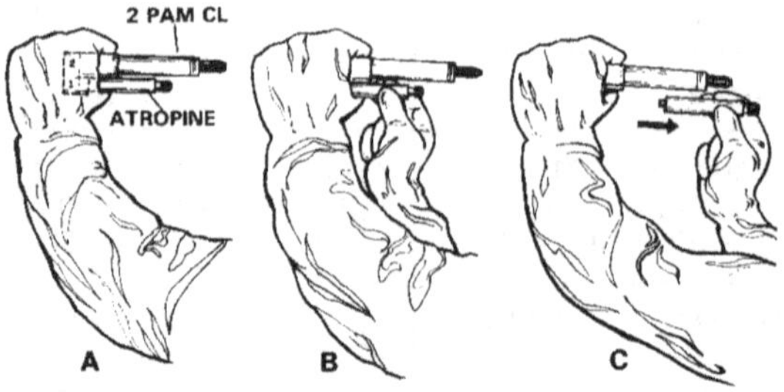

Figure 7-5. Removing the atropine autoinjector from the MARK I clip.

> **CAUTION**
>
> **DO NOT** cover or hold the needle end with your hand, thumb, or fingers—you might accidentally inject yourself. An accidental injection into the hand **WILL NOT** deliver an effective dose of the antidote, especially if the needle goes through the hand.

Figure 7-6. Thigh injection site for self-aid.

NOTE

If you are thinly built, inject yourself into the upper outer quadrant of the buttock (Figure 7-7). There is a nerve that crosses the buttocks; hitting this nerve can cause paralysis. Therefore, you must only inject into the *upper outer quadrant* of the buttock.

Figure 7-7. Buttocks injection site for self-aid.

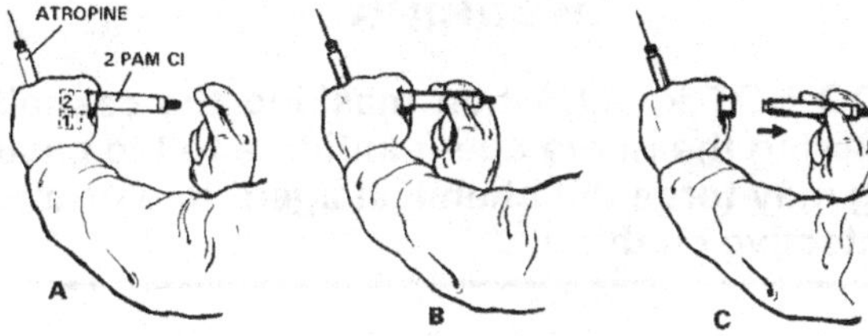

Figure 7-8. Removing the 2 PAM Cl autoinjector from the MARK I clip.

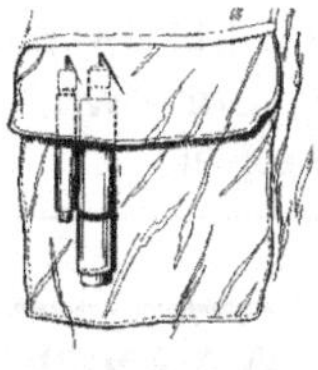

Figure 7-9. One set of used MARK I autoinjectors attached to pocket flap.

NOTES

1. **DO NOT** give yourself another set of injections. If you are able to walk without assistance, know who you are, and where you are, you **WILL NOT** need the second set of injections. (If not needed, giving yourself a second set of MARK I injections or ATNAA may create a nerve agent antidote overdose, which could cause incapacitation [inability to perform mission or defend yourself].)

2. If you continue to have symptoms of nerve agent poisoning, seek someone else (a buddy) to check your symptoms and administer the additional sets of injections, if required.

c. Buddy Evaluation and Buddy Aid. Service members may seek assistance after self-aid (self-administering one MARK I or ATNAA) or may become incapacitated after self-aid. A buddy must evaluate the individual to determine if additional antidotes are required to counter the effects of the nerve agent. Also, service members may experience **SEVERE** symptoms of nerve agent poisoning (paragraph 7-7*b*); they will not be able to treat themselves. In either case, other service members must perform buddy aid as quickly as possible. Before initiating buddy aid, determine if one set of MARK I autoinjectors has already been used so that no more than three sets of the antidote are administered. Buddy aid also includes administering the CANA with the third MARK I or ATNAA to prevent convulsions. Follow the procedures indicated in Table 7-2.

WARNING

Squat, DO NOT kneel, when masking the casualty or administering the nerve agent antidote to the casualty. Kneeling may force the chemical agent into or through your protective clothing.

CAUTION

DO NOT use your own MARK I, ATNAA, or CANA on a casualty. If you use your own, you may not have any antidote if needed for self-aid.

WARNING

DO NOT inject into areas close to the hip, knee, or thigh-bone.

Table 7-2. Buddy Aid/Combat Lifesaver Aid for Nerve Agent Casualty.

MARK I*	ATNAA*	CANA**
STEP 1. MASK THE CASUALTY AND POSITION HIM ON HIS SIDE (SWIMMER'S POSITION).	STEP 1. MASK THE CASUALTY AND POSITION HIM ON HIS SIDE (SWIMMER'S POSITION).	STEP 1. OBTAIN BUDDY'S CANA.
STEP 2. POSITION YOURSELF NEAR THE CASUALTY'S THIGH.	STEP 2. POSITION YOURSELF NEAR THE CASUALTY'S THIGH.	STEP 2. CHECK INJECTION SITE.

Table 7-2. Buddy Aid/Combat Lifesaver Aid for Nerve Agent Casualty (Continued).

MARK I*	ATNAA*	CANA**
STEP 3. OBTAIN BUDDY'S THREE OR REMAINING MARK Is.	STEP 3. OBTAIN BUDDY'S THREE OR REMAINING ATNAAs.	STEP 3. HOLD CANA IN A CLOSED FIST WITH DOMINANT HAND (FIGURE 7-12A).
STEP 4. CHECK INJECTION SITE.	STEP 4. CHECK INJECTION SITE.	STEP 4. GRASP SAFETY CAP WITH NONDOMINANT HAND AND REMOVE FROM INJECTOR (FIGURE 7-12B).
STEP 5. HOLD MARK I WITH NONDOMINANT HAND (FIGURE 7-5A).	STEP 5. HOLD ATNAA IN A CLOSED FIST WITH DOMINANT HAND (FIGURE 7-12A).	STEP 5. CLEAR HARD OBJECTS FROM INJECTION SITE.
STEP 6. GRASP SMALL INJECTOR (ATROPINE) AND REMOVE FROM CLIP (FIGURE 7-5B).	STEP 6. GRASP SAFETY CAP WITH NONDOMINANT HAND AND REMOVE FROM INJECTOR (FIGURE 7-12B).	STEP 6. INJECT CANA AT INJECTION SITE BY APPLYING EVEN PRESSURE TO THE INJECTOR, NOT A JABBING MOTION (FIGURE 7-14 OR 7-15). HOLD IN PLACE FOR 10 SECONDS.
STEP 7. CLEAR HARD OBJECTS FROM INJECTION SITE.	STEP 7. CLEAR HARD OBJECTS FROM INJECTION SITE.	STEP 7. BEND NEEDLE OF INJECTOR BY PRESSING ON A HARD SURFACE TO FORM A HOOK.
STEP 8. INJECT ATROPINE AT INJECTION SITE BY APPLYING EVEN PRESSURE TO THE INJECTOR, NOT A JABBING MOTION (FIGURE 7-10 OR 7-11). HOLD IN PLACE FOR 10 SECONDS.	STEP 8. INJECT ATNAA AT INJECTION SITE BY APPLYING EVEN PRESSURE TO THE INJECTOR, NOT A JABBING MOTION (FIGURE 7-14 OR 7-15). HOLD IN PLACE FOR 10 SECONDS.	STEP 8. ATTACH USED INJECTOR TO BLOUSE POCKET FLAP OF BDO/JSLIST (FIGURE 7-16).
STEP 9. HOLD USED INJECTOR BETWEEN LITTLE FINGER AND RING FINGER OF NONDOMINANT HAND (FIGURE 7-5A).	STEP 9. BEND NEEDLE OF INJECTOR BY PRESSING ON A HARD SURFACE TO FORM A HOOK.	STEP 9. MASSAGE INJECTION SITE, MISSION PERMITTING.
STEP 10. PULL LARGE INJECTOR (2 PAM CI) FROM CLIP (FIGURE 7-5C). DROP CLIP TO GROUND.	STEP 10. ATTACH ALL USED INJECTORS TO BLOUSE POCKET FLAP OF BDO/JSLIST (FIGURE 7-16).	

Table 7-2. Buddy Aid/Combat Lifesaver Aid for Nerve Agent Casualty (Continued).

MARK I*	ATNAA*	CANA**
STEP 11. INJECT 2 PAM CI AT INJECTION SITE BY APPLYING EVEN PRESSURE TO THE INJECTOR, NOT A JAB-BING MOTION (FIGURE 7-10 OR 7-11). HOLD IN PLACE FOR 10 SECONDS.	STEP 11. MASSAGE INJECTION SITE, MISSION PERMITTING.	
STEP 12. REPEAT STEPS ABOVE FOR REMAINING MARK Is.		
STEP 13. BEND THE NEEDLES OF ALL USED INJECTORS BY PRESSING ON A HARD SURFACE TO FORM A HOOK.		
STEP 14. ATTACH ALL USED INJECTORS TO BLOUSE POCKET FLAP OF BDO/JSLIST (FIGURE 7-13).		
STEP 15. MASSAGE INJECTION SITE, MISSION PERMITTING.		

* USE STEPS LISTED FOR TYPE OF ANTIDOTE DEVICE ISSUED.
** CANA IS USED IN BUDDY AID/CLS AID ONLY. DO NOT USE IN SELF-AID.

NOTE

If the casualty is thinly built, inject the antidote into the buttock. Only inject the antidote into the upper outer portion of the casualty's buttock (Figure 7-11). This avoids hitting the nerve that crosses the buttocks (Figure 7-4). Hitting this nerve can cause paralysis.

Figure 7-10. Injecting the casualty's thigh (Mark I or CANA).

Figure 7-11. Injecting the casualty's buttocks (Mark I or CANA).

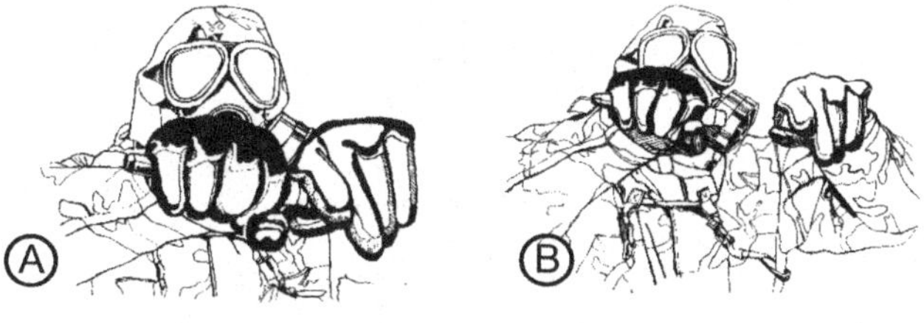

Figure 7-12. Preparing CANA or ATNAA for injection.

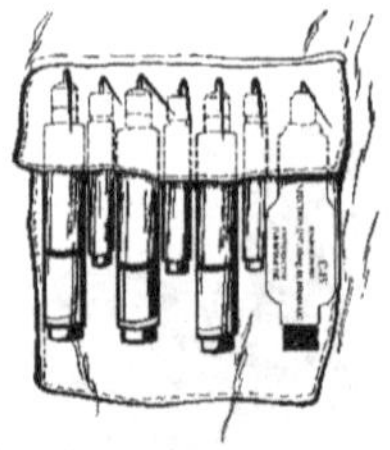

Figure 7-13. Three sets of used MARK I autoinjectors and one CANA autoinjector attached to pocket flap.

d. Self-Administer Antidote Treatment Nerve Agent Autoinjector. If you experience any or all of the nerve agent **MILD** symptoms (paragraph 7-7*b*), you must **IMMEDIATELY** self-administer one ATNAA following the procedure given Table 7-1.

NOTE

If you are thinly-built, inject yourself into the upper outer quarter (quadrant) of the buttock (Figure 7-15). There is a nerve that crosses the buttocks; hitting this nerve can cause paralysis. Therefore, you must only inject into the upper outer quarter (quadrant) of the buttocks.

Figure 7-14. Self-administration of ATNAA (thigh).

Figure 7-15. Self-administration of ATNAA (buttock).

NOTE

If you continue to have symptoms of nerve agent poisoning, seek someone else (a buddy) to check your symptoms and administer your remaining sets of injections, if required.

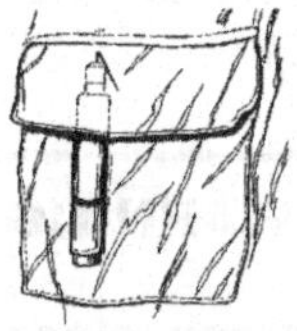

Figure 7-16. Used ATNAA attached to clothing.

e. Buddy Assistance. Service members may seek assistance after self-aid (self-administering one ATNAA) or may become incapacitated after self-aid. A buddy must evaluate the individual to determine if additional antidotes are required to counter the effects of the nerve agent. Also, service members may experience **SEVERE** symptoms of nerve agent poisoning (paragraph 7-7*b*); they will not be able to treat themselves. In either case, other service members must perform buddy aid as quickly as possible. Before initiating buddy aid, determine if one ATNAA has already been used so that no more than three ATNAA are administered. Buddy aid also includes administering the CANA with the third ATNAA to prevent convulsions. Follow the procedures indicated in Table 7-2.

WARNING

Squat, DO NOT kneel, when masking the casualty or administering the nerve agent antidotes to the casualty. Kneeling may force any chemical agent on your overgarment into or through your protective clothing.

Figure 7-17. Buddy injecting casualty's outer thigh (ATNAA or CANA).

NOTE

If the casualty is thinly built, inject the antidote into the buttocks (Figure 7-18). Only inject the antidote into the upper outer portion of the casualty's buttocks. This avoids hitting the nerve that crosses the buttocks (Figure 7-4). Hitting this nerve can cause paralysis.

WARNING

DO NOT inject into areas close to the hip, knee, or thighbone.

Figure 7-18. Buddy injecting casualty's buttocks (ATNAA or CANA).

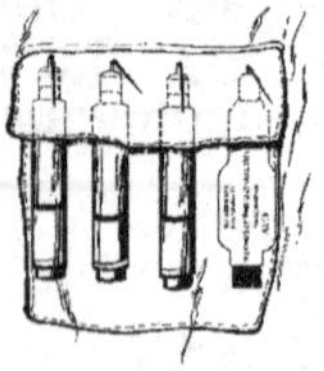

Figure 7-19. Three used ATNAAs and one CANA autoinjector attached to clothing.

f. Combat Lifesaver.

(1) The combat lifesaver must check to verify if the individual has received three sets of MARK I or ATNAAs. If not, the combat lifesaver performs first aid as described for buddy aid above. If the

individual has received the initial three sets of MARK I, then the combat lifesaver may administer additional atropine injections at approximately 15 minute intervals until atropinization is achieved (that is a heart rate above 90 beats per minute, reduced bronchial secretions, and reduced salivations). Administer additional atropine at intervals of 30 minutes to 4 hours to maintain atropinization or until the casualty is placed under the care of medical personnel. Check the heart rate by lifting the casualty's mask hood and feeling for a pulse at the carotid artery. Request medical assistance as soon as the tactical situation permits.

(2) The CLS should administer additional CANA to casualties suffering convulsions. Administer a second, and if needed, a third CANA at 5 to 10 minute intervals for a maximum of three injections (30 milligrams diazepam). Follow the steps and procedures described in buddy aid for administering the CANA. **DO NOT** give more than two additional injections for a total of three (one self-aid plus two by the CLS).

7-9. Blister Agents

Blister agents (vesicants) include mustard (H and HD), nitrogen mustards (HN), lewisite (L), and other arsenicals, mixtures of mustards and arsenicals, and phosgene oxime (CX). Blister agents may act on the eyes, mucous membranes, lungs, and skin. They burn and blister the skin or any other body parts they contact. Even relatively low doses may cause serious injury. Blister agents damage the respiratory tract (nose, sinuses, and windpipe) when inhaled and cause vomiting and diarrhea when absorbed. Lewisite and CX cause immediate pain on contact. However, mustard agents are deceptive as there is little or no pain at the time of exposure. Thus, in some cases, signs of injury may not appear for several hours after exposure.

a. Protective Measures. Your protective mask with hood and protective overgarment provide protection against blister agents. If it is known or suspected that blister agents are being used, **STOP BREATHING**, put on your mask and your protective overgarment.

CAUTION

Large drops of liquid vesicants on the protective overgarment ensemble may penetrate it if allowed to stand for an extended period. Remove large drops as soon as possible.

b. Signs and Symptoms of Blister Agent Poisoning.

(1) Immediate and intense pain upon contact with L, LH (lewisite and mustard) mixture, and CX. No initial pain upon contact with mustard.

(2) Inflammation and blisters (burns) resulting in tissue destruction. The severity of a chemical burn is directly related to the concentration of the agent and the duration of contact with the skin. The longer the agent is in contact with the tissue, the more serious the injury will be.

(3) Vomiting and diarrhea. Exposure to high concentrations of vesicants may cause vomiting or diarrhea.

(4) Death. The blister agent vapors absorbed during ordinary field exposure will probably not cause enough internal body (systemic) damage to result in death. However, death may occur from prolonged exposure to high concentrations of vapor or from extensive liquid contamination over wide areas of the skin, particularly *when decontamination is neglected or delayed.*

c. First Aid Measures.

(1) Use your M291 Skin Decontaminating Kit to decontaminate your skin and use water to flush contaminated eyes. Decontamination of vesicants must be done immediately (within 1 minute is best).

(2) If blisters form, cover them loosely with a field dressing and secure the dressing.

CAUTION

Blisters are actually burns. **DO NOT** attempt to decontaminate the skin where blisters have formed, as the agent has already been absorbed.

(3) If you receive blisters over a wide area of the body, you are considered seriously burned. Seek medical assistance immediately.

(4) If vomiting occurs, the mask should be lifted momentarily and drained—while the eyes are closed and the breath is held—and replaced, cleared, and sealed.

(5) Remember, if vomiting or diarrhea occurs after having been exposed to blister agents, seek medical assistance immediately.

7-10. Choking Agents (Lung-Damaging Agents)

Chemical agents that attack lung tissue, primarily causing fluid buildup (pulmonary edema), are classified as choking agents (lung-damaging agents). This group includes phosgene (CG), diphosgene (DP), chlorine (Cl), and chloropicrin (PS). Of these four agents, CG is the most dangerous and is more likely to be employed by the enemy in future conflict.

a. Protective Measures. Your protective mask gives adequate protection against choking agents.

b. Signs and Symptoms. During and immediately after exposure to choking agents (depending on agent concentration and length of exposure), you may experience some or all of the following signs and symptoms:

- Tears (lacrimation).
- Coughing.
- Choking.
- Tightness of chest.
- Nausea and vomiting.
- Headaches.

c. Self-Aid.

(1) The protective mask should be put on immediately when any of the conditions described in *b* above exist. Another indication of a CG attack is an odor like newly mown hay; however, **DO NOT** rely upon odor as indication of a chemical attack.

(2) If some CG is inhaled, normal combat duties should be continued unless there is difficulty in breathing, nausea, vomiting, or more than the usual shortness of breath during exertion. If any of the above symptoms occur and the mission permits, remain at quiet rest until medical evacuation is accomplished.

d. Death. With ordinary field exposure to choking agents, death will probably not occur. *However, prolonged exposure to high concentrations of the vapor and neglect or delay in masking can be fatal.*

7-11. Cyanogen (Blood) Agents

Cyanogen agents interfere with proper oxygen utilization in the body. Hydrogen cyanide (AC) and cyanogen chloride (CK) are the primary agents in this group.

a. Protective Measures. Your protective mask with a fresh filter gives adequate protection against field concentrations of cyanogen agent vapor. The protective overgarments, as well as the mask, are needed when exposed to liquid AC.

b. Signs and Symptoms. During and immediately after exposure to cyanogen agents (depending on agent concentration and length of exposure), you may experience some or all of the following signs and symptoms:

- Tearing (lacrimation).
- Eye, nose, and throat irritation.
- Sudden stimulation of breathing (unable to hold breath).
- Nausea.
- Coughing.
- Tightness of chest.
- Headache.
- Light-headedness (dizziness).
- Unconsciousness.

c. First Aid.

(1) *Hydrogen cyanide.* During any chemical attack, if you get a sudden stimulation of breath or detect an odor like bitter almonds, **PUT ON YOUR MASK IMMEDIATELY**. Speed is absolutely essential since this agent acts so rapidly that within a few seconds its effects will make it impossible for service members to put on their mask by themselves. Stop breathing until the mask is on, if at all possible. This may be very difficult since the agent strongly stimulates respiration.

(2) *Cyanogen chloride.* Put your mask on immediately if you experience any irritation of the eyes, nose, or throat. Service members

who are unable to mask should be masked by the nearest service member (buddy).

d. Medical Assistance. If you suspect that you have been exposed to blood agents, seek medical assistance immediately.

7-12. Incapacitating Agents

An incapacitating agent is a chemical agent which produces temporary, disabling conditions which persist for hours to days after exposure. Unlike riot control agents, which usually are momentary or fleeting in action, incapacitating agents have a persistent effect. It is likely that smoke-producing munitions or aerosols will disseminate such agents, thus making breathing their means of entry into the body. The protective mask is, therefore, essential.

a. There are no specific first aid measures to relieve the symptoms of incapacitating agents. Supportive first aid and physical restraint may be indicated. If the casualty is stuporous or comatose, be sure that respiration is unobstructed; then turn him on his side in case vomiting should occur. Complete cleansing of the skin with soap and water should be done as soon as possible; or, the M291 Skin Decontaminating Kit can be used if washing is impossible. Remove weapons and other potentially harmful items from service members who are suspected of having these symptoms. Harmful items include cigarettes, matches, medications, and small items that might be swallowed accidentally. Delirious (confused) persons have been known to attempt to eat items bearing only a superficial resemblance to food.

b. Incapacitating agents (anticholinergic drugs BZ type) may produce alarming dryness and coating of the lips and tongue; however, there is usually no danger of immediate dehydration. Fluids should be given sparingly, if at all, because of the danger of vomiting and because of the likelihood of temporary urinary retention due to paralysis of bladder muscles.

c. If the body temperature is elevated and mucous membranes are dry, immediate and vigorous cooling (as for heatstroke) is indicated. Methods that can be used to cool the skin are spraying with cool water and air circulation (fanning); applying alcohol soaked cloths and air circulation; and providing maximum exposure to air in a shaded area, along with maximum air circulation. Such cases are usually a result of anticholinergic poisoning. Rapid evacuation should be accomplished since medical treatment with the appropriate medication may be lifesaving.

CAUTION

DO NOT use **ice** for cooling the skin.

d. Reassurance and a firm, but friendly, attitude by individuals providing first aid will be beneficial if the casualty appears to comprehend what is being said. Conversation is a waste of time if the service member is incoherent or cannot understand what is being said. In such cases, the less said, the better it is—these casualties will benefit more from prompt and vigorous restraint and evacuation to an MTF.

7-13. Incendiaries

Incendiaries can be grouped as WP, thickened gasoline, metal, and oil and metal. You must learn to protect yourself against these incendiaries.

a. White phosphorus is used primarily as a smoke producer but can be used for its incendiary effect to ignite field expedients and combustible materials. The burns from WP are usually multiple, deep, and variable in size. When particles of WP get on the skin or clothing, they continue to burn until deprived of air. They also have a tendency to stick to a surface and must be brushed off or picked out.

(1) If burning particles of WP strike and stick to your clothing, quickly take off the contaminated clothing before the WP burns through to the skin.

(2) If burning WP strikes your skin, smother the flame with water, a wet cloth, or mud.

NOTE

Since WP is soluble in oil, **DO NOT** use grease, oily ointments, or eye ointments to smother the flame.

(3) Keep the WP particles covered with a wet material to exclude air until you can remove them or have them removed from your skin.

(4) Remove the WP particles from the skin by brushing them with a wet cloth and by picking them out with a knife, bayonet, stick, or other available object.

(5) Seek medical assistance when the mission permits.

b. Thickened fuel mixtures (napalm) have a tendency to cling to clothing and body surfaces, thereby producing prolonged exposure and severe burns. The first aid for these burns is the same as for other heat burns. The heat and irritating gases given off by these combustible mixtures may cause lung damage, which must be treated by medical personnel.

c. Metal incendiaries pose special problems. Thermite particles on the skin should be immediately cooled with water and then removed. The first aid for these burns is the same as for other heat burns. Particles of magnesium on the skin burn quickly and deeply. Like other metal incendiaries, they must be removed. Ordinarily, medical personnel should do the complete removal of these particles as soon as possible. Immediate medical treatment is required.

d. Oil and metal incendiaries have much the same effect on contact with the skin and clothing as those discussed (*b* and *c* above). First aid measures for burns are discussed in Chapter 3.

7-14. Biological Agents and First Aid

a. Biological attacks can result in combat ineffectiveness by introducing disease-causing organisms into a troop population.

b. Once a disease is identified, first aid or medical treatment is initiated, depending on the seriousness of the disease. First aid measures are concerned with observable symptoms of the disease such as diarrhea or vomiting.

7-15. Toxins

Toxins are alleged to have been used in past conflicts. Witnesses and victims have described the agent as toxic rain (or yellow rain) because it was reported to have been released from aircraft as a yellow powder or liquid that covered ground, structures, vegetation, and people.

a. *Signs and Symptoms.* The occurrence of the symptoms from toxins may appear in a period of a few minutes to several hours depending on the particular toxin, the service member's susceptibility, and the amount of toxin inhaled, ingested, or deposited on the skin. Symptoms from toxins usually involve the central nervous system but are often preceded by less prominent symptoms, such as nausea, vomiting, diarrhea, cramps, or stomach irritation and burning sensation. Typical neurological symptoms often develop rapidly in severe cases; for example, visual disturbances, inability to swallow, speech difficulty, lack of muscle coordination, and sensory abnormalities (numbness of mouth, throat, or extremities). Yellow rain (mycotoxins) also may have hemorrhagic symptoms, which could include any or all of the following:

- Dizziness.
- Severe itching or tingling of the skin.
- Formation of multiple, small, hard blisters.
- Coughing up blood.
- Shock (which could result in death).

b. *Self-Aid.* Upon recognition of an attack employing toxins, you must immediately take the following actions:

(1) Stop breathing, put on your protective mask with hood, and then resume breathing. Next, put on your protective clothing.

(2) Should severe itching of the face become unbearable, quickly—

- Loosen the cap on your canteen.
- Take and hold a deep breath and lift your mask.
- While holding your breath, close your eyes and flush your face with generous amounts of water.

CAUTION

DO NOT rub or scratch your eyes. Try not to let the water run onto your clothing or protective overgarment.

- Put your protective mask back on, seat it properly, clear it, and check it for a seal; then resume breathing.
- Decontaminate your skin by bathing with soap and water as soon as the mission permits.

- Change clothing and decontaminate your protective mask using soap and water. Replace the filters if directed.

(3) If vomiting occurs, the mask should be lifted momentarily and drained—while the eyes are closed and the breath is held—and replaced, cleared, and sealed.

c. *Medical Assistance.* If you suspect that you have been exposed to toxins, you should seek medical assistance immediately.

7-16. Nuclear Detonation

a. Three types of injuries may result from a nuclear detonation. These are thermal, blast, and radiation injuries. Many times the casualty will have a combination of these types of injuries. First aid for thermal and blast injuries is provided based on observable injuries, such as burns, hemorrhage, or fractures.

b. The signs and symptoms of radiation illness in the initial phase include the rapid onset of nausea, vomiting, and malaise (tiredness). The only first aid procedure for radiological casualties is decontamination.

CHAPTER 8

FIRST AID FOR PSYCHOLOGICAL REACTIONS

8-1. General

Psychological first aid is as natural and reasonable as physical first aid and is just as familiar. When you were hurt as a child, the understanding attitude of your parents did as much as the psychological effect of a bandage. Later, your disappointment or grief was eased by supportive words from a friend. Certainly, taking a walk and talking things out with a friend are familiar ways of dealing with an emotional crisis. The same natural feelings that make us want to help a person who is injured make us want to give a helping hand to a buddy who is upset. *Psychological first aid* really means nothing more complicated than assisting people with emotional distress whether it results from physical injury, disease, or excessive stress. Emotional distress is not always as visible as a wound or a broken bone. However, overexcitement, severe fear, excessive worry, deep depression, misdirected irritability, and anger are signs that stress has reached the point of interfering with effective coping. The more noticeable the symptoms become, the more urgent the need for you to be of help and the more important it is for you to know *how* to help.

8-2. Importance of Psychological First Aid

You must know how to give psychological first aid to be able to help yourself, your buddies, and your unit in order to keep performing the mission. Psychological first aid measures are simple and easy to understand. Your decision of what to do depends upon your ability to observe the service member and understand his needs. Making the best use of resources requires ingenuity on your part. A stress reaction resulting in poor judgment can cause injury or even death to yourself or others on the battlefield. It can be even more dangerous if other persons are affected by the judgment of an emotionally upset service member. If it is detected early enough, the affected service member stands a good chance of remaining in his unit as an effective member. If it is not detected early and if the service member becomes more emotionally upset, he may become a threat to himself and to others.

8-3. Situations Requiring Psychological First Aid

- Psychological first aid (buddy aid) is most needed at the first sign that a service member cannot perform the mission because of emotional distress. Stress is inevitable in combat, in hostage and terrorist situations, and in civilian disasters such as floods, hurricanes, or industrial accidents. Most emotional reactions to such situations are temporary, and the service member can still carry on with encouragement. Painful or disruptive symptoms may last for minutes, hours, or days. However, if the stress symptoms are seriously disabling, they may be psychologically contagious and endanger not only the emotionally upset service member but also the entire unit.

- Sometimes people continue to function well during a disastrous event, but suffer from emotional scars which impair their job performance or quality of life at a later time. Painful memories and dreams may recur for months and years and still be considered a normal reaction. However, if the memories are so painful that the person must avoid all situations which arouse them, becomes socially withdrawn, or shows symptoms of anxiety, depression, or substance abuse, he needs treatment. Experience with police, firemen, emergency medical technicians, and others who deal with disasters has proved that the routine application of psychological first aid to all the participants, including those who have functioned well, greatly reduces the likelihood of future serious post-traumatic stress disorders (PTSDs).

8-4. Interrelationship of Psychological and Physical First Aid

Psychological first aid should go hand in hand with physical first aid. The discovery of a physical injury or cause for an inability to function does not rule out the possibility of a psychological injury (or vice versa). The person suffering from pain, shock, fear of serious injury, or fear of death does not respond well to joking, indifference, or fearful-tearful attention. Fear and anxiety may take as high a toll of the service member's strength as does the loss of blood.

8-5. Goals of Psychological First Aid

The goals of psychological first aid are to—

- Be supportive; assist the service member in dealing with his stress reaction.
- Prevent, and if necessary control, behavior harmful to himself and to others.
- Return the service member to duty as soon as possible after dealing with the stress reaction.

8-6. Respect for Others' Feelings

a. Accept the service member you are trying to help without censorship or ridicule. Respect his right to his own feelings. Even though your feelings, beliefs, and behavior are different, DO NOT blame or make light of him for the way he feels or acts. Your purpose is to help him in this tough situation, not to be his critic. A person DOES NOT WANT to be upset and worried. When he seeks help, he needs and expects consideration of his fears, not abrupt dismissal or ridicule.

b. Realize that people are the products of a wide variety of factors. All people DO NOT react the same way to the same situations. Each individual has complex needs and motivations, both conscious and unconscious, that are uniquely his own. Often the one thing that finally causes the person to become overloaded by a stressful situation is not the stressor itself, but some other problem.

8-7. Emotional and Physical Disability

a. Accept emotional disability as being just as real as physical disability. If a service member's ankle is seriously sprained in a fall, no one expects him to run right away. A service member's emotions may be temporarily strained by the overwhelming stress of battle or other traumatic incident. DO NOT demand that he pull himself together immediately and carry on without a break. Some individuals can pull themselves together immediately, but others cannot. The service member whose emotional stability has been disrupted has a disability just as real as the service member who has sprained his ankle. There is an unfortunate tendency in many people to regard as real only what they can see, such as a wound or bleeding. Some people tend to assume that damage involving a person's mind and emotions is just imagined, that he is not really sick or injured, and that he could overcome his trouble by using his will power.

b. The terms *it's all in your head, snap out of it,* and *get control of yourself* are often used by people who believe they are being helpful. Actually, these terms are expressions of hostility because they show lack of understanding. They only emphasize weakness and inadequacy. Such terms are of no use in psychological first aid.

c. Every physically injured person has some emotional reaction to the fact that he is injured.

(1) It is normal for an injured person to feel upset. The more severe the injury, the more insecure and fearful he becomes, especially

if the injury is to a body part which is highly valued. For example, an injury to the eyes or the genitals, even though relatively minor, is likely to be extremely upsetting. An injury to some other part of the body may be especially disturbing to an individual for his own particular reason. For example, an injury of the hand may be a terrifying blow to a surgeon or an injury to the eye of a pilot.

(2) An injured service member always feels less secure, more anxious, and more afraid not only because of what has happened to him but because of what he imagines may happen as a result of his injury. This fear and insecurity may cause him to be irritable, uncooperative, or unreasonable. As you help him, always keep in mind that such behavior has little or nothing to do with you personally. He needs your patience, reassurance, encouragement, and support.

8-8. Combat and Other Operational Stress Reactions

Stress reaction is a temporary emotional disorder or inability to function, experienced by a previously normal service member as a reaction to the overwhelming or cumulative stress of combat. Stress reaction gets better with reassurance, rest, physical replenishment, and activities that restore confidence. All service members are likely to feel stress reaction under conditions of intense and/or prolonged stress. They may even become stress reaction casualties, unable to perform their mission for hours or days. Other combat and operational stress reactions (COSRs) may result in negative behavior, but are not termed *stress reaction,* as they need more intensive treatment. These negative COSRs may result in misconduct stress behaviors such as drug and alcohol abuse, criminal acts, looting, desertion, and self-inflicted wounds. These harmful COSRs can often be prevented by good psychological first aid. Service members who commit misconduct stress behaviors may require disciplinary action rather than medical treatment.

8-9. Reactions to Stress

Most service members react to stressful incidents after the situation has passed. All service members feel some fear. This fear may be greater than they have experienced at any other time, or they may be more aware of their fear. In such a situation, they should not be surprised if they feel shaky or become sweaty, nauseated, or confused. These reactions are normal and are not a cause for concern. However, some reactions, either short- or long-term, will cause problems if left unchecked. See paragraph 8-13 for more information.

a. Emotional Reactions.

(1) The most obvious combat stress reaction (CSR) is inefficient performance. This can be demonstrated by—

- Slow thinking (or reaction time).
- Difficulty recognizing priorities and seeing what needs to be done.
- Difficulty getting started.
- Indecisiveness and having trouble focusing attention.
- Tendency to do familiar tasks and be preoccupied with familiar details. (This can reach the point where the person is very passive, such as just sitting or wandering about not knowing what to do.)

(2) A less common reaction may be uncontrolled emotional outbursts; this can be demonstrated by crying, screaming, or laughing. Some service members will react in the opposite way. They will be very withdrawn and silent and try to isolate themselves from everyone. These service members should be encouraged to remain with their assigned unit. Uncontrolled reactions may appear by themselves or in any combination (the person may be crying uncontrollably one minute and then laughing the next). In this state, the person is restless and cannot keep still. He may run about, apparently without purpose. Inside, he feels a great rage or fear and his physical acts may show this. In his anger he may indiscriminately strike out at others.

b. Loss of Adaptability.

(1) In a desperate attempt to get away from the danger, which has overwhelmed him, a service member may panic and become confused. His mental ability may be so impaired he cannot think clearly or even follow simple commands. His judgment may be faulty and he may not be aware of his actions, such as standing up in his fighting position during an attack.

(2) In other cases, overwhelming stress may produce symptoms that are often associated with head injuries. For example, the service member may appear dazed or be found wandering around aimlessly. He may appear confused and disoriented and may seem to have a complete or partial loss of memory. In such cases, especially when no eyewitnesses can provide evidence that the service member has NOT suffered a head injury, it is necessary for him to be rapidly medically evacuated. **DO NOT** allow the

service member to expose himself to further personal danger until the cause of the problem has been determined.

c. Sleep Disturbance and Repetition of Dreams. A person who has been overwhelmed by stress often has difficulty sleeping. The service member may experience nightmares related to the stressors. Remember that nightmares, in themselves, are not considered abnormal when they occur soon after a period of intensive stress. As time passes, the nightmares usually become less frequent and less intense. In extreme cases, a service member, even when awake, may think repeatedly of the incident, feel as though it is happening again, and act out parts of his stress over and over again. For some persons, this repetitious reexperiencing of the stressful event may be necessary for eventual recovery; therefore, it should not be discouraged or viewed as abnormal. For the person reexperiencing the event, such reaction may be disruptive. The service member needs to be encouraged to *ventilate* about the incident. Ventilation is a technique where the service member is given the opportunity to talk extensively, often repetitiously about the experience.

8-10. Severe Stress or Stress Reaction

You do not need specialized training to recognize severe stress or stress reaction that will cause problems for the service member, the unit, or the mission. Reactions that are less severe, however, are more difficult to detect. To determine whether a person needs help, you must observe him to see whether he is doing something meaningful, performing his duties, taking care of himself, behaving in an unusual fashion, or acting out of character.

8-11. Application of Psychological First Aid

The emotionally disturbed service member has built a barrier against fear. He does this for his own protection, although he is probably not aware that he is doing it. If he finds that he does not have to be afraid and that there are normal, understandable things about him, he will feel safer in dropping this barrier. Persistent efforts to make him realize that you want to understand him will be reassuring, especially if you remain calm. Nothing can cause an emotionally disturbed person to become even more fearful than feeling that others are afraid of him. Try to remain calm. Familiar things, such as a cup of coffee, the use of his name, attention to a minor wound, being given a simple job to do, or the sight of familiar people and activities, will add to his ability to overcome his fear. He may not respond well if you get excited, angry, or abrupt.

a. *Ventilation.* After the service member becomes calmer, he is likely to have dreams about the stressful event. He also may think about it when he is awake or even repeat his personal reaction to the event. One benefit of this natural pattern is that it helps him master the stress by going over it just as one masters the initial fear of parachuting from an aircraft by doing it over and over again. Eventually, it is difficult to remember how frightening the event was initially. In giving first aid to the emotionally disturbed service member, you should let him follow this natural pattern. Encourage him to talk. Be a good listener. Let him tell, in his own words, what actually happened. If home front problems or worries have contributed to the stress, it will help him to talk about them. Your patient listening will prove to him that you are interested in him, and by describing his personal problem, he can work at mastering his fear. If he becomes overwhelmed in the telling, suggest a cup of coffee or a break. Whatever you do, assure him that you will listen again as soon as he is ready. Do try to help put the service member's perception of what happened back into realistic perspective; but DO NOT argue about it.

b. *Activity.*

(1) A person who is emotionally disturbed as the result of a combat action is a casualty of anxiety and fear. He is disabled because he has become temporarily overwhelmed by his anxiety. A good way to control fear is through activity. Almost all service members, for example, experience a considerable sense of anxiety and fear while they are poised, awaiting the opening of a big offensive; but this is normally relieved, and they actually feel better once they begin to move into action. They take pride in effective performance and pleasure in knowing that they are good service members, perhaps being completely unaware that overcoming their initial fear was their first major accomplishment.

(2) Useful activity is very beneficial to the emotionally disturbed service member who is not physically incapacitated. After you help a service member get over his initial fear, help him to regain some self-confidence. Make him realize his job is continuing by finding him something useful to do. Encourage him to be active. Get him to help load trucks, clean up debris, or dig fighting positions. If possible, get him back to his usual duty. Seek out his strong points and help him apply them. Avoid having him just sit around. You may have to provide direction by telling him what to do and where to do it. The instructions should be clear and simple and should be repeated. A person who has panicked is likely to argue. Respect his feelings, but point out more immediate, obtainable, and demanding needs. Channel his excessive energy and, above all, DO NOT argue. If you cannot get him interested in doing more profitable work, it may be necessary to enlist aid in controlling his overactivity before it spreads to the group and

results in more panic. Prevent the spread of such infectious feelings by restraining and segregating if necessary.

(3) Involvement in activity helps a service member in three ways; he—

- Forgets himself.
- Has an outlet for his excessive tensions.
- Proves to himself he can do something useful.

c. *Rest.* There are times, particularly in combat, when physical exhaustion is a principal cause for emotional reactions. A unit sleep plan should be established and implemented. When possible, service members should be given a safe and relatively comfortable area in which to sleep. Examples would be an area away from heavy traffic, noise, and congestion or a place that is clean and dry and protected from environmental conditions. The more uninterrupted sleep a service member gets the better he will be able to function in the tactical environment.

d. *Hygiene.* Field hygiene is an important ingredient in a service member's morale. A service member who is dirty and unkempt will not function as well as a service member who has had the opportunity to bathe and put on clean, dry clothing. During combat, unit leaders should stress the importance of personal hygiene. Good personal hygiene not only improves morale, it also is a preventive measure against disease and nonbattle injury (DNBI).

e. *Group Activity.* You have probably already noticed that a person works, faces danger, and handles serious problems better if he is a member of a closely-knit group. Each service member in the team supports the other team members. Esprit de corps is built because the service members have the same interests, goals, and mission, and as a result they are more productive; furthermore, they are less worried because everyone is involved. It is this spirit that takes a strategic hill in battle. It is so powerful that it is one of the most effective tools you have in your *psychological first aid bag*. Getting the service member back into the team or squad activities will reestablish his sense of belonging and security and will go far toward making him a useful member of the unit.

8-12. Reactions and Limitations

Up to this point the discussion has been primarily about the feelings of the emotionally distressed service member. What about your feelings toward him? Whatever the situation, you will have emotional reactions (conscious or unconscious) toward this service member. Your reactions can either help or hinder your ability to help him. When you are tired or worried, you may very easily become impatient with him if he is unusually slow or exaggerates. You may even feel resentful toward him. At times when many physically wounded lie about you, it will be especially natural for you to resent disabilities that you cannot see. Physical wounds can be seen and easily accepted. Emotional reactions are more difficult to accept as injuries. On the other hand, will you tend to be overly sympathetic? Excessive sympathy for an incapacitated person can be as harmful as negative feelings in your relationship with him. He needs strong help, but not your sorrow. To overwhelm him with pity will make him feel even more inadequate. You must expect your buddy to recover, to be able to return to duty, and to become a useful service member again. This expectation should be displayed in your behavior and attitude as well as in what you say. If he can see your calmness, confidence, and competence, he will be reassured and will feel a sense of greater security.

8-13. Stress Reactions

See Tables 8-1, 8-2, and 8-3 for more information.

Table 8-1. Mild Stress Reaction

PHYSICAL SIGNS*	EMOTIONAL SIGNS*
1. TREMBLING, TEARFUL 2. JUMPINESS, NERVOUSNESS 3. COLD SWEAT, DRY MOUTH 4. POUNDING HEART, DIZZINESS 5. INSOMNIA, NIGHTMARES 6. NAUSEA, VOMITING, DIARRHEA 7. FATIGUE 8. THOUSAND-YARD STARE 9. DIFFICULTY THINKING, SPEAKING, AND COMMUNICATING	1. ANXIETY, INDECISIVENESS 2. IRRITABLE, COMPLAINING 3. FORGETFUL, UNABLE TO CONCENTRATE 4. EASILY STARTLED BY NOISE, MOVEMENT 5. GRIEF, TEARFUL 6. ANGER, BEGINNING TO LOSE CONFIDENCE IN SELF AND UNIT

SELF- AND BUDDY AID
1. CONTINUE MISSION PERFORMANCE, FOCUS ON IMMEDIATE MISSION. 2. EXPECT SERVICE MEMBER TO PERFORM ASSIGNED DUTIES. 3. REMAIN CALM AT ALL TIMES; BE DIRECTIVE AND IN CONTROL. 4. LET SERVICE MEMBER KNOW HIS REACTION IS NORMAL, AND THAT THERE IS NOTHING SERIOUSLY WRONG WITH HIM. 5. KEEP SERVICE MEMBER INFORMED OF THE SITUATION, OBJECTIVES, EXPECTATIONS, AND SUPPORT. CONTROL RUMORS. 6. BUILD SERVICE MEMBER'S CONFIDENCE, TALK ABOUT SUCCEEDING. 7. KEEP SERVICE MEMBER PRODUCTIVE (WHEN NOT RESTING) THROUGH RECREATIONAL ACTIVITIES, EQUIPMENT MAINTENANCE.

8. ENSURE SERVICE MEMBER MAINTAINS GOOD PERSONAL HYGIENE.
9. ENSURE SERVICE MEMBER EATS, DRINKS, AND SLEEPS AS SOON AS POSSIBLE.
10. LET SERVICE MEMBER TALK ABOUT HIS FEELINGS. DO NOT "PUT DOWN" HIS FEELINGS OF GRIEF OR WORRY. GIVE PRACTICAL ADVICE AND PUT EMOTIONS INTO PERSPECTIVE.

* MOST OR ALL OF THESE SIGNS ARE PRESENT IN MILD STRESS REACTION. THEY CAN BE PRESENT IN ANY NORMAL SERVICE MEMBER IN COMBAT YET HE CAN STILL DO HIS JOB.

Table 8-2. More Serious Stress Reaction

PHYSICAL SIGNS*	EMOTIONAL SIGNS*
1. CONSTANTLY MOVES AROUND	1. RAPID AND/OR INAPPROPRIATE TALKING
2. FLINCHING OR DUCKING AT SUDDEN SOUNDS	2. ARGUMENTATIVE, RECKLESS MOVEMENTS/ACTIONS
3. SHAKING, TREMBLING (WHOLE BODY OR ARMS)	3. INATTENTIVE TO PERSONAL HYGIENE
4. CANNOT USE PART OF BODY, NO PHYSICAL REASON (HAND, ARM, LEGS)	4. INDIFFERENT TO DANGER
5. CANNOT SEE, HEAR, OR FEEL (PARTIAL OR COMPLETE LOSS)	5. MEMORY LOSS
6. PHYSICAL EXHAUSTION, CRYING	6. SEVERE STUTTERING, MUMBLING, OR CANNOT SPEAK AT ALL
7. FREEZING UNDER FIRE, OR TOTAL IMMOBILITY	7. INSOMNIA, NIGHTMARES
8. VACANT STARES, STAGGERS, SWAYS WHEN STANDS	8. SEEING OR HEARING THINGS THAT DO NOT EXIST
9. PANIC RUNNING UNDER FIRE	9. RAPID EMOTIONAL SHIFTS
	10. SOCIAL WITHDRAWAL
	11. APATHETIC
	12. HYSTERICAL OUTBURSTS
	13. FRANTIC OR STRANGE BEHAVIOR

TREATMENT PROCEDURES**

1. IF A SERVICE MEMBER'S BEHAVIOR ENDANGERS THE MISSION, SELF, OR OTHERS, DO WHATEVER IS NECESSARY TO CONTROL HIM.
2. IF THE SERVICE MEMBER IS UPSET, CALMLY TALK HIM INTO COOPERATING.
3. IF CONCERNED ABOUT THE SERVICE MEMBER'S RELIABILITY:
   - UNLOAD HIS WEAPON.
   - TAKE WEAPON IF SERIOUSLY CONCERNED.
   - PHYSICALLY RESTRAIN HIM ONLY WHEN NECESSARY FOR SAFETY OR TRANSPORTATION.
4. REASSURE EVERYONE THAT THE SIGNS ARE PROBABLY JUST STRESS REACTION AND WILL QUICKLY IMPROVE.
5. IF STRESS REACTION SIGNS CONTINUE:
   - GET THE SERVICE MEMBER TO A SAFER PLACE.
   - DO NOT LEAVE THE SERVICE MEMBER ALONE, KEEP SOMEONE HE KNOWS WITH HIM.
   - NOTIFY SENIOR NONCOMMISSIONED OFFICER (NCO) OR OFFICER.
   - HAVE THE SERVICE MEMBER EXAMINED BY MEDICAL PERSONNEL.

Table 8-2. More Serious Stress Reaction (Continued)

TREATMENT PROCEDURES**
6. GIVE THE SERVICE MEMBER EASY TASKS TO DO WHEN NOT SLEEPING, EATING, OR RESTING. 7. ASSURE THE SERVICE MEMBER HE WILL RETURN TO FULL DUTY IN 24 HOURS; AND, RETURN HIM TO NORMAL DUTIES AS SOON AS HE IS READY.
* THESE SIGNS ARE PRESENT IN ADDITION TO THE SIGNS OF MILD STRESS REACTION. ** DO THESE PROCEDURES IN ADDITION TO THE SELF- AND BUDDY AID CARE.

Table 8-3. Preventive Measures to Combat Stress Reaction

1. WELCOME NEW MEMBERS INTO YOUR TEAM, GET TO KNOW THEM QUICKLY. IF YOU ARE NEW, BE ACTIVE IN MAKING FRIENDS.
2. BE PHYSICALLY FIT (STRENGTH, ENDURANCE, AND AGILITY).
3. KNOW AND PRACTICE LIFESAVING SELF- AND BUDDY AID.
4. PRACTICE RAPID RELAXATION TECHNIQUES (FM 22-51).
5. HELP EACH OTHER OUT WHEN THINGS ARE TOUGH AT HOME OR IN THE UNIT.
6. KEEP INFORMED; ASK YOUR LEADER QUESTIONS, IGNORE RUMORS.
7. WORK TOGETHER TO GIVE EVERYONE FOOD, WATER, SHELTER, HYGIENE, AND SANITATION.
8. SLEEP WHEN MISSION AND SAFETY PERMIT; LET EVERYONE GET TIME TO SLEEP.
   - SLEEP ONLY IN SAFE PLACES AND BY STANDING OPERATING PROCEDURE (SOP).
   - IF POSSIBLE, SLEEP 6 TO 9 HOURS PER DAY.
   - TRY TO GET AT LEAST 4 HOURS SLEEP PER DAY.
   - GET GOOD SLEEP BEFORE GOING ON SUSTAINED OPERATIONS.
   - CATNAP WHEN YOU CAN, BUT ALLOW TIME TO WAKE UP FULLY.
   - CATCH UP ON SLEEP AFTER GOING WITHOUT.

APPENDIX A

FIRST AID CASE AND KITS, DRESSINGS, AND BANDAGES

A-1. First Aid Case with Field Dressings and Bandages

Every service member is issued a first aid case (Figure A-1A) with a field first aid dressing encased in a plastic wrapper (Figure A-1B). He carries it at all times for his use. The field first aid dressing is a standard sterile (germ-free) compress or pad with bandages attached (Figure A-1C). This dressing is used to cover the wound, to protect against further contamination, and to stop bleeding (pressure dressing). When a service member administers first aid to another person, he must remember to use the wounded person's dressing; he may need his own later. The service member must check his first aid case regularly and replace any used or missing dressing. The field first aid dressing may normally be obtained from his unit supply.

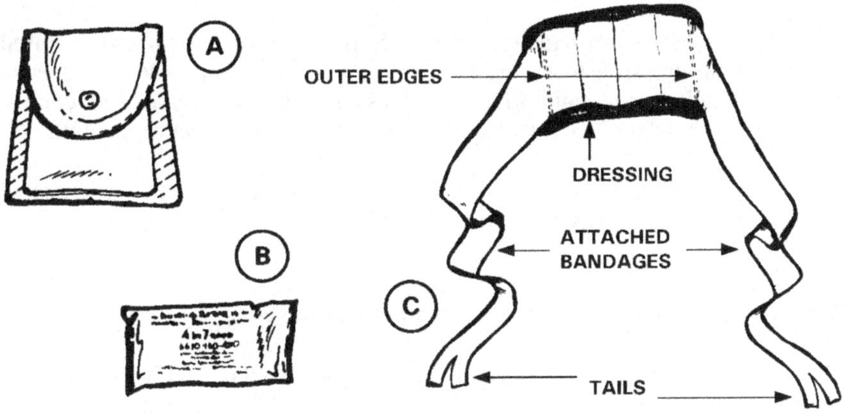

Figure A-1. Field first aid case and dressing (Illustrated A—C).

A-2. General Purpose First Aid Kits

General-purpose first aid kits are listed in the common table of allowances (CTA) 8-100. The operators, crew, and passengers carry these kits on Army vehicles, aircraft, and boats for use. Individuals designated by unit TSOP to be responsible for these kits are required to check them regularly and replace all items used. The general-purpose kit and its contents can be obtained through the unit supply system.

NOTE

> Periodically check the dressings (for holes or tears in the packaging) and the medicines (for expiration date) that are in the first aid kits. If necessary, replace defective or outdated items.

A-3. Dressings

Dressings are sterile pads or compresses used to cover wounds. They usually are made of gauze or cotton wrapped in gauze (Figure A-1C). In addition to the standard field first aid dressing, other dressings such as sterile gauze compresses and small sterile compresses on adhesive strips may be available under CTA 8-100.

A-4. Standard Bandages

a. Standard bandages are made of gauze or muslin and are used over a sterile dressing to secure the dressing in place, to close off its edge from dirt and germs, and to create pressure on the wound and control bleeding. A bandage can also support an injured part or secure a splint.

b. Tailed bandages may be attached to the dressing as indicated on the field first aid dressing (Figure A-1C).

A-5. Triangular and Cravat (Swathe) Bandages

a. Triangular and cravat (or swathe) bandages (Figure A-2) are fashioned from a triangular piece of muslin (37 by 37 by 52 inches) provided in the general-purpose first aid kit. If it is folded into a strip, it is called a cravat. Two safety pins are packaged with each bandage. These bandages are valuable in an emergency since they are easily applied.

b. To improvise a triangular bandage, cut a square of available material, slightly larger than 3 feet by 3 feet, and *fold it diagonally*. If two bandages are needed, cut the material along the diagonal fold.

c. A cravat can be improvised from such common items as T-shirts, other shirts, bed linens, trouser legs, scarfs, or any other item made of pliable and durable material that can be folded, torn, or cut to the desired size.

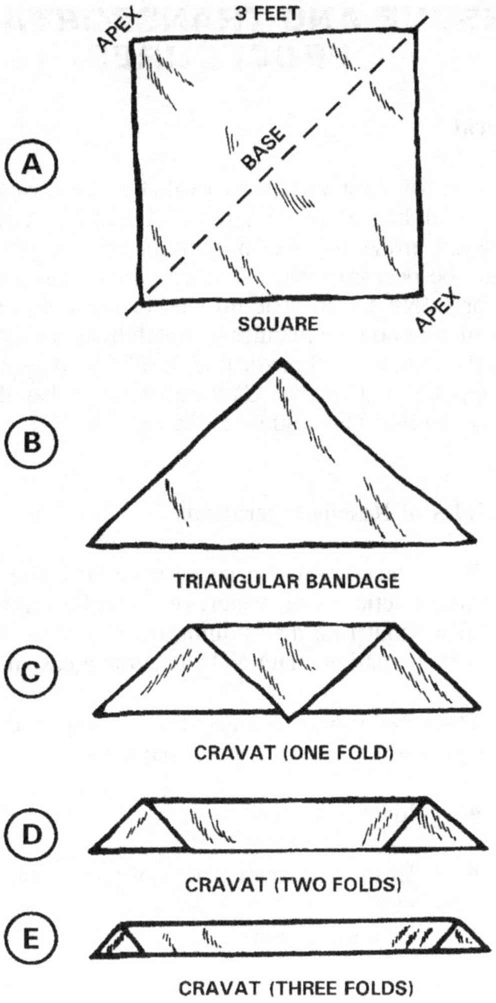

Figure A-2. Triangular and cravat bandages (Illustrated A—E).

APPENDIX B

RESCUE AND TRANSPORTATION PROCEDURES

B-1. General

A basic principle of first aid is to evaluate the casualty's injuries and administer first aid before moving him. However, adverse situations or conditions may jeopardize the lives of both the rescuer and the casualty if this is done. It may be necessary *first to rescue* the casualty before first aid can be effectively or safely given. The life and/or the well-being of the casualty will depend as much upon the manner in which he is *rescued and transported,* as it will upon the first aid and medical treatment he receives. Rescue actions must be done quickly and safely. Careless or rough handling of the casualty during rescue operations can aggravate his injuries.

B-2. Principles of Rescue Operations

a. When faced with the necessity of rescuing a casualty who is threatened by hostile action, fire, water, or any other immediate hazard, DO NOT take action without first determining the extent of the hazard and your ability to handle the situation. DO NOT become a casualty.

b. The rescuer must evaluate the situation and analyze the factors involved. This evaluation involves three major steps:

- Identify the task.
- Evaluate circumstances of the rescue.
- Plan the action.

B-3. Considerations

a. First determine if a rescue attempt is actually needed. It is a waste of time, equipment, and personnel to rescue someone not in need of rescuing. It is also a waste to look for someone who is not lost or needlessly risk the lives of the rescuer(s). In planning a rescue, attempt to obtain the following information:

- Who, what, where, when, why, and how the situation happened?

- How many casualties are involved and the nature of their injuries?
- What is the tactical situation?
- What are the terrain features and the location of the casualties?
- Will there be adequate assistance available to aid in the rescue/evacuation?
- Can first aid and/or medical treatment be provided at the scene; will the casualties require movement to a safer location?
- What specialized equipment will be required for the rescue operation?
- Is the rescue area contaminated? Will decontamination equipment and materiel be required for casualties, rescue personnel, and rescue equipment?
- How much time is available?

b. The time element can play a significant role in how the rescue is attempted. If the casualties are in imminent danger of losing their lives (such as near a burning vehicle or in a burning building) the time available will be relatively short and will sometimes cause a rescuer to compromise planning stages and/or the first aid which can be given. However, if the casualty is in a relatively secure area and his physical condition is strong, more deliberate planning can take place. A realistic estimate of time available must be made as quickly as possible to determine action time remaining. The key elements are the casualty's physical and mental condition, the tactical situation, and the environment.

B-4. Plan of Action

a. The casualty's ability to endure is of primary importance in estimating the time available. Age, physical condition, and extent of wounds and/or injuries will differ from casualty to casualty. Therefore, to determine the time available, you will have to consider—

- Endurance time of the casualty.
- Extent of injuries.

- Type of situation.
- Personnel and/or equipment availability.
- Weather.
- Terrain (natural and man-made).
- Environment (contaminated or uncontaminated).

b. In respect to terrain, you must consider altitude and visibility. In some cases, the casualty may be of assistance because he knows more about the particular terrain or situation than you do. Maximum use of secure/reliable trails or roads is essential.

c. When taking weather into account, ensure that blankets and/or rain gear are available. Even a mild rain can complicate a normally simple rescue. In high altitudes and/or extreme cold and gusting winds, the time available is critically shortened. Be prepared to provide shelter and warmth for the casualty as well as the rescuers.

B-5. Proper Handling of Casualties

a. You may have saved the casualty's life through the application of appropriate first aid measures. However, his life can be lost through rough handling or careless transportation procedures. Before you attempt to move the casualty—

- Evaluate the type and extent of his injuries.
- Ensure that dressings over wounds are adequately reinforced.
- Ensure that fractured bones are properly immobilized and supported to prevent them from cutting through muscle, blood vessels, and skin.

b. Based upon your evaluation of the type and extent of the casualty's injury and your knowledge of the various manual carries, you must select the best possible method of manual transportation. If the casualty is conscious, tell him how he is to be transported. This will help allay his fear of movement and gain his cooperation and confidence.

c. Buddy aid for chemical agent casualties includes those actions required to prevent an incapacitated casualty from receiving additional injury from the effects of chemical hazards. If a casualty is physically unable to decontaminate himself or administer the proper chemical agent antidote, the casualty's buddy assists him and assumes responsibility for his care. Buddy-aid includes—

- Administering the proper chemical agent antidote.
- Decontaminating the incapacitated casualty's exposed skin.
- Ensuring that his protective ensemble remains correctly emplaced.
- Maintaining respiration.
- Controlling bleeding.
- Providing other standard first aid measures
- Transporting the casualty out of the contaminated area.

B-6. Positioning the Casualty

The first step in any manual carry is to position the casualty to be lifted. If he is conscious, he should be told how he is to be positioned and transported. This helps lessen his fear of movement and to gain his cooperation. It may be necessary to roll the casualty onto his abdomen, or his back, depending upon the position in which he is lying and the particular carry to be used.

a. To roll a casualty onto his abdomen, kneel at the casualty's uninjured side.

(1) Place his arms above his head; cross his ankle which is farther from you over the one that is closer to you.

(2) Place your hands on the shoulder which is farther from you; place your other hand in the area of his hip or thigh (Figure B-1).

(3) Roll him gently toward you onto his abdomen (Figure B-2).

b. To roll a casualty onto his back, follow the same procedure described in *a* above, except gently roll the casualty onto his back, rather than onto his abdomen.

Figure B-1. Positioning the casualty.

Figure B-2. Rolling casualty onto his abdomen.

B-7. Medical Evacuation and Transportation of Casualties

a. Medical evacuation of the sick and wounded (with en route medical care) is the responsibility of medical personnel who have been provided special training and equipment. Therefore, unless a good reason for you to transport a casualty arises, wait for some means of medical evacuation to be provided. When the situation is urgent and you are unable to obtain medical assistance or know that no medical evacuation assets are available, you will have to transport the casualty. For this reason, you must know how to transport him without increasing the seriousness of his condition.

b. Transporting a casualty by litter (FM 8-10-6) is safer and more comfortable for him than by manual means; it is also easier for you.

Manual transportation, however, may be the only feasible method because of the terrain or the combat situation; or it may be necessary to save a life. In these situations, the casualty should be transferred to a litter as soon as one can be made available or improvised.

B-8. Manual Carries

Casualties carried by manual means must be carefully and correctly handled, otherwise their injuries may become more serious or possibly fatal. Situation permitting, transport of a casualty should be organized and unhurried. Each movement should be performed as deliberately and gently as possible. Casualties should not be moved before the type and extent of injuries are evaluated and the required first aid is administered. The exception to this occurs when the situation dictates immediate movement for safety purposes (for example, it may be necessary to remove a casualty from a burning vehicle); that is, the situation dictates that the urgency of casualty movement outweighs the need to administer first aid. Manual carries are tiring for the bearers and involve the risk of increasing the severity of the casualty's injury. In some instances, however, they are essential to save the casualty's life. Although manual carries are accomplished by one or two bearers, the two-man carries are used whenever possible. They provide more comfort to the casualty, are less likely to aggravate his injuries, and are also less tiring for the bearers. The distance a casualty can be carried depends on many factors, such as—

- Nature of the casualty's injuries.
- Strength and endurance of the bearer(s).
- Weight of the casualty.
- Obstacles encountered during transport (natural or manmade).
- Type of terrain.

a. One-man Carries. These carries should be used when only one bearer is available to transport the casualty.

(1) The *fireman's carry* (Figure B-3) is one of the easiest ways for one individual to carry another. After an unconscious or disabled casualty has been properly positioned, he is raised from the ground, then supported and placed in the carrying position.

(*a*) After rolling the casualty onto his abdomen, straddle him. Extend your hands under his chest and lock them together.

(*b*) Lift the casualty to his knees as you move backward.

(*c*) Continue to move backward, thus straightening the casualty's legs and locking his knees.

(*d*) Walk forward, bringing the casualty to a standing position; tilt him slightly backward to prevent his knees from buckling.

(*e*) As you maintain constant support of the casualty with one arm, free your other arm, quickly grasp his wrist, and raise his arm high. Instantly pass your head under his raised arm, releasing it as you pass under it.

(*f*) Move swiftly to face the casualty and secure your arms around his waist. Immediately place your foot between his feet and spread them apart (approximately 6 to 8 inches).

(*g*) Grasp the casualty's wrist and raise his arm high over your head.

(*h*) Bend down and pull the casualty's arm over and down on your shoulder, bringing his body across your shoulders. At the same time, pass your arm between his legs.

(*i*) Grasp the casualty's wrist with one hand, and place your other hand on your knee for support.

(*j*) Rise with the casualty positioned correctly. Your other hand is free for use.

Figure B-3. Fireman's carry (Illustrated A—J).

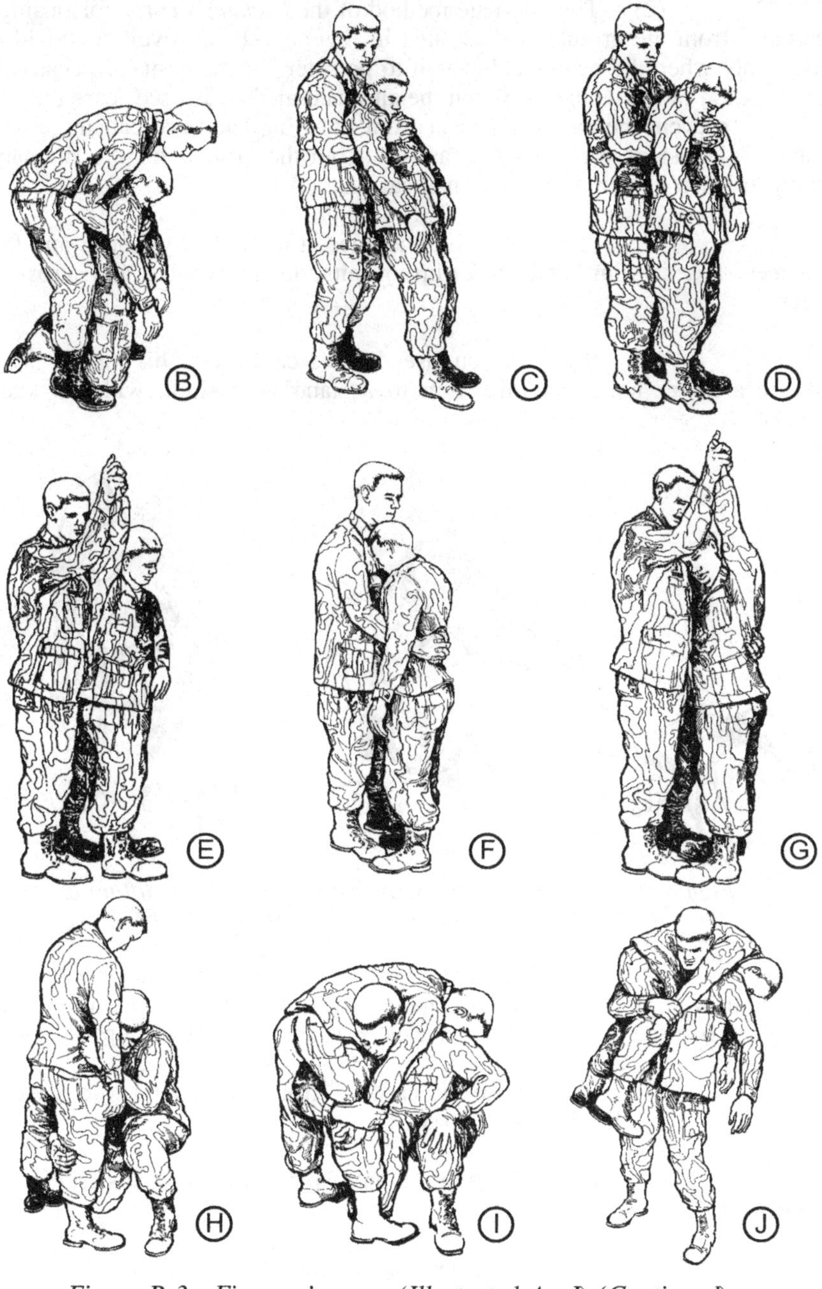

Figure B-3. Fireman's carry (Illustrated A—J) (Continued).

(2) The alternate method of the *fireman's carry* for raising a casualty from the ground is illustrated in Figure B-4; however, it should be used only when the bearer believes it to be safer for the casualty because of the location of his wounds. When the alternate method is used, care must be taken to prevent the casualty's head from snapping back and causing a neck injury. The steps for raising a casualty from the ground for the fireman's carry are also used in other one-man carries.

(*a*) Kneel on one knee at the casualty's head and face his feet. Extend your hands under his armpits, down his sides, and across his back.

(*b*) As you rise, lift the casualty to his knees. Then secure a lower hold and raise him to a standing position with his knees locked.

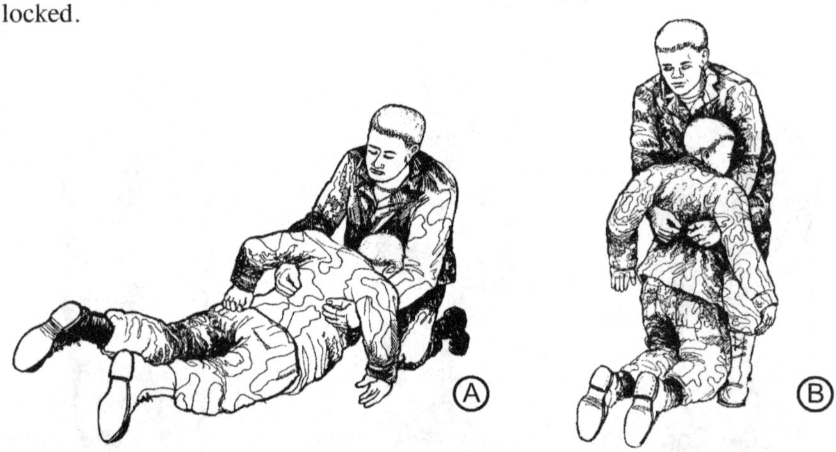

Figure B-4. Fireman's carry (alternate method) for lifting a casualty to a standing position (Illustrated A—B).

(3) In the *supporting carry* (Figure B-5), the casualty must be able to walk or at least hop on one leg, using the bearer as a crutch. This carry can be used to assist him as far as he is able to walk or hop.

(*a*) Raise the casualty from the ground to a standing position by using the fireman's carry.

(*b*) Grasp the casualty's wrist and draw his arm around your neck.

(*c*) Place your arm around his waist. The casualty is now able to walk or hop using you as a support.

Figure B-5. Supporting carry.

(4) The *arms carry* (Figure B-6) is useful in carrying a casualty for a short distance (up to 50 meters) and for placing him on a litter.

(*a*) Raise or lift the casualty from the ground to a standing position, as in the fireman's carry.

(*b*) Place one arm under the casualty's knees and your other arm around his back.

(*c*) Lift the casualty.

(*d*) Carry the casualty high to lessen fatigue.

Figure B-6. Arms carry.

(5) Only a conscious casualty can be transported by the *saddleback carry* (Figure B-7), because he must be able to hold onto the bearer's neck. To use this technique—

(*a*) Raise the casualty to an upright position, as in the fireman's carry.

(*b*) Support the casualty by placing an arm around his waist. Move to the casualty's side. Have the casualty put his arm around your neck and move in front of him with your back to support him.

(*c*) Have the casualty encircle his arms around your neck

(*d*) Stoop, raise him on your back and clasp your hands together beneath his thighs, if possible.

Figure B-7. Saddleback carry.

(6) In the *pack-strap carry* (Figure B-8), the casualty's weight rests high on the your back. This makes it easier for you to carry the casualty a moderate distance (50 to 300 meters). To eliminate the possibility of injury to the casualty's arms, you must hold his arms in a palms-down position.

(*a*) Lift the casualty from the ground to a standing position, as in the fireman's carry.

(*b*) Support the casualty with your arms around him and grasp his wrist closer to you.

(*c*) Place his arm over your head and across your shoulders.

(*d*) Move in front of him while still supporting his weight against your back.

(*e*) Grasp his other wrist and place this arm over your shoulder.

(*f*) Bend forward and raise or hoist the casualty as high on your back as possible so that his weight is resting on your back.

NOTE

Once the casualty is positioned on the bearer's back, the bearer remains as erect as possible to prevent straining or injuring his back.

Figure B-8. Pack-strap carry.

(7) The *pistol-belt carry* (Figure B-9) is the best one-man carry for a long distance (over 300 meters). The casualty is securely supported upon your shoulders by a belt. Both your hands and the casualty's (if conscious) are free for carrying a weapon or equipment, or climbing obstacles. With your hands free and the casualty secured in place, you are also able to creep through shrubs and under low-hanging branches.

(*a*) Link two pistol belts (or three, if necessary) together to form a sling. Place the sling under the casualty's thighs and lower back so that a loop extends from each side.

NOTE

If pistol belts are not available for use, other items such as a rifle sling, two cravat bandages, two litter straps, or any other suitable material, which will not cut or bind the casualty may be used.

(*b*) Lie face up between the casualty's outstretched legs. Thrust your arms through the loops and grasp his hands and trouser leg on his injured side.

(*c*) Roll toward the casualty's uninjured side onto your abdomen, bringing him onto your back. Adjust the sling, if necessary.

(*d*) Rise to a kneeling position. The belt will hold the casualty in place.

(*e*) Place one hand on your knee for support and rise to an upright position. (The casualty is supported on your shoulders.)

(*f*) Carry the casualty with your hands free for use in rifle firing, climbing, or surmounting obstacles.

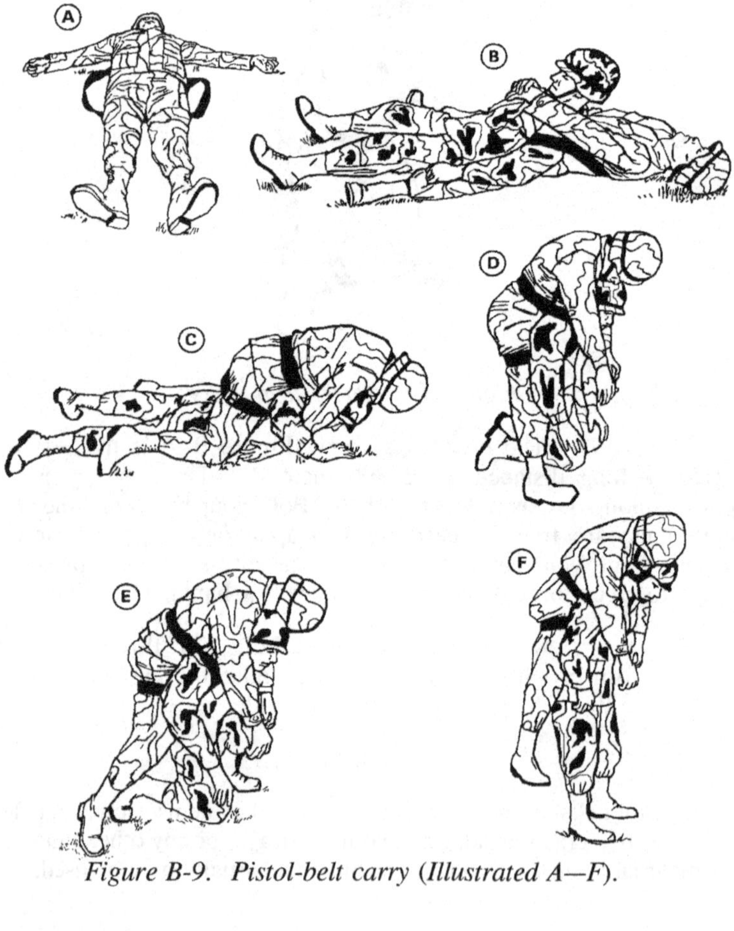

Figure B-9. Pistol-belt carry (Illustrated A—F).

(8) The *pistol-belt drag* (Figure B-10), as well as other drags, is generally used for short distances (up to 50 meters). This drag is useful in combat, since both the bearer and the casualty can remain closer to the ground than in any other drags.

(*a*) Extend two pistol belts or similar objects to their full length and join them together to make a continuous loop.

(*b*) Roll the casualty onto his back, as in the fireman's carry.

(*c*) Pass the loop over the casualty's head, and position it across his chest and under his armpits. Then cross the remaining portion of the loop, thus forming a figure eight. Keep tension on the belts so they do not come unhooked.

(*d*) Lie on your side facing the casualty.

(*e*) Slip the loop over your head and turn onto your abdomen. This enables you to drag the casualty as you crawl.

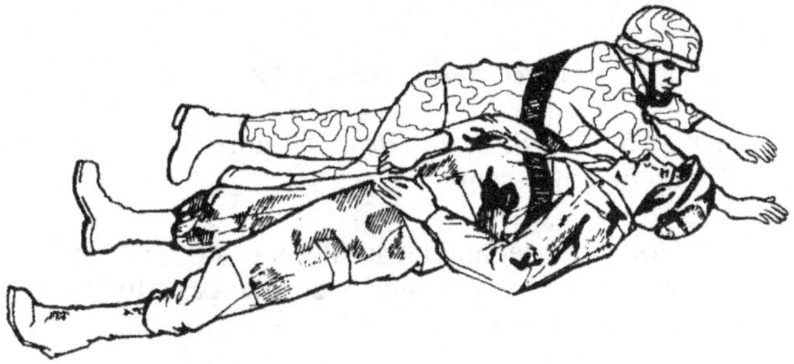

Figure B-10. Pistol-belt drag.

(9) The *neck drag* (Figure B-11) is useful in combat because the bearer can transport the casualty as he creeps behind a low wall or shrubbery, under a vehicle, or through a culvert. If the casualty is unconscious, his head must be protected from the ground. The neck drag cannot be used if the casualty has a broken arm.

NOTE

If the casualty is conscious, he may clasp his hands together around your neck.

(*a*) Tie the casualty's hands together at the wrists.

(*b*) Straddle the casualty in a kneeling face-to-face position.

(*c*) Loop the casualty's tied hands over and around your neck.

(*d*) Crawl forward dragging the casualty with you.

NOTE

If the casualty is unconscious, protect his head from the ground.

Figure B-11. Neck drag.

(10) The *cradle drop drag* (Figure B-12) is effective in moving a casualty up or down steps.

(*a*) Kneel at the casualty's head (with him lying on his back). Slide your hands, with palms up, under the casualty's shoulders and get a firm hold under his armpits.

(*b*) Rise (partially), supporting the casualty's head on one of your forearms. (You may bring your elbows together and let the casualty's head rest on both of your forearms.)

(*c*) Rise and drag the casualty backward. (The casualty is in a semisitting position.)

(*d*) Back down the steps, supporting the casualty's head and body and letting his hips and legs drop from step to step.

NOTE

If the casualty needs to be moved up the steps, you should back up the steps, using the same procedure.

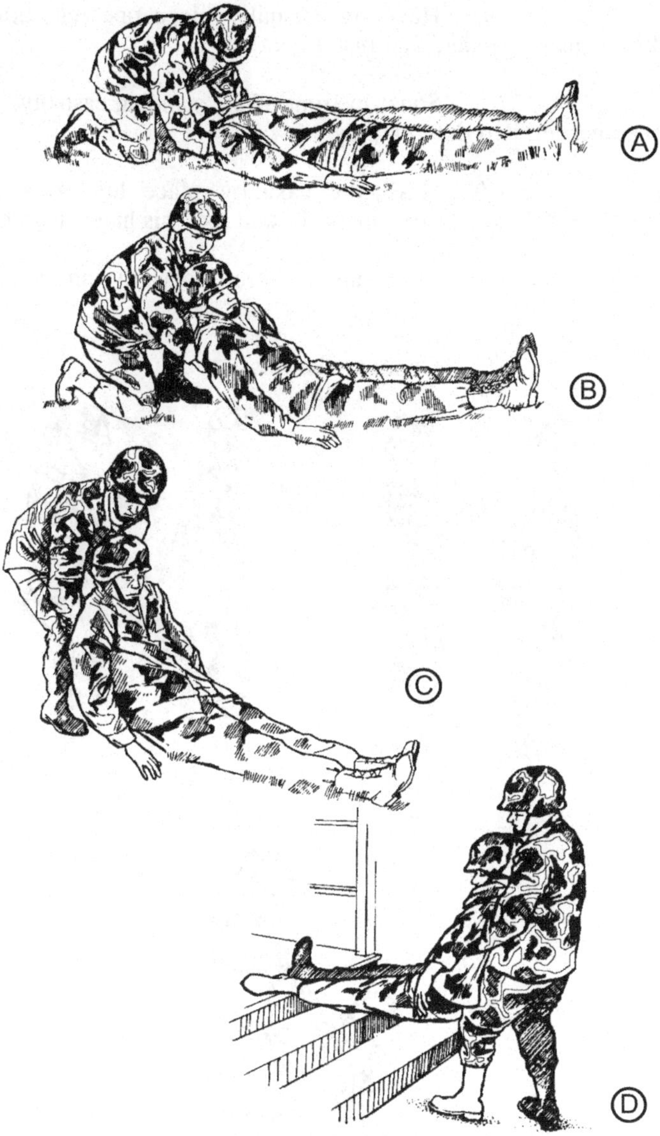

Figure B-12. Cradle-drop drag (Illustrated A—D).

(11) The *LBE carry using the bearer's LBE* can be used with a conscious casualty (Figure B-13).

(*a*) Loosen all suspenders on your LBE.

(*b*) Have the casualty place one leg into the loop formed by your suspenders and pistol belt.

(*c*) Squat in front of the standing casualty. Have him place his other leg into the loop, also.

(*d*) Have the casualty place his arms over your shoulders, lean forward onto your back, and lock his hands together.

(*e*) Stand up and lean forward into a comfortable position.

(*f*) Continue the mission.

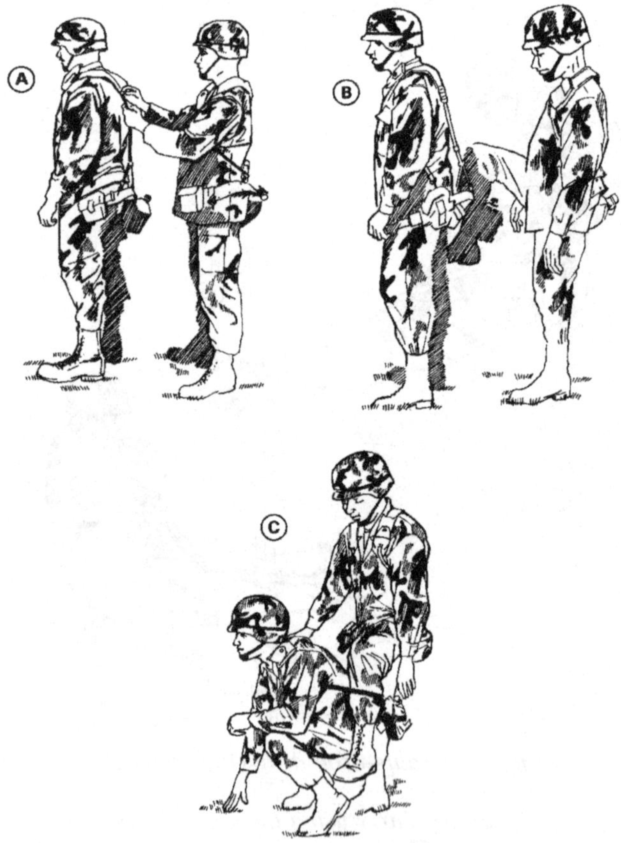

Figure B-13. Load bearing equipment carry using bearer's LBE (conscious casualty) (Illustrated A—F).

Figure B-13. Load bearing equipment carry using bearer's LBE (conscious casualty) (Illustrated A—F) (Continued).

(12) The *LBE carry using the bearer's LBE* can be used with an unconscious casualty or one who cannot stand (Figure B-14).

(*a*) Position the casualty on the flat of his back.

(*b*) Remove your LBE and loosen all suspender straps.

(*c*) Lift the casualty's leg and place it through the loop formed by your suspenders and pistol belt. Then place the other leg through the same loop. The LBE is moved up until the pistol belt is behind the casualty's thighs.

(*d*) Lay between the casualty's legs; work your arms through the LBE suspenders.

(*e*) Grasp the casualty's hand (on the injured side), and roll the casualty (on his uninjured side) onto your back.

(*f*) Rise to one knee and then push into a standing position.

(*g*) Bring the casualty's arms over your shoulders. Grasp his hands and secure them if the casualty is unconscious. If the casualty is conscious, have him lock his hands in front if he is able to do so.

(*h*) Lean forward into a comfortable position and continue the mission.

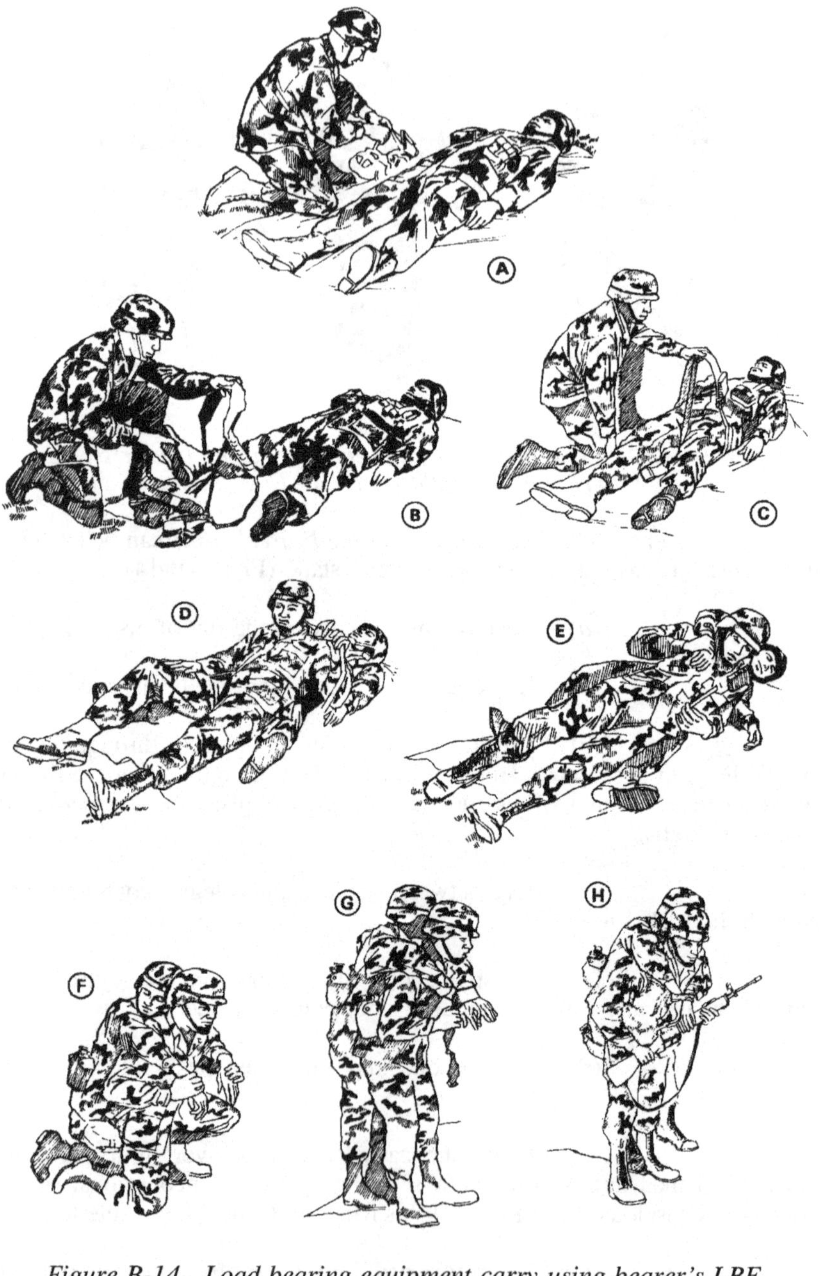

Figure B-14. Load bearing equipment carry using bearer's LBE (unconscious casualty or one that cannot stand) (Illustrated A—H).

(13) The *LBE carry using the casualty's LBE* (Figure B-15) can be used with a conscious or unconscious casualty.

(*a*) Position the casualty on his back with his LBE on.

(*b*) Loosen the casualty's two front suspenders.

(*c*) Position yourself between the casualty's legs, and slip your arms into the casualty's two front suspenders (up to your shoulders).

(*d*) Work his arms out of his LBE suspenders.

(*e*) Grasp the casualty's hand (on the injured side), and roll him (on his uninjured side) onto your back.

(*f*) Rise to one knee, then into a standing position.

(*g*) Grasp the casualty's hands and secure them, if the casualty is unconscious. Have the casualty lock his hands in front of you, if he is conscious.

(*h*) Lean forward into a comfortable position and continue the mission.

Figure B-15. Load bearing equipment carry using casualty's LBE (Illustrated A—G).

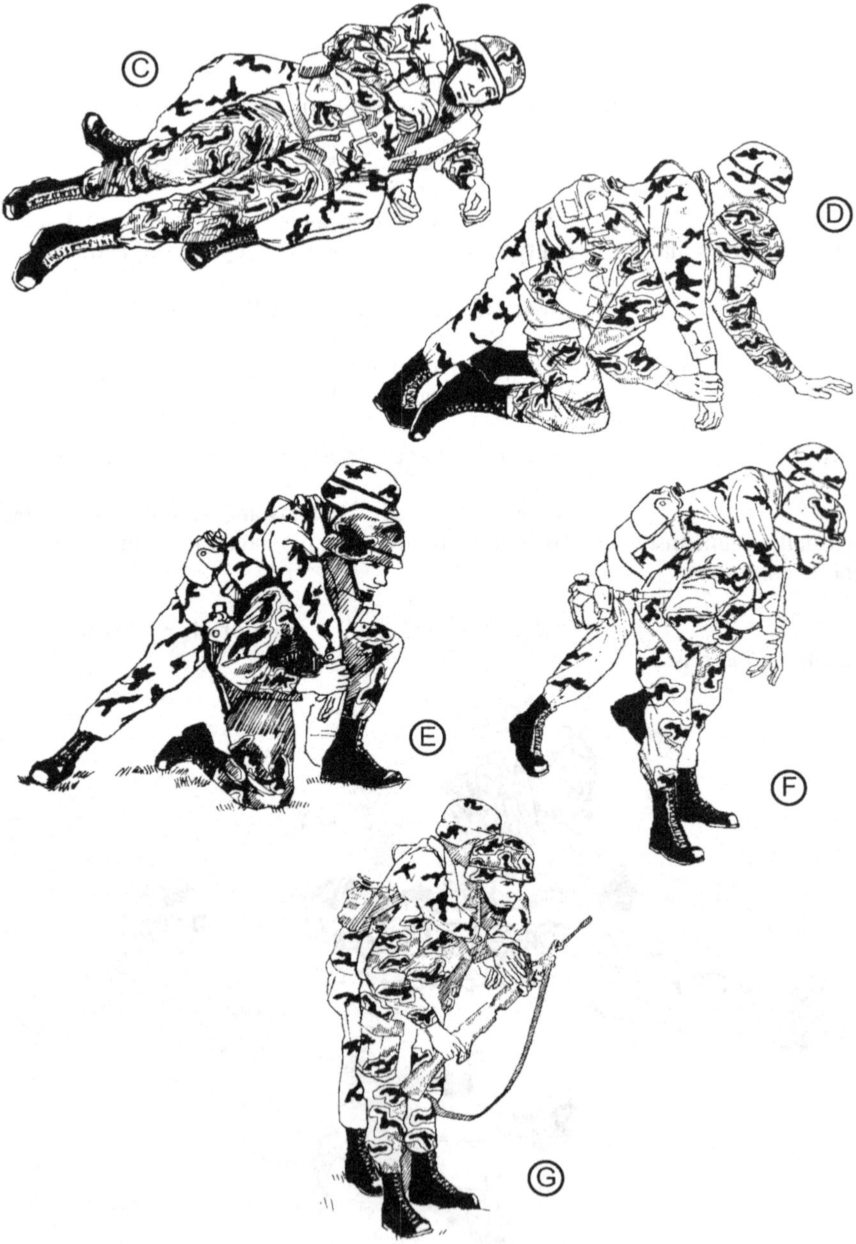

Figure B-15. Load bearing equipment carry using casualty's LBE (Illustrated A—G) (Continued).

b. *Two-man Carries.* These carries should be used whenever possible. They provide more casualty comfort, are less likely to aggravate injuries, and are less tiring for the bearers. Five different two-man carries can be used.

(1) The *two-man support carry* (Figure B-16) can be used in transporting either conscious or unconscious casualties. If the casualty is taller than the bearers, it may be necessary for the bearers to lift the casualty's legs and let them rest on their forearms. The bearers—

(*a*) Help the casualty to his feet and support him with their arms around his waist.

(*b*) Grasp the casualty's wrists and draw his arms around their necks.

Figure B-16. Two-man supporting carry.

(2) The *two-man arms carry* (Figure B-17) is useful in carrying a casualty for a moderate distance (50 to 300 meters) and placing him on a litter. To lessen fatigue, the bearers should carry the casualty high and as close to their chests as possible. In extreme emergencies when there is no time to obtain a spine board, this carry is the safest one for transporting a casualty with a back injury. If possible, two additional bearers should be used to keep the casualty's head and legs in alignment with his body. The bearers—

(*a*) Kneel at one side of the casualty; then they place their arms beneath the casualty's back, waist, hips, and knees.

(*b*) Lift the casualty while rising to their knees.

(*c*) Turn the casualty toward their chests, while rising to a standing position. Carry the casualty high to lessen fatigue.

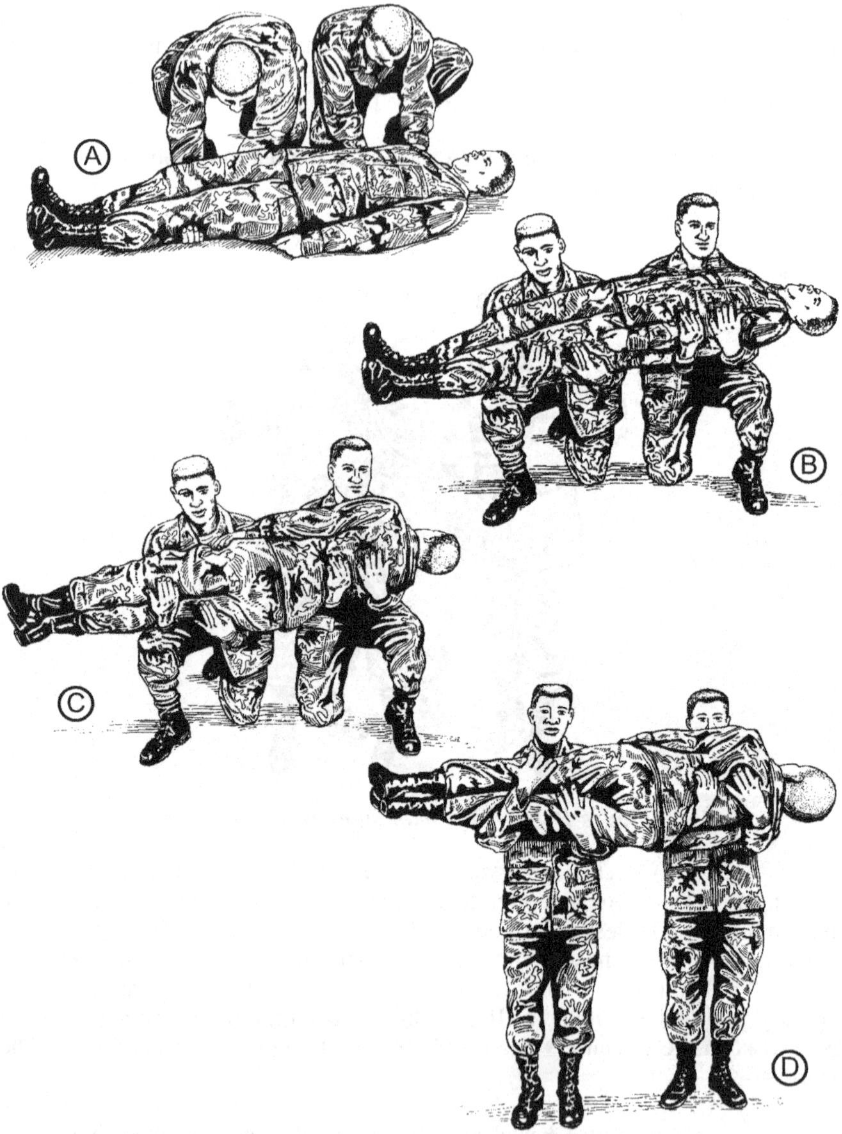

Figure B-17. Two-man arms carry (Illustrated A—D).

(3) The *two-man fore-and aft-carry* (Figure B-18) is a useful *two-man carry* for transporting a casualty for a long distance (over 300 meters). The taller of the two bearers should position himself at the casualty's head. By altering this carry so that both bearers face the casualty, it is useful for placing a casualty on a litter.

(*a*) The shorter bearer spreads the casualty's legs and kneels between them with his back to the casualty. He positions his hands behind the casualty's knees. The other bearer kneels at the casualty's head, slides his hands under the arms, across the chest, and locks his hands together.

(*b*) The two bearers rise together, lifting the casualty.

Figure B-18. Two-man fore-and-aft carry (Illustrated A—B).

(4) Only a conscious casualty can be transported with the *four-hand seat carry* (Figure B-19) because he must help support himself by placing his arms around the bearers' shoulders. This carry is especially useful in transporting a casualty with a head or foot injury for a moderate distance (50 to 300 meters). It is also useful for placing a casualty on a litter.

(*a*) Each bearer grasps one of his wrists and one of the other bearer's wrists, thus forming a packsaddle.

(*b*) The two bearers lower themselves sufficiently for the casualty to sit on the packsaddle; then, they have the casualty place his arms around their shoulders for support. The bearers then rise to an upright position.

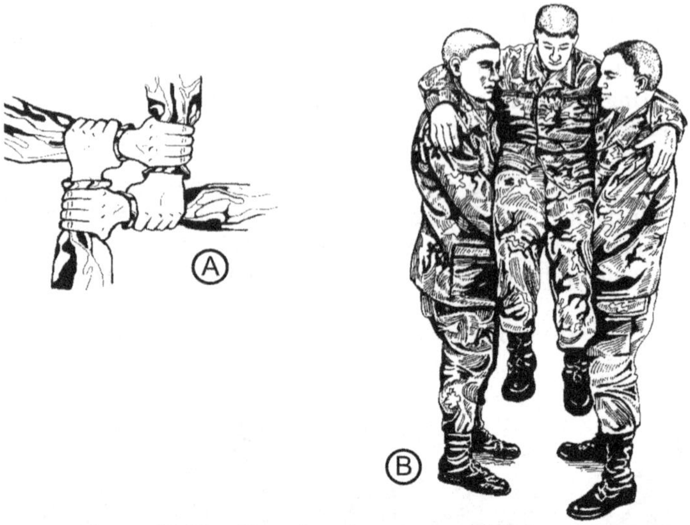

Figure B-19. Four-hand seat carry (Illustrated A—B).

(5) The *two-hand seat carry* (Figure B-20) is used when carrying a casualty for a short distance or for placing him on a litter. With the casualty lying on his back, a bearer kneels on each side of the casualty at his hips. Each bearer passes his arms under the casualty's thighs and back, and grasps the other bearer's wrists. The bearers rise lifting the casualty.

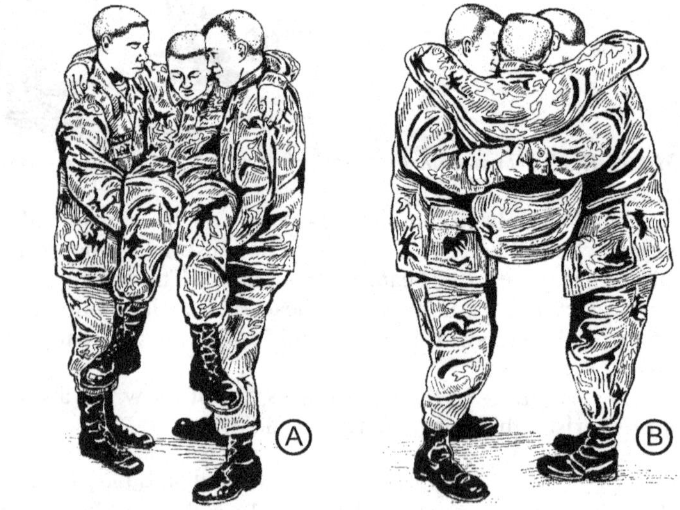

Figure B-20. Two-hand seat carry (Illustrated A—B).

B-9. Improvised Litters

Two men can support or carry a casualty without equipment for only short distances. By using available materials to improvise equipment, the casualty can be transported greater distances by two or more rescuers.

a. There are times when a casualty may have to be moved and a standard litter is not available. The distance may be too great for manual carries or the casualty may have an injury (such as a fractured neck, back, hip, or thigh) that would be aggravated by manual transportation. In these situations, litters can be improvised from materials at hand. Improvised litters must be as well constructed as possible to avoid risk of dropping or further injuring the casualty. Improvised litters are emergency measures and must be replaced by standard litters at the first opportunity.

b. Many different types of litters can be improvised, depending upon the materials available. A satisfactory litter can be made by securing poles inside such items as a blanket, poncho, shelter half, tarpaulin, mattress cover, jacket, shirt, or bed ticks, bags, and sacks (Figure B-18). Poles can be improvised from strong branches, tent supports, skis, lengths of pipe or other objects. If objects for improvising poles are not available, a blanket, poncho, or similar item can be rolled from both sides toward the center so the rolls can be gripped for carrying a patient. Most flat-surface objects of suitable size can be used as litters. Such objects include doors, boards, window shutters, benches, ladders, cots, and chairs. If possible, these objects should be padded for the casualty's comfort.

(1) To improvise a litter using a blanket and poles (Figure B-21), the following steps should be used.

Figure B-21. Litter made with blanket and poles.

(*a*) Open the blanket and lay one pole lengthwise across the center; then fold the blanket over the pole.

(*b*) Place the second pole across the center of the folded blanket.

(*c*) Fold the free edges of the blanket over the second pole and across the first pole.

(2) To improvise a litter using shirts or jackets (Figure B-22), button the shirt or jacket and turn it inside out, leaving the sleeves inside, (more than one shirt or jacket may be required), then pass the pole through the sleeves.

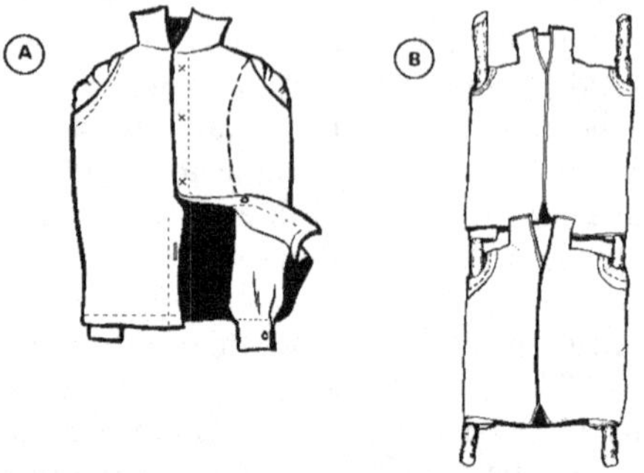

Figure B-22. Litter improvised from jackets and poles (Illustrated A—B).

(3) To improvise a litter from bed sacks and poles (Figure B-23), rip open the corners of bed ticks, bags, or sacks; then pass the poles through them.

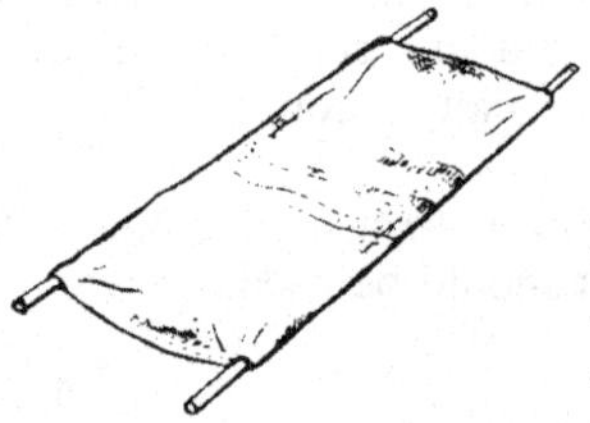

Figure B-23. Litter improvised from bed sacks and poles.

(4) If no poles are available, roll a blanket, shelter half, tarpaulin, or similar item from both sides toward the center (Figure B-24). Grip the rolls to carry the casualty.

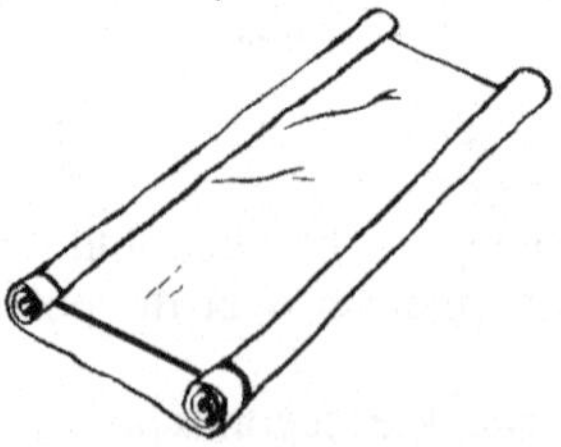

Figure B-24. Rolled blanket used as a litter.

c. Any of the appropriate carries may be used to place a casualty on a litter. These carries are:

- The one-man arms carry (Figure B-6).
- The two-man arms carry (Figure B-17).
- The two-man fore-and-aft carry (Figure B-18).
- The two-hand seat carry (Figure B-20).
- The four-hand seat carry (Figure B-19).

WARNING

Unless there is an immediate life-threatening situation (such as fire, explosion), DO NOT move a casualty with a suspected back or neck injury. Seek medical personnel for guidance on how to transport.

d. Either two or four service members (head/foot) may be used to lift a litter. To lift the litter, follow the procedure below.

(1) Raise the litter at the same time as the other carriers/ bearers.

(2) Keep the casualty as level as possible.

NOTE

Use caution when transporting on a sloping incline/hill.

GLOSSARY

ACRONYMS, ABBREVIATIONS, AND DEFINITIONS

AC hydrogen cyanide
AFMAN Air Force Manual
AOC area of concentration
AR Army regulation
ATM advanced trauma management
ATNAA Antidote Treatment, Nerve Agent, Autoinjector
attn attention

BDO battle dress overgarment
BDU battle dress uniform
BZ anticholinergic drugs

C Celsius
CANA Convulsant Antidote for Nerve Agent
CASEVAC casualty evacuation
cc cubic centimeter
CG phosgene
CHS combat health support
CK cyanogen chloride
Cl chlorine
CLS Combat Lifesaver
CNS central nervous system
CO_2 carbon dioxide
COSR combat and operational stress reactions
CSR combat stress reaction
CTA common table of allowance
CX phosgene oxime

DA Department of the Army
DD Department of Defense
DM diphenylaminochloroarsine (adamsite)
DNBI disease and nonbattle injury
DOD Department of Defense
DP diphosgene
DS direct support

EMT emergency medical treatment

F Fahrenheit
FM field manual

H mustard

HD mustard
HM Hospital Corpsman
HN nitrogen mustard
HSS health service support

IPE individual protective equipment
IV intravenous

JSLIST Joint Services Light Weight Integrated Suit Technology

L lewisite
lasers laser means Light Amplification by Stimulated Emission of Radiation and sources include range finders, weapons/guidance, communication systems, and weapons simulations such as MILES [Multiple Integrated Laser Engagement System].
LBE load bearing equipment
LX lewisite and mustard

MCRP Marine Corps Reference Publication
MILES Multiple Integrated Laser Engagement System
ml milliliter
MOPP mission-oriented protective posture
MOS military occupational specialty
MTF medical treatment facility

NAPP Nerve Agent Pyridostigmine Pretreatment
NATO North Atlantic Treaty Organization
NBC nuclear, biological, and chemical
NCO noncommissioned officer
NTRP Navy Tactical Reference Publication

occlusive dressing air tight transparent dressing used to seal and cover wounds
oz ounce

PAM pamphlet
PS chloropicrin
PTSD post-traumatic stress disorder

QSTAG Quadripartite Standardization Agreement

SOP standing operating procedure
STANAG standardization agreement
STP soldier training publication

2 PAM Cl pralidoxime chloride
TB MED technical bulletin medical
TM technical manual
TSOP tactical standing operating procedure

US United States

WP white phosphorus

REFERENCES

DOCUMENTS NEEDED

These documents must be available to the intended users of this publication.

NATO STANAGs

These agreements are available on request using DD Form 1425 from Standardization Document Order Desk, 700 Robin Avenue, Building 4, Section D, Philadelphia, Pennsylvania 19111-5094.

2122. *Medical Training in First Aid, Basic Hygiene and Emergency Care*. 10 December 1975.
2126. *First Aid Kits and Emergency Medical Care Kits*. 27 September 1983.
2358. *First Aid and Hygiene Training in NBC Operations*. 3 March 1989.
2871. *First Aid Material for Chemical Injuries*. 10 March 1986.

ABCA QSTAGs

These agreements are available on request using DD Form 1425 from Standardization Document Order Desk, 700 Robin Avenue, Building 4, Section D, Philadelphia, Pennsylvania 19111-5094.

535. *Medical Training in First Aid, Basic Hygiene and Emergency Care.* 12 November 1979.

Joint and Multiservice Publications

FM 21-10. *Field Hygiene and Sanitation*. MCRP 4-11.1D. 21 June 2000.

Army Publications

AR 350-41. *Training In Units*. 19 March 1993.
DA PAM 350-59. *Army Correspondence Course Program Catalog*. 26 October 2001.
FM 3-4. *NBC Protection*. FMFM 11-9. 29 May 1992 (Reprinted with basic including Change 1, 28 October 1992; Change 2, 26 February 1996.)
FM 3-5. *NBC Decontamination*. MCWP 3-37.3. 28 July 2000. (Change 1, 31 January 2002.)

FM 3-100. *Chemical Operations Principles and Fundamentals.* MCWP 3-3.7.1. 8 May 1996.
FM 4-02.33 (8-33). *Control of Communicable Diseases Manual* (17th Edition). 3 January 2000.
FM 8-10-6 (4-02.2). *Medical Evacuation in a Theater of Operations—Tactics, Techniques, and Procedures.* 14 April 2000.
FM 4-02.7 (8-10-7). *Health Service Support in a Nuclear, Biological, and Chemical Environment.* 1 October 2002.
FM 8-284 (4-02.284). *Treatment of Biological Warfare Agent Casualties.* NAVMED P-5042; AFMAN (I) 44-156; MCRP 4-11.1C. 17 July 2000. (Change 1, 8 July 2002.)
FM 8-285 (4-02.285). *Treatment of Chemical Agent Casualties and Conventional Military Chemical Injuries.* NAVMED P-5041; AFJMAN 44-149; FMFM 11-11. 22 December 1995.
FM 22-51 (4-02.22). *Leaders' Manual for Combat Stress Control.* 29 September 1994.
CTA 8-100. *Army Medical Department Expendable/Durable Items.* 31 August 1994.
CTA 50-900. *Clothing and Individual Equipment.* 1 September 1994.
STP 21-1-SMCT. *Soldier's Manual of Common Tasks Skill Level I.* 1 October 2001.

Department of Defense Forms

DD Form 1425. *Specifications and Standards Requisition.* March 1986.

READINGS RECOMMENDED

These readings contain relevant supplemental information.

Joint and Multiservice Publications

FM 8-9 (4-02.11). *NATO Handbook on the Medical Aspects of NBC Defensive Operations AMEDP-6 (B), Part I—Nuclear, Part II—Biological, Part III—Chemical.* NAVMED P-5059; AFJMAN 44-151V1V2V3. 1 February 1996.
TB MED 81. *Cold Injury.* NAVMED P-5052-29; AFP 161-11. 30 September 1976.
TB MED 507. *Occupational and Environmental Health Prevention, Treatment and Control of Heat Injury.* NAVMED P-5052-5; AFP 160-1. 25 July 1980.

Army Publications

AR 310-25. *Dictionary of United States Army Terms (Short Title: AD)*. 15 October 1983. (Reprinted with basic including Change 1, 21 May 1986.)

AR 310-50. *Authorized Abbreviations and Brevity Codes*. 15 November 1985.

TM 3-4230-216-10. *Operator's Manual for Decontaminating Kit, Skin: M258A1 (NSN 4230-01-101-3984) and Training Aid, Skin Decontaminating: M58A1 (6910-01-101-1768)*. 17 May 1985. (Change 1, 22 January 1997.)

INDEX

References are to paragraph numbers except where specified otherwise.

PREFACE

Mountains exist in almost every country in the world and almost every war has included some type of mountain operations. This pattern will not change; therefore, soldiers will fight in mountainous terrain in future conflicts. Although mountain operations have not changed, several advancements in equipment and transportation have increased the soldiers' capabilities. The helicopter now allows access to terrain that was once unreachable or could be reached only by slow methodical climbing. Inclement weather, however, may place various restrictions on the capabilities of air assets available to a commander. The unit must then possess the necessary mountaineering skills to overcome adverse terrain to reach an objective.

This field manual details techniques soldiers and leaders must know to cope with mountainous terrain. These techniques are the foundation upon which the mountaineer must build. They must be applied to the various situations encountered to include river crossings, glaciers, snow-covered mountains, ice climbing, rock climbing, and urban vertical environments. The degree to which this training is applied must be varied to conform to known enemy doctrine, tactics, and actions. This FM also discusses basic and advanced techniques to include acclimatization, illness and injury, equipment, anchors, evacuation, movement on glaciers, and training.

This field manual is a training aid for use by *qualified* personnel in conjunction with FM 3-97.6, Mountain Operations, which is used for planning operations in mountainous terrain. Personnel using FM 3-97.61 should attend a recognized Department of Defense Mountain Warfare School for proper training. **Improper use of techniques and procedures by untrained personnel may result in serious injury or death.** Personnel should be certified as Level I, Basic Mountaineer; Level II, Assault Climber; or Level III, Mountain Leader before using FM 3-97.61 for training (see Appendix A).

The measurements in this manual are stated as they are used in training (either metric or standard). Appendix B contains a measurement conversion chart for your convenience.

The proponent of this publication is HQ TRADOC. Submit changes for improving this publication to doctrine@benning.army.mil or on DA Form 2028 (Recommended Changes to Publications and Blank Forms) and forward to the Commander, U.S. Army Infantry School, ATTN: ATSH-RBO, Fort Benning, GA 31905-5593.

Unless otherwise stated, whenever the masculine gender is used, both men and women are included.

FIELD MANUAL
No. 3-97.61

HEADQUARTERS
DEPARTMENT OF THE ARMY
WASHINGTON, DC, 26 August 2002

MILITARY MOUNTAINEERING

CONTENTS

DISTRIBUTION RESTRICTION: Approved for public release; distribution is unlimited.

*This publication supersedes TC 90-6-1, dated 26 April 1989.

CHAPTER 1

MOUNTAIN TERRAIN, WEATHER, AND HAZARDS

Commanders must consider the effects terrain and weather will have on their operations, mainly on their troops and logistics efforts. Weather and terrain combine to challenge efforts in moving supplies to forward areas. Spring storms, which may deposit a foot of snow on dry roads, combined with unprepared vehicles create hazardous situations. Helicopters are a valuable asset for use in moving men and supplies, but commanders should not plan to use them as the only means of movement and resupply. Alternate methods must be planned due to the variability of weather. Units scheduled for deployment in mountainous terrain should become self-sufficient and train under various conditions. Commanders must be familiar with the restraints that the terrain can place on a unit.

Section I. MOUNTAIN TERRAIN

Operations in the mountains require soldiers to be physically fit and leaders to be experienced in operations in this terrain. Problems arise in moving men and transporting loads up and down steep and varied terrain in order to accomplish the mission. Chances for success in this environment are greater when a leader has experience operating under the same conditions as his men. Acclimatization, conditioning, and training are important factors in successful military mountaineering.

1-1. DEFINITION

Mountains are land forms that rise more than 500 meters above the surrounding plain and are characterized by steep slopes. Slopes commonly range from 4 to 45 degrees. Cliffs and precipices may be vertical or overhanging. Mountains may consist of an isolated peak, single ridges, glaciers, snowfields, compartments, or complex ranges extending for long distances and obstructing movement. Mountains usually favor the defense; however, attacks can succeed by using detailed planning, rehearsals, surprise, and well-led troops.

1-2. COMPOSITION

All mountains are made up of rocks and all rocks of minerals (compounds that cannot be broken down except by chemical action). Of the approximately 2,000 known minerals, seven rock-forming minerals make up most of the earth's crust: quartz and feldspar make up granite and sandstone; olivene and pyroxene give basalt its dark color; and amphibole and biotite (mica) are the black crystalline specks in granitic rocks. Except for calcite, found in limestone, they all contain silicon and are often referred to as silicates.

1-3. ROCK AND SLOPE TYPES

Different types of rock and different slopes present different hazards. The following paragraphs discuss the characteristics and hazards of the different rocks and slopes.

a. **Granite.** Granite produces fewer rockfalls, but jagged edges make pulling rope and raising equipment more difficult. Granite is abrasive and increases the danger of ropes or accessory cords being cut. Climbers must beware of large loose boulders. After a rain, granite dries quickly. Most climbing holds are found in cracks. Face climbing can be found, however, it cannot be protected.

b. **Chalk and Limestone.** Chalk and limestone are slippery when wet. Limestone is usually solid; however, conglomerate type stones may be loose. Limestone has pockets, face climbing, and cracks.

c. **Slate and Gneiss.** Slate and gneiss can be firm and or brittle in the same area (red coloring indicates brittle areas). Rockfall danger is high, and small rocks may break off when pulled or when pitons are emplaced.

d. **Sandstone.** Sandstone is usually soft causing handholds and footholds to break away under pressure. Chocks placed in sandstone may or may not hold. Sandstone should be allowed to dry for a couple of days after a rain before climbing on it—wet sandstone is extremely soft. Most climbs follow a crack. Face climbing is possible, but any outward pull will break off handholds and footholds, and it is usually difficult to protect.

e. **Grassy Slopes.** Penetrating roots and increased frost cracking cause a continuous loosening process. Grassy slopes are slippery after rain, new snow, and dew. After long, dry spells clumps of the slope tend to break away. Weight should be distributed evenly; for example, use flat hand push holds instead of finger pull holds.

f. **Firm Spring Snow (Firn Snow).** Stopping a slide on small, leftover snow patches in late spring can be difficult. Routes should be planned to avoid these dangers. Self-arrest should be practiced before encountering this situation. Beginning climbers should be secured with rope when climbing on this type surface. Climbers can glissade down firn snow if necessary. Firn snow is easier to ascend than walking up scree or talus.

g. **Talus.** Talus is rocks that are larger than a dinner plate, but smaller than boulders. They can be used as stepping-stones to ascend or descend a slope. However, if a talus rock slips away it can produce more injury than scree because of its size.

h. **Scree.** Scree is small rocks that are from pebble size to dinner plate size. Running down scree is an effective method of descending in a hurry. One can run at full stride without worry—the whole scree field is moving with you. Climbers must beware of larger rocks that may be solidly planted under the scree. Ascending scree is a tedious task. The scree does not provide a solid platform and will only slide under foot. If possible, avoid scree when ascending.

1-4. ROCK CLASSIFICATIONS

Rock is classified by origin and mineral composition.

a. **Igneous Rocks.** Deep within the earth's crust and mantle, internal heat, friction and radioactive decay creates magmas (melts of silicate minerals) that solidify into igneous rocks upon cooling. When the cooling occurs at depth, under pressure, and over time, the minerals in the magma crystallize slowly and develop well, making coarse-grained plutonic rock. The magma may move upward, propelled by its own lower density, either melting and combining with the overlying layers or forcing them aside. This results in an intrusive rock. If the melt erupts onto the surface it cools rapidly and the minerals form little or no crystal matrix, creating a volcanic or extrusive rock.

(1) ***Plutonic or Intrusive Rocks.*** Slow crystallization from deeply buried magmas generally means good climbing, since the minerals formed are relatively large and interwoven into a solid matrix. Weathering develops protrusions of resistant minerals, which makes for either a rough-surfaced rock with excellent friction, or, if the resistant

crystals are much larger than the surrounding matrix, a surface with numerous knobby holds. Pieces of foreign rock included in the plutonic body while it was rising and crystallizing, or clusters of segregated minerals, may weather differently than the main rock mass and form "chicken heads."

(a) Intrusions are named according to location and size. Large (100 square kilometers or larger) masses of plutonic rock are called "batholiths" and small ones "stocks." Most plutonic rock is in the granite family, differing only in the amounts of constituent minerals contained. A core of such batholiths is in every major mountain system in the world. In the Alps, Sierras, North Cascades, Rockies, Adirondacks, and most other ranges this core is at least partly exposed.

(b) Small plutonic intrusions are stocks, forced between sedimentary strata, and dikes, which cut across the strata. Many of these small intrusive bodies are quickly cooled and thus may look like extrusive rock.

(2) ***Volcanic or Extrusive Rocks.*** Explosive eruptions eject molten rock so quickly into the air that it hardens into loose aerated masses of fine crystals and uncrystallized glass (obsidian). When this ash consolidates while molten or after cooling, it is called "tuff," a weak rock that breaks down quickly and erodes easily. Quieter eruptions, where widespread lava flows from large fissures, produce basalt. Basaltic rocks are fine-grained and often sharp-edged.

(3) ***Jointing Rocks.*** In plutonic rocks, joints or cracks are caused by internal stresses such as contraction during cooling or expansion when overlying rock erodes or exfoliates. Some joints tend to follow a consistent pattern throughout an entire mountain and their existence can often be predicted. Therefore, when a ledge suddenly ends, the joint—and thus the ledge—may begin again around the corner. When molten rock extrudes onto the surface as a lava flow or intrudes into a cold surrounding mass as a dike or sill, the contraction from rapid cooling usually causes so much jointing that climbing can be extremely hazardous. Occasionally, this jointing is regular enough to create massed pillars with usable vertical cracks such as Devil's Tower in Wyoming.

b. **Sedimentary Rocks.** Sedimentary rocks are born high in the mountains, where erosion grinds down debris and moves it down to rivers for transportation to its final deposition in valleys, lakes, or oceans. As sediments accumulate, the bottom layers are solidified by pressure and by mineral cements precipitated from percolating groundwater. Gravel and boulders are transformed into conglomerates; sandy beaches into sandstone; beds of mud into mudstone or shale; and shell beds and coral reefs into limestone or dolomite.

(1) Though in general sedimentary rocks are much more friable than those cooled from molten magmas, pressure and cementing often produce solid rocks. In fact, by sealing up internal cracks cementing can result in flawless surfaces, especially in limestone.

(2) Most high mountain ranges have some sedimentary peaks. Ancient seafloor limestone can be found on the summits of the Himalayas and the Alps. The Canadian Rockies are almost exclusively limestone. With the exception of the Dolomites, in general sedimentary rocks do not offer high-angle climbing comparable to that of granite.

c. **Metamorphic Rocks.** These are igneous or sedimentary rocks that have been altered physically and or chemically by the tremendous heat and pressures within the earth. After sediments are solidified, high heat and pressure can cause their minerals to recrystallize. The bedding planes (strata) may also be distorted by folding and squeezing. Shale changes to slate or schist, sandstone and conglomerate into quartzite, and limestone to marble. These changes may be minimal, only slightly altering the sediments, or extensive enough to produce gneiss, which is almost indistinguishable from igneous rock.

(1) Metamorphic rocks may have not only joints and bedding, but cleavage or foliation, a series of thinly spaced cracks caused by the pressures of folding. Because of this cleavage, lower grades of metamorphic rocks may be completely unsuitable for climbing because the rock is too rotten for safe movement.

(2) Higher degrees of metamorphism or metamorphism of the right rocks provide a solid climbing surface. The Shawangunks of New York are an excellent example of high-grade conglomerate quartzite, which offers world class climbing. The center of the Green Mountain anticline contains heavily metamorphosed schist, which also provides solid climbing.

1-5. MOUNTAIN BUILDING

The two primary mechanisms for mountain-building are volcanic and tectonic activity. Volcanoes are constructed from lava and ash, which begin within the earth as magma. Tectonic activity causes plates to collide, heaving up fold mountains, and to pull apart and crack, forming fault-block mountain ranges.

a. **Plate Tectonics.** The massive slabs composing the outer layer are called tectonic plates. These plates are made up of portions of lighter, granitic continental crust, and heavier, basaltic oceanic crust attached to slabs of the rigid upper mantle. Floating slowly over the more malleable asthenosphere, their movement relative to each other creates earthquakes, volcanoes, ocean trenches, and mountain ridge systems.

b. **Mountain Structure.** The different horizontal and vertical stresses that create mountains usually produce complex patterns. Each type of stress produces a typical structure, and most mountains can be described in terms of these structures.

(1) ***Dome Mountains.*** A simple upward bulge of the crust forms dome mountains such as the Ozarks of Arkansas and Missouri, New York's Adirondacks, the Olympics of Washington, and the High Uintahs of Utah. They are usually the result of the upward movement of magma and the folding of the rock layers overhead. Erosion may strip away the overlying layers, exposing the central igneous core.

(2) ***Fault-Block Mountains.*** Faulting, or cracking of the crust into large chunks, often accompanies upwarp, which results in fault-block mountains. Many forms are created by the motion of these chunks along these faults.

(a) The ranges of the desert country of California, Nevada, and Utah provide the clearest display of faulting. The breakage extends to the surface and often during earthquakes—caused by slippage between the blocks—fresh scarps many feet high develop.

(b) Sometimes a block is faulted on both sides and rises or falls as a unit. More often, however, it is faulted on one side only. The Tetons of Wyoming and the Sierra Nevada display this—along the single zone of faults the range throws up impressive steep scarps, while on the other side the block bends but does not break, leaving a gentler slope from the base of the range to the crest. An example of a dropped block is California's Death Valley, which is below sea level and could not have been carved by erosion.

(3) ***Fold Mountains.*** Tectonic forces, in which continental plates collide or ride over each other, have given rise to the most common mountain form—fold mountains. Geologists call folds geosynclines. Upward folded strata are anticlines and downward folds are synclines. When erosion strips down the overburden of rock from folded mountain ranges, the oldest, central core is all that remains. The Alps and the Appalachians are examples of fold mountains. When the squeezing of a range is intense the rocks of the mountain mass first fold but then may break, and parts of the rocks are pushed sideways and override neighboring formations. This explains why older rocks are often found perched on top of younger ones. Isolated blocks of the over thrust mass may form when erosion strips away links connecting them with their place of origin. Almost every range of folded mountains in the world exhibits an over thrust of one sort or another.

(4) ***Volcanic Mountains.*** Along convergent plate boundaries volcanic activity increases. As it is forced underneath an overriding neighbor, continental crust melts and turns to magma within the mantle. Since it is less dense than the surrounding material it rises and erupts to form volcanoes.

(a) These volcanoes are found in belts, which correspond to continental margins around the world. The best known is the "Ring of Fire" encircling the Pacific Ocean from Katmai in Alaska through the Cascades (Mount Rainier and Mount Saint Helens) down through Mexico's Popocatepetl to the smokes of Tierra del Fuego. This belt then runs west down the Aleutian chain to Kamchatka, south to the volcanoes of Japan and the Philippines, and then east through New Guinea into the Pacific. Smaller volcanic belts are found along the Indonesian-SE Asian arc, the Caucasus region, and the Mediterranean.

(b) Volcanic activity also arises at boundaries where two plates are moving away from each other, creating deep rifts and long ridges where the crust has cracked apart and magma wells up to create new surface material. Examples of this are the Mid-Atlantic Ridge, which has created Iceland and the Azores, and the Rift Valley of East Africa with Kilimanjaro's cone.

(5) ***Complex Mountains.*** Most ranges are complex mountains with portions that have been subject to several processes. A block may have been simply pushed upward without tilting with other portions folded, domed, and faulted, often with a sprinkling of volcanoes. In addition, these processes occur both at the macro and the micro level. One massive fold can make an entire mountain peak; however, there are folds measured by a rope length, and tiny folds found within a handhold. A mountain front may be formed from a single fault, but smaller faults that form ledges and gullies may also be present.

1-6. ROUTE CLASSIFICATION

Military mountaineers must be able to assess a vertical obstacle, develop a course of action to overcome the obstacle, and have the skills to accomplish the plan. Assessment of a vertical obstacle requires experience in the classifications of routes and understanding the levels of difficulty they represent. Without a solid understanding of the difficulty of a chosen route, the mountain leader can place his life and the life of other soldiers in extreme danger. Ignorance is the most dangerous hazard in the mountain environment.

a. In North America the Yosemite Decimal System (YDS) is used to rate the difficulty of routes in mountainous terrain. The YDS classes are:

- Class 1—Hiking trail.
- Class 2—Off-trail scramble.
- Class 3—Climbing, use of ropes for beginners (moderate scrambling).
- Class 4—Belayed climbing. (This is moderate to difficult scrambling, which may have some exposure.)
- Class 5—Free climbing. (This class requires climbers to be roped up, belay and emplace intermediate protection.)

b. Class 5 is further subdivided into the following classifications:

(1) Class 5.0-5.4—Little difficulty. This is the simplest form of free climbing. Hands are necessary to support balance. This is sometimes referred to as advanced rock scrambling.

(2) Class 5.5—Moderate difficulty. Three points of contact are necessary.

(3) Class 5.6—Medium difficulty. The climber can experience vertical position or overhangs where good grips can require moderate levels of energy expenditure.

(4) Class 5.7—Great difficulty. Considerable climbing experience is necessary. Longer stretches of climbing requiring several points of intermediate protection. Higher levels of energy expenditure will be experienced.

(5) Class 5.8—Very great difficulty. Increasing amount of intermediate protection is the rule. High physical conditioning, climbing technique, and experience required.

(6) Class 5.9—Extremely great difficulty. Requires well above average ability and excellent condition. Exposed positions, often combined with small belay points. Passages of the difficult sections can often be accomplished under good conditions. Often combined with aid climbing (A0-A4).

(7) Class 5.10—Extraordinary difficulty. Climb only with improved equipment and intense training. Besides acrobatic climbing technique, mastery of refined security technique is indispensable. Often combined with aid climbing (A0-A4).

(8) Class 5.11-5.14—Greater increases of difficulty, requiring more climbing ability, experience, and energy expenditure. Only talented and dedicated climbers reach this level.

c. Additional classifications include the following.

(1) Classes are further divided into a, b, c, and d categories starting from 5.10 to 5.14 (for example, 5.10d).

(2) Classes are also further divided from 5.9 and below with +/- categories (for example, 5.8+).

(3) All class 5 climbs can also be designated with "R" or "X," which indicates a run-out on a climb. This means that placement of intermediate protection is not possible on portions of the route. (For example, in a classification of 5.8R, the "R" indicates periods of run-out where, if a fall was experienced, ground fall would occur.) Always check the local guidebook to find specific designation for your area.

(4) All class 5 climbs can also be designated with "stars." These refer to the popularity of the climb to the local area. Climbs are represented by a single "star" up to five "stars;" a five-star climb is a classic climb and is usually aesthetically pleasing.

d. Aid climb difficulty classification includes:

(1) A0—"French-free." This technique involves using a piece of gear to make progress; for example, clipping a sling into a bolt or piece of protection and then pulling up on it or stepping up in the sling. Usually only needed to get past one or two more difficult moves on advanced free climbs.

(2) A1—Easy aid. The placement of protection is straight forward and reliable. There is usually no high risk of any piece of protection pulling out. This technique requires etriers and is fast and simple.

(3) A2—Moderate aid. The placement of protection is generally straight forward, but placement can be awkward and strenuous. Usually A2 involves one or two moves that are difficult with good protection placement below and above the difficult moves. No serious fall danger.

(4) A3—Hard aid. This technique requires testing your protection. It involves several awkward and strenuous moves in a row. Generally solid placements which will hold a fall and are found within a full rope length. However, long fall potential does exist, with falls of 40 to 60 feet and intermediate protection on the awkward placements failing. These falls, however, are usually clean and with no serious bodily harm.

(5) A4—Serious aid. This technique requires lots of training and practice. More like walking on eggs so none of them break. Leads will usually take extended amounts of time which cause the lead climber to doubt and worry about each placement. Protection placed will usually only hold a climber's weight and falls can be as long as two-thirds the rope length.

(6) A5—Extreme aid. All protection is sketchy at best. Usually no protection placed on the entire route can be trusted to stop a fall.

(7) A6—Extremely severe aid. Continuous A5 climbing with A5 belay stations. If the leader falls, the whole rope team will probably experience ground fall.

(8) Aid climbing classes are also further divided into +/- categories, such as A3+ or A3-, which would simply refer to easy or hard.

e. Grade ratings (commitment grades) inform the climber of the approximate time a climber trained to the level of the climb will take to complete the route.

- I—Several hours.
- II—Half of a day.
- III—About three-fourths of a day.
- IV—Long hard day (usually not less than 5.7).
- V—1 1/2 to 2 1/2 days (usually not less than 5.8).
- VI—Greater than 2 days.

f. Climbing difficulties are rated by different systems. Table 1-1 shows a comparison of these systems.

- YDS (Yosemite Decimal System)—Used in the United States.
- UIAA (Union des International Alpine Association)—Used in Europe.
- British—The British use adjectives and numbers to designate the difficulty of climbs. This system can be confusing if the climber is not familiar with it.
- French—The French use numbers and letters to designate the difficulty of climbs.
- Brazilian—Brazil uses Roman Numerals and adjectives to designate difficulty.
- Australian—Australia uses only numbers to designate difficulty.

YDS	UIAA	BRITISH	FRENCH	BRAZIL	AUSTRALIA
Class 1	I	easy (E)			
Class 2	II	easy (E)			
Class 3	III	easy (E)	1a, b, c		
Class 4	III-	moderate (MOD)	1a, b, c		
5.0	III	moderate (MOD)	2a, b		4
5.1	III+	difficult (DIFF)	2a, b		5
5.2	IV-	hard difficult	2c, 3a		6
5.3	IV	very difficult	3b, c, 4a		7
5.4	IV+	hard very difficult	3b, c, 4a	II	8, 9
5.5	V-	mild severe	3b, c, 4a	IIsup	10, 11
5.6	V	severe, hard severe, 4a	4a, b, c	III	12, 13
5.7	V+	severe, hard severe, 4b	4a, b, c	IIIsup	14
5.8	VI-	hard severe, hard very severe, 4c	5a, b	IV	15
5.9	VI	5a	5b, c	IVsup	16, 17
5.10a	VII-	E1, 5b	5b, c	V	18
5.10b	VII	E1, 5b	5b, c	Vsup	19
5.10c	VII	E1, 5b	5b, c	VI	20
5.10d	VII+	E1/E2, 5b-5c	5b, c	VIsup	21
5.11a	VIII-	E3, 6a	6a, b, c	VII	22
5.11b	VIII	E3/E4, 6a	6a, b, c	VII	23
5.11c	VIII	E4, 6b	6a, b, c	VIIsup	24
5.12a	IX-	E5, E6/7, 6c	7a	VIII	26
5.12b	IX	E5, E6/7, 6c	7a	VIIIsup	27
5.12c	IX	E5, E6/7, 6c	7a		28
5.12d	IX+	E6/7, 7a	7a		29

Table 1-1. Rating systems.

g. Ice climbing ratings can have commitment ratings and technical ratings. The numerical ratings are often prefaced with WI (waterfall ice), AI (alpine ice), or M (mixed rock and ice).

(1) ***Commitment Ratings.*** Commitment ratings are expressed in Roman numerals.

- I—A short, easy climb near the road, with no avalanche hazard and a straightforward descent.
- II—A route of one or two pitches within a short distance of rescue assistance, with little objective hazard.
- III—A multipitch route at low elevation, or a one-pitch climb with an approach that takes about an hour. The route requires anywhere from a few hours to a long day to complete. The descent may require building rappel anchors, and the route might be prone to avalanche.
- IV—A multipitch route at higher elevations; may require several hours of approach on skis or foot. This route is subject to objective hazards, possibly with a hazardous descent.
- V—A long climb in a remote setting, requiring all day to complete the climb itself. Requires many rappels off anchors for the descent. This route has sustained exposure to avalanche or other objective hazards.

- VI—A long ice climb in an alpine setting, with sustained technical climbing. Only elite climbers will complete it in a day. A difficult and involved approach and descent, with objective hazards ever-present, all in a remote area.
- VII—Everything a grade VI has, and more of it. Possibly days to approach the climb, and objective hazards rendering survival as questionable. Difficult physically and mentally.

(2) ***Technical Ratings.*** Technical ratings are expressed as Arabic numerals.

- 1—A frozen lake or stream bed.
- 2—A pitch with short sections of ice up to 80 degrees; lots of opportunity for protection and good anchors.
- 3—Sustained ice up to 80 degrees; the ice is usually good, with places to rest, but it requires skill at placing protection and setting anchors.
- 4—A sustained pitch that is vertical or slightly less than vertical; may have special features such as chandeliers and run-outs between protection.
- 5—A long, strenuous pitch, possibly 50 meters of 85- to 90-degree ice with few if any rests between anchors. The pitch may be shorter, but on featureless ice. Good skills at placing protection are required.
- 6—A full 50-meter pitch of dead vertical ice, possibly of poor quality; requires efficiency of movement and ability to place protection while in awkward stances.
- 7—A full rope length of thin vertical or overhanging ice of dubious adhesion. An extremely tough pitch, physically and mentally, requiring agility and creativity.
- 8—Simply the hardest ice climbing ever done; extremely bold and gymnastic.

1-7. CROSS-COUNTRY MOVEMENT

Soldiers must know the terrain to determine the feasible routes for cross-country movement when no roads or trails are available.

a. A pre-operations intelligence effort should include topographic and photographic map coverage as well as detailed weather data for the area of operations. When planning mountain operations, additional information may be needed about size, location, and characteristics of landforms; drainage; types of rock and soil; and the density and distribution of vegetation. Control must be decentralized to lower levels because of varied terrain, erratic weather, and communication problems inherent to mountainous regions.

b. Movement is often restricted due to terrain and weather. The erratic weather requires that soldiers be prepared for wide variations in temperature, types, and amounts of precipitation.

(1) Movement above the timberline reduces the amount of protective cover available at lower elevations. The logistical problem is important; therefore, each man must be self-sufficient to cope with normal weather changes using materials from his rucksack.

(2) Movement during a storm is difficult due to poor visibility and bad footing on steep terrain. Although the temperature is often higher during a storm than during clear weather, the dampness of rain and snow and the penetration of wind cause soldiers to chill quickly. Although climbers should get off the high ground and seek shelter and

warmth, if possible, during severe mountain storms, capable commanders may use reduced visibility to achieve tactical surprise.

c. When the tactical situation requires continued movement during a storm, the following precautions should be observed:

- Maintain visual contact.
- Keep warm. Maintain energy and body heat by eating and drinking often; carry food that can be eaten quickly and while on the move.
- Keep dry. Wear wet-weather clothing when appropriate, but do not overdress, which can cause excessive perspiration and dampen clothing. As soon as the objective is reached and shelter secured, put on dry clothing.
- Do not rush. Hasty movement during storms leads to breaks in contact and accidents.
- If lost, stay warm, dry, and calm.
- Do not use ravines as routes of approach during a storm as they often fill with water and are prone to flash floods.
- Avoid high pinnacles and ridgelines during electrical storms.
- Avoid areas of potential avalanche or rock-fall danger.

1-8. COVER AND CONCEALMENT

When moving in the mountains, outcroppings, boulders, heavy vegetation, and intermediate terrain can provide cover and concealment. Digging fighting positions and temporary fortifications is difficult because soil is often thin or stony. The selection of dug-in positions requires detailed planning. Some rock types, such as volcanic tuff, are easily excavated. In other areas, boulders and other loose rocks can be used for building hasty fortifications. In alpine environments, snow and ice blocks may be cut and stacked to supplement dug-in positions. As in all operations, positions and routes must be camouflaged to blend in with the surrounding terrain to prevent aerial detection.

1-9. OBSERVATION

Observation in mountains varies because of weather and ground cover. The dominating height of mountainous terrain permits excellent long-range observation. However, rapidly changing weather with frequent periods of high winds, rain, snow, sleet, hail, and fog can limit visibility. The rugged nature of the terrain often produces dead space at midranges.

a. Low cloud cover at higher elevations may neutralize the effectiveness of OPs established on peaks or mountaintops. High wind speeds and sound often mask the noises of troop movement. Several OPs may need to be established laterally, in depth, and at varying altitudes to provide visual coverage of the battle area.

b. Conversely, the nature of the terrain can be used to provide concealment from observation. This concealment can be obtained in the dead space. Mountainous regions are subject to intense shadowing effects when the sun is low in relatively clear skies. The contrast from lighted to shaded areas causes visual acuity in the shaded regions to be considerably reduced. These shadowed areas can provide increased concealment when combined with other camouflage and should be considered in maneuver plans.

1-10. FIELDS OF FIRE

Fields of fire, like observation, are excellent at long ranges. However, dead space is a problem at short ranges. When forces cannot be positioned to cover dead space with direct fires, mines and obstacles or indirect fire must be used. Range determination is deceptive in mountainous terrain. Soldiers must routinely train in range estimation in mountainous regions to maintain their proficiency.

Section II. MOUNTAIN WEATHER

Most people subconsciously "forecast" the weather. If they look outside and see dark clouds they may decide to take rain gear. If an unexpected wind strikes, people glance to the sky for other bad signs. A conscious effort to follow weather changes will ultimately lead to a more accurate forecast. An analysis of mountain weather and how it is affected by mountain terrain shows that such weather is prone to patterns and is usually severe, but patterns are less obvious in mountainous terrain than in other areas. Conditions greatly change with altitude, latitude, and exposure to atmospheric winds and air masses. Mountain weather can be extremely erratic. It varies from stormy winds to calm, and from extreme cold to warmth within a short time or with a minor shift in locality. The severity and variance of the weather causes it to have a major impact on military operations.

1-11. CONSIDERATIONS FOR PLANNING

Mountain weather can be either a dangerous obstacle to operations or a valuable aid, depending on how well it is understood and to what extent advantage is taken of its peculiar characteristics.

a. Weather often determines the success or failure of a mission since it is highly changeable. Military operations plans must be flexible, especially in planning airmobile and airborne operations. The weather must be anticipated to allow enough time for planning so that the leaders of subordinate units can use their initiative in turning an important weather factor in their favor. The clouds that often cover the tops of mountains and the fogs that cover valleys are an excellent means of concealing movements that normally are made during darkness or in smoke. Limited visibility can be used as a combat multiplier.

b. The safety or danger of almost all high mountain regions, especially in winter, depends upon a change of a few degrees of temperature above or below the freezing point. Ease and speed of travel depend mainly on the weather. Terrain that can be crossed swiftly and safely one day may become impassable or highly dangerous the next due to snowfall, rainfall, or a rise in temperature. The reverse can happen just as quickly. The prevalence of avalanches depends on terrain, snow conditions, and weather factors.

c. Some mountains, such as those found in desert regions, are dry and barren with temperatures ranging from extreme heat in the summer to extreme cold in the winter. In tropical regions, lush jungles with heavy seasonal rains and little temperature variation often cover mountains. High rocky crags with glaciated peaks can be found in mountain ranges at most latitudes along the western portion of the Americas and Asia.

d. Severe weather may decrease morale and increase basic survival problems. These problems can be minimized when men have been trained to accept the weather by being self-sufficient. Mountain soldiers properly equipped and trained can use the weather to their advantage in combat operations.

1-12. MOUNTAIN AIR

High mountain air is dry and may be drier in the winter. Cold air has a reduced capacity to hold water vapor. Because of this increased dryness, equipment does not rust as quickly and organic material decomposes slowly. The dry air also requires soldiers to increase consumption of water. The reduced water vapor in the air causes an increase in evaporation of moisture from the skin and in loss of water through transpiration in the respiratory system. Due to the cold, most soldiers do not naturally consume the quantity of fluids they would at higher temperatures and must be encouraged to consciously increase their fluid intake.

a. Pressure is low in mountainous areas due to the altitude. The barometer usually drops 2.5 centimeters for every 300 meters gained in elevation (3 percent).

b. The air at higher altitudes is thinner as atmospheric pressure drops with the increasing altitude. The altitude has a natural filtering effect on the sun's rays. Rays are absorbed or reflected in part by the molecular content of the atmosphere. This effect is greater at lower altitudes. At higher altitudes, the thinner, drier air has a reduced molecular content and, consequently, a reduced filtering effect on the sun's rays. The intensity of both visible and ultraviolet rays is greater with increased altitude. These conditions increase the chance of sunburn, especially when combined with a snow cover that reflects the rays upward.

1-13. WEATHER CHARACTERISTICS

The earth is surrounded by an atmosphere that is divided into several layers. The world's weather systems are in the lower of these layers known as the "troposphere." This layer reaches as high as 40,000 feet. Weather is a result of an atmosphere, oceans, land masses, unequal heating and cooling from the sun, and the earth's rotation. The weather found in any one place depends on many things such as the air temperature, humidity (moisture content), air pressure (barometric pressure), how it is being moved, and if it is being lifted or not.

a. Air pressure is the "weight" of the atmosphere at any given place. The higher the pressure, the better the weather will be. With lower air pressure, the weather will more than likely be worse. In order to understand this, imagine that the air in the atmosphere acts like a liquid. Areas with a high level of this "liquid" exert more pressure on an area and are called high-pressure areas. Areas with a lower level are called low-pressure areas. The average air pressure at sea level is 29.92 inches of mercury (hg) or 1,013 millibars (mb). The higher in altitude, the lower the pressure.

(1) ***High Pressure.*** The characteristics of a high-pressure area are as follows:

- The airflow is clockwise and out.
- Otherwise known as an "anticyclone".
- Associated with clear skies.
- Generally the winds will be mild.
- Depicted as a blue "H" on weather maps.

(2) ***Low Pressure.*** The characteristics of a low-pressure area are as follows:

- The airflow is counterclockwise and in.
- Otherwise known as a "cyclone".
- Associated with bad weather.
- Depicted as a red "L" on weather maps.

b. Air from a high-pressure area is basically trying to flow out and equalize its pressure with the surrounding air. Low pressure, on the other hand, is building up vertically by pulling air in from outside itself, which causes atmospheric instability resulting in bad weather.

c. On a weather map, these differences in pressure are depicted as isobars. Isobars resemble contour lines and are measured in either millibars or inches of mercury. The areas of high pressure are called "ridges" and lows are called "troughs."

1-14. WIND

In high mountains, the ridges and passes are seldom calm; however, strong winds in protected valleys are rare. Normally, wind speed increases with altitude since the earth's frictional drag is strongest near the ground. This effect is intensified by mountainous terrain. Winds are accelerated when they converge through mountain passes and canyons. Because of these funneling effects, the wind may blast with great force on an exposed mountainside or summit. Usually, the local wind direction is controlled by topography.

a. The force exerted by wind quadruples each time the wind speed doubles; that is, wind blowing at 40 knots pushes four times harder than a wind blowing at 20 knots. With increasing wind strength, gusts become more important and may be 50 percent higher than the average wind speed. When wind strength increases to a hurricane force of 64 knots or more, soldiers should lay on the ground during gusts and continue moving during lulls. If a hurricane- force wind blows where there is sand or snow, dense clouds fill the air. The rocky debris or chunks of snow crust are hurled near the surface. During the winter season, or at high altitudes, commanders must be constantly aware of the wind-chill factor and associated cold-weather injuries (see Chapter 2).

b. Winds are formed due to the uneven heating of the air by the sun and rotation of the earth. Much of the world's weather depends on a system of winds that blow in a set direction.

c. Above hot surfaces, air expands and moves to colder areas where it cools and becomes denser, and sinks to the earth's surface. The results are a circulation of air from the poles along the surface of the earth to the equator, where it rises and moves to the poles again.

d. Heating and cooling together with the rotation of the earth causes surface winds. In the Northern Hemisphere, there are three prevailing winds:

(1) ***Polar Easterlies.*** These are winds from the polar region moving from the east. This is air that has cooled and settled at the poles.

(2) ***Prevailing Westerlies.*** These winds originate from approximately 30 degrees north latitude from the west. This is an area where prematurely cooled air, due to the earth's rotation, has settled to the surface.

(3) ***Northeast Tradewinds.*** These are winds that originate from approximately 30° north from the northeast.

e. The jet stream is a long meandering current of high-speed winds often exceeding 250 miles per hour near the transition zone between the troposphere and the stratosphere known as the tropopause. These winds blow from a generally westerly direction dipping down and picking up air masses from the tropical regions and going north and bringing down air masses from the polar regions.

f. The patterns of wind mentioned above move air. This air comes in parcels called "air masses." These air masses can vary from the size of a small town to as large as a country. These air masses are named from where they originate:

- Maritime—over water.
- Continental—over land
- Polar—north of 60° north latitude.
- Tropical—south of 60° north latitude.

Combining these parcels of air provides the names and description of the four types of air masses:

- Continental Polar—cold, dry air mass.
- Maritime Polar—cold, wet air mass.
- Maritime Tropical—warm, wet air mass.
- Continental Tropical—warm, dry air mass.

g. Two types of winds are peculiar to mountain environments, but do not necessarily affect the weather.

(1) ***Anabatic Wind (Valley Winds).*** These winds blow up mountain valleys to replace warm rising air and are usually light winds.

(2) ***Katabatic Wind (Mountain Wind).*** These winds blow down mountain valley slopes caused by the cooling of air and are occasionally strong winds.

1-15. HUMIDITY

Humidity is the amount of moisture in the air. All air holds water vapor even if it cannot be seen. Air can hold only so much water vapor; however, the warmer the air, the more moisture it can hold. When air can hold all that it can the air is "saturated" or has 100 percent relative humidity.

a. If air is cooled beyond its saturation point, the air will release its moisture in one form or another (clouds, fog, dew, rain, snow, and so on). The temperature at which this happens is called the "condensation point". The condensation point varies depending on the amount of water vapor contained in the air and the temperature of the air. If the air contains a great deal of water, condensation can occur at a temperature of 68 degrees Fahrenheit, but if the air is dry and does not hold much moisture, condensation may not form until the temperature drops to 32 degrees Fahrenheit or even below freezing.

b. The adiabatic lapse rate is the rate at which air cools as it rises or warms as it descends. This rate varies depending on the moisture content of the air. Saturated (moist) air will warm and cool approximately 3.2 degrees Fahrenheit per 1,000 feet of elevation gained or lost. Dry air will warm and cool approximately 5.5 degrees Fahrenheit per 1,000 feet of elevation gained or lost.

1-16. CLOUD FORMATION

Clouds are indicators of weather conditions. By reading cloud shapes and patterns, observers can forecast weather with little need for additional equipment such as a barometer, wind meter, and thermometer. Any time air is lifted or cooled beyond its saturation point (100 percent relative humidity), clouds are formed. The four ways air gets lifted and cooled beyond its saturation point are as follows.

a. **Convective Lifting.** This effect happens due to the sun's heat radiating off the Earth's surface causing air currents (thermals) to rise straight up and lift air to a point of saturation.

b. **Frontal Lifting.** A front is formed when two air masses of different moisture content and temperature collide. Since air masses will not mix, warmer air is forced aloft over the colder air mass. From there it is cooled and then reaches its saturation point. Frontal lifting creates the majority of precipitation.

c. **Cyclonic Lifting.** An area of low pressure pulls air into its center from all over in a counterclockwise direction. Once this air reaches the center of the low pressure, it has nowhere to go but up. Air continues to lift until it reaches the saturation point.

d. **Orographic Lifting.** This happens when an air mass is pushed up and over a mass of higher ground such as a mountain. Air is cooled due to the adiabatic lapse rate until the air's saturation point is reached.

1-17. TYPES OF CLOUDS

Clouds are one of the signposts to what is happening with the weather. Clouds can be described in many ways. They can be classified by height or appearance, or even by the amount of area covered vertically or horizontally. Clouds are classified into five categories: low-, mid-, and high-level clouds; vertically-developed clouds; and less common clouds.

a. **Low-Level Clouds.** Low-level clouds (0 to 6,500 feet) are either cumulus or stratus (Figures 1-1 and 1-2, page 1-16). Low-level clouds are mostly composed of water droplets since their bases lie below 6,500 feet. When temperatures are cold enough, these clouds may also contain ice particles and snow.

(1) The two types of precipitating low-level clouds are nimbostratus and stratocumulus (Figures 1-3 and 1-4, page 1-17).

(a) Nimbostratus clouds are dark, low-level clouds accompanied by light to moderately falling precipitation. The sun or moon is not visible through nimbostratus clouds, which distinguishes them from mid-level altostratus clouds. Because of the fog and falling precipitation commonly found beneath and around nimbostratus clouds, the cloud base is typically extremely diffuse and difficult to accurately determine.

(b) Stratocumulus clouds generally appear as a low, lumpy layer of clouds that is sometimes accompanied by weak precipitation. Stratocumulus vary in color from dark gray to light gray and may appear as rounded masses with breaks of clear sky in between. Because the individual elements of stratocumulus are larger than those of altocumulus, deciphering between the two cloud types is easier. With your arm extended toward the sky, altocumulus elements are about the size of a thumbnail while stratocumulus are about the size of a fist.

Figure 1-1. Cumulus clouds.

Figure 1-2. Stratus clouds.

Figure 1-3. Nimbostratus clouds.

Figure 1-4. Stratocumulus clouds.

(2) Low-level clouds may be identified by their height above nearby surrounding relief of known elevation. Most precipitation originates from low-level clouds because rain or snow usually evaporate before reaching the ground from higher clouds. Low-level clouds usually indicate impending precipitation, especially if the cloud is more than 3,000 feet thick. (Clouds that appear dark at their bases are more than 3,000 feet thick.)

b. **Mid-Level Clouds.** Mid-level clouds (between 6,500 to 20,000 feet) have a prefix of alto. Middle clouds appear less distinct than low clouds because of their height. Alto clouds with sharp edges are warmer because they are composed mainly of water droplets. Cold clouds, composed mainly of ice crystals and usually colder than -30 degrees F, have distinct edges that grade gradually into the surrounding sky. Middle clouds usually indicate fair weather, especially if they are rising over time. Lowering middle clouds indicate potential storms, though usually hours away. There are two types of mid-level clouds, altocumulus and altostratus clouds (Figures 1-5 and 1-6).

(1) Altocumulus clouds can appear as parallel bands or rounded masses. Typically a portion of an altocumulus cloud is shaded, a characteristic which makes them distinguishable from high-level cirrocumulus. Altocumulus clouds usually form in advance of a cold front. The presence of altocumulus clouds on a warm humid summer morning is commonly followed by thunderstorms later in the day. Altocumulus clouds that are scattered rather than even, in a blue sky, are called "fair weather" cumulus and suggest arrival of high pressure and clear skies.

(2) Altostratus clouds are often confused with cirrostratus. The one distinguishing feature is that a halo is not observed around the sun or moon. With altostratus, the sun or moon is only vaguely visible and appears as if it were shining through frosted glass.

Figure 1-5. Altocumulus.

Figure 1-6. Altostratus.

c. **High-Level Clouds.** High-level clouds (more than 20,000 feet above ground level) are usually frozen clouds, indicating air temperatures at that elevation below -30 degrees Fahrenheit, with a fibrous structure and blurred outlines. The sky is often covered with a thin veil of cirrus that partly obscures the sun or, at night, produces a ring of light around the moon. The arrival of cirrus indicates moisture aloft and the approach of a traveling storm system. Precipitation is often 24 to 36 hours away. As the storm approaches, the cirrus thickens and lowers, becoming altostratus and eventually stratus. Temperatures are warm, humidity rises, and winds become southerly or south easterly. The two types of high-level clouds are cirrus and cirrostratus (Figure 1-7 and Figure 1-8, page 1-20).

(1) Cirrus clouds are the most common of the high-level clouds. Typically found at altitudes greater than 20,000 feet, cirrus are composed of ice crystals that form when super-cooled water droplets freeze. Cirrus clouds generally occur in fair weather and point in the direction of air movement at their elevation. Cirrus can be observed in a variety of shapes and sizes. They can be nearly straight, shaped like a comma, or seemingly all tangled together. Extensive cirrus clouds are associated with an approaching warm front.

(2) Cirrostratus clouds are sheet-like, high-level clouds composed of ice crystals. They are relatively transparent and can cover the entire sky and be up to several thousand feet thick. The sun or moon can be seen through cirrostratus. Sometimes the only indication of cirrostratus clouds is a halo around the sun or moon. Cirrostratus clouds tend to thicken as a warm front approaches, signifying an increased production of ice crystals. As a result, the halo gradually disappears and the sun or moon becomes less visible.

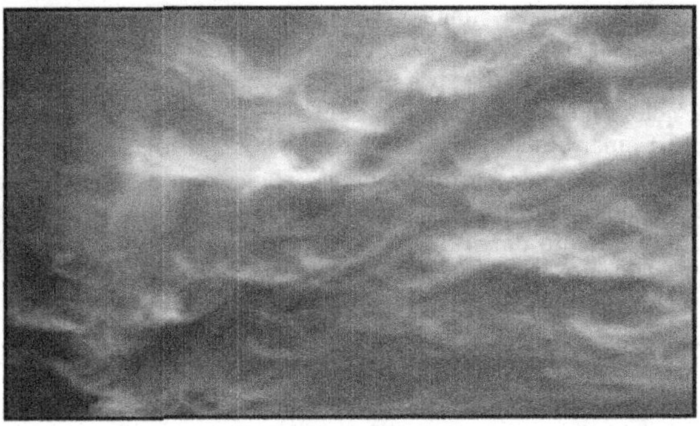

Figure 1-7. Cirrus.

Figure 1-8. Cirrostratus.

d. **Vertical-Development Clouds.** Clouds with vertical development can grow to heights in excess of 39,000 feet, releasing incredible amounts of energy. The two types of clouds with vertical development are fair weather cumulus and cumulonimbus.

(1) Fair weather cumulus clouds have the appearance of floating cotton balls and have a lifetime of 5 to 40 minutes. Known for their flat bases and distinct outlines, fair weather cumulus exhibit only slight vertical growth, with the cloud tops designating the limit of the rising air. Given suitable conditions, however, these clouds can later develop into towering cumulonimbus clouds associated with powerful thunderstorms. Fair weather cumulus clouds are fueled by buoyant bubbles of air known as thermals that rise up from the earth's surface. As the air rises, the water vapor cools and condenses forming water droplets. Young fair weather cumulus clouds have sharply defined edges and bases while the edges of older clouds appear more ragged, an artifact of erosion. Evaporation along the cloud edges cools the surrounding air, making it heavier and producing sinking motion outside the cloud. This downward motion inhibits further convection and growth of additional thermals from down below, which is why fair weather cumulus typically have expanses of clear sky between them. Without a continued supply of rising air, the cloud begins to erode and eventually disappears.

(2) Cumulonimbus clouds are much larger and more vertically developed than fair weather cumulus (Figure 1-9). They can exist as individual towers or form a line of towers called a squall line. Fueled by vigorous convective updrafts, the tops of cumulonimbus clouds can reach 39,000 feet or higher. Lower levels of cumulonimbus clouds consist mostly of water droplets while at higher elevations, where the temperatures are well below freezing, ice crystals dominate the composition. Under favorable conditions, harmless fair weather cumulus clouds can quickly develop into large cumulonimbus associated with powerful thunderstorms known as super-cells. Super-cells are large thunderstorms with deep rotating updrafts and can have a lifetime of several hours. Super-cells produce frequent lightning, large hail, damaging winds, and

tornadoes. These storms tend to develop during the afternoon and early evening when the effects of heating from the sun are the strongest.

Figure 1-9. Cumulonimbus.

e. **Other Cloud Types.** These clouds are a collection of miscellaneous types that do not fit into the previous four groups. They are orographic clouds, lenticulars, and contrails.

(1) Orographic clouds develop in response to the forced lifting of air by the earth's topography. Air passing over a mountain oscillates up and down as it moves downstream. Initially, stable air encounters a mountain, is lifted upward, and cools. If the air cools to its saturation temperature during this process, the water vapor condenses and becomes visible as a cloud. Upon reaching the mountain top, the air is heavier than the environment and will sink down the other side, warming as it descends. Once the air returns to its original height, it has the same buoyancy as the surrounding air. However, the air does not stop immediately because it still has momentum carrying it downward. With continued descent, the air becomes warmer then the surrounding air and accelerates back upwards towards its original height. Another name for this type of cloud is the lenticular cloud.

(2) Lenticular clouds are cloud caps that often form above pinnacles and peaks, and usually indicate higher winds aloft (Figure 1-10, page 1-22). Cloud caps with a lens shape, similar to a "flying saucer," indicate extremely high winds (over 40 knots). Lenticulars should always be watched for changes. If they grow and descend, bad weather can be expected.

Figure 1-10. Lenticular.

(3) Contrails are clouds that are made by water vapor being inserted into the upper atmosphere by the exhaust of jet engines (Figure 1-11). Contrails evaporate rapidly in fair weather. If it takes longer than two hours for contrails to evaporate, then there is impending bad weather (usually about 24 hours prior to a front).

Figure 1-11. Contrails.

f. **Cloud Interpretation.** Serious errors can occur in interpreting the extent of cloud cover, especially when cloud cover must be reported to another location. Cloud cover always appears greater on or near the horizon, especially if the sky is covered with cumulus clouds, since the observer is looking more at the sides of the clouds rather than between them. Cloud cover estimates should be restricted to sky areas more than 40 degrees above the horizon—that is, to the local sky. Assess the sky by dividing the 360 degrees of sky around you into eighths. Record the coverage in eighths and the types of clouds observed.

1-18. FRONTS

Fronts occur when two air masses of different moisture and temperature contents meet. One of the indicators that a front is approaching is the progression of the clouds. The four types of fronts are warm, cold, occluded, and stationary.

a. **Warm Front.** A warm front occurs when warm air moves into and over a slower or stationary cold air mass. Because warm air is less dense, it will rise up and over the cooler air. The cloud types seen when a warm front approaches are cirrus, cirrostratus, nimbostratus (producing rain), and fog. Occasionally, cumulonimbus clouds will be seen during the summer months.

b. **Cold Front.** A cold front occurs when a cold air mass overtakes a slower or stationary warm air mass. Cold air, being more dense than warm air, will force the warm air up. Clouds observed will be cirrus, cumulus, and then cumulonimbus producing a short period of showers.

c. **Occluded Front.** Cold fronts generally move faster than warm fronts. The cold fronts eventually overtake warm fronts and the warm air becomes progressively lifted from the surface. The zone of division between cold air ahead and cold air behind is called a "cold occlusion." If the air behind the front is warmer than the air ahead, it is a warm occlusion. Most land areas experience more occlusions than other types of fronts. The cloud progression observed will be cirrus, cirrostratus, altostratus, and nimbostratus. Precipitation can be from light to heavy.

d. **Stationary Front.** A stationary front is a zone with no significant air movement. When a warm or cold front stops moving, it becomes a stationary front. Once this boundary begins forward motion, it once again becomes a warm or cold front. When crossing from one side of a stationary front to another, there is typically a noticeable temperature change and shift in wind direction. The weather is usually clear to partly cloudy along the stationary front.

1-19. TEMPERATURE

Normally, a temperature drop of 3 to 5 degrees Fahrenheit for every 1,000 feet gain in altitude is encountered in motionless air. For air moving up a mountain with condensation occurring (clouds, fog, and precipitation), the temperature of the air drops 3.2 degrees Fahrenheit with every 1,000 feet of elevation gain. For air moving up a mountain with no clouds forming, the temperature of the air drops 5.5 degrees Fahrenheit for every 1,000 feet of elevation gain.

a. An expedient to this often occurs on cold, clear, calm mornings. During a troop movement or climb started in a valley, higher temperatures may often be encountered as altitude is gained. This reversal of the normal cooling with elevation is called temperature inversion. Temperature inversions are caused when mountain air is cooled by ice, snow, and heat loss through thermal radiation. This cooler, denser air settles into the valleys and low areas. The inversion continues until the sun warms the surface of the earth or a moderate wind causes a mixing of the warm and cold layers. Temperature inversions are common in the mountainous regions of the arctic, subarctic, and mid-latitudes.

b. At high altitudes, solar heating is responsible for the greatest temperature contrasts. More sunshine and solar heat are received above the clouds than below. The important effect of altitude is that the sun's rays pass through less of the atmosphere and more direct heat is received than at lower levels, where solar radiation is absorbed and reflected by dust and water vapor. Differences of 40 to 50 degrees Fahrenheit may occur between surface temperatures in the shade and surface temperatures in the sun. This is particularly true for dark metallic objects. The difference in temperature felt on the skin between the sun and shade is normally 7 degrees Fahrenheit. Special care must be taken to avoid sunburn and snow blindness. Besides permitting rapid heating, the clear air at high altitudes also favors rapid cooling at night. Consequently, the temperature rises fast after sunrise and drops quickly after sunset. Much of the chilled air drains downward, due to convection currents, so that the differences between day and night temperatures are greater in valleys than on slopes.

c. Local weather patterns force air currents up and over mountaintops. Air is cooled on the windward side of the mountain as it gains altitude, but more slowly (3.2 degrees Fahrenheit per 1,000 feet) if clouds are forming due to heat release when water vapor becomes liquid. On the leeward side of the mountain, this heat gained from the condensation on the windward side is added to the normal heating that occurs as the air descends and air pressure increases. Therefore, air and winds on the leeward slope are considerably warmer than on the windward slope, which is referred to as Chinook winds. The heating and cooling of the air affects planning considerations primarily with regard to the clothing and equipment needed for an operation.

1-20. WEATHER FORECASTING

The use of a portable aneroid barometer, thermometer, wind meter, and hygrometer help in making local weather forecasts. Reports from other localities and from any weather service, including USAF, USN, or the National Weather Bureau, are also helpful. Weather reports should be used in conjunction with the locally observed current weather situation to forecast future weather patterns.

a. Weather at various elevations may be quite different because cloud height, temperature, and barometric pressure will all be different. There may be overcast and rain in a lower area, with mountains rising above the low overcast into warmer clear weather.

b. To be effective, a forecast must reach the small-unit leaders who are expected to utilize weather conditions for assigned missions. Several different methods can be used to create a forecast. The method a forecaster chooses depends upon the forecaster's experience, the amount of data available, the level of difficulty that the forecast situation presents, and the degree of accuracy needed to make the forecast. The five ways to forecast weather are:

(1) ***Persistence Method.*** "Today equals tomorrow" is the simplest way of producing a forecast. This method assumes that the conditions at the time of the forecast will not

change; for example, if today was hot and dry, the persistence method predicts that tomorrow will be the same.

(2) ***Trends Method.*** "Nowcasting" involves determining the speed and direction of fronts, high- and low-pressure centers, and clouds and precipitation. For example, if a cold front moves 300 miles during a 24-hour period, we can predict that it will travel 300 miles in another 24-hours.

(3) ***Climatology Method.*** This method averages weather statistics accumulated over many years. This only works well when the pattern is similar to the following years.

(4) ***Analog Method.*** This method examines a day's forecast and recalls a day in the past when the weather looked similar (an analogy). This method is difficult to use because finding a perfect analogy is difficult.

(5) ***Numerical Weather Prediction.*** This method uses computers to analyze all weather conditions and is the most accurate of the five methods.

1-21. RECORDING DATA

An accurate observation is essential in noting trends in weather patterns. Ideally, under changing conditions, trends will be noted in some weather parameters. However, this may not always be the case. A minor shift in the winds may signal an approaching storm.

a. **Wind Direction.** Assess wind direction as a magnetic direction from which the wind is blowing.

b. **Wind Speed.** Assess wind speed in knots.

(1) If an anemometer is available, assess speed to the nearest knot.

(2) If no anemometer is available, estimate the speed in knots. Judge the wind speed by the way objects, such as trees, bushes, tents, and so forth, are blowing.

c. **Visibility in Meters.** Observe the farthest visible major terrain or man-made feature and determine the distance using any available map.

d. **Present Weather.** Include any precipitation or obscuring weather. The following are examples of present weather:

- Rain—continuous and steady liquid precipitation that will last at least one hour.
- Rain showers—short-term and potentially heavy downpours that rarely last more than one hour.
- Snow—continuous and steady frozen precipitation that will last at least one hour.
- Snow showers—short-term and potentially heavy frozen downpours that rarely last more than one hour.
- Fog, haze—obstructs visibility of ground objects.
- Thunderstorms—a potentially dangerous storm. Thunderstorms will produce lightning, heavy downpours, colder temperatures, tornadoes (not too frequently), hail, and strong gusty winds at the surface and aloft. Winds commonly exceed 35 knots.

e. **Total Cloud Cover.** Assess total cloud cover in eighths. Divide the sky into eight different sections measuring from horizon to horizon. Count the sections with cloud cover, which gives the total cloud cover in eighths. (For example, if half of the sections are covered with clouds, total cloud cover is 4/8.)

f. **Ceiling Height.** Estimate where the cloud base intersects elevated terrain. Note if bases are above all terrain. If clouds are not touching terrain, then estimate to the best of your ability.

g. **Temperature.** Assess temperature with or without a thermometer.

(1) With a thermometer, assess temperature in degrees Celsius (use Fahrenheit only if Celsius conversion is not available). To convert Fahrenheit to Celsius: C = F minus 32 times .55. To convert Celsius to Fahrenheit: F = 1.8 times C plus 32.

Example: 41 degrees F – 32 x .55 = 5 degrees C.
5 degrees C x 1.8 + 32 = 41 degrees F.

(2) Without a thermometer, estimate temperature as above or below freezing (0°C), as well as an estimated temperature.

h. **Pressure Trend.** With a barometer or altimeter, assess the pressure trend.

(1) A high pressure moving in will cause altimeters to indicate lower elevation.

(2) A low pressure moving in will cause altimeters to indicate higher elevation.

i. **Observed Weather.** Note changes or trends in observed weather conditions.

(1) Deteriorating trends include:

- Marked wind direction shifts. A high pressure system wind flows clockwise. A low pressure system wind flows counterclockwise. The closer the isometric lines are, the greater the differential of pressure (greater wind speeds).
- Marked wind speed increases.
- Changes in obstructions to visibility.
- Increasing cloud coverage.
- Increase in precipitation. A steady drizzle is usually a long-lasting rain.
- Lowering cloud ceilings.
- Marked cooler temperature changes, which could indicate that a cold front is passing through.
- Marked increase in humidity.
- Decreasing barometric pressure, which indicates a lower pressure system is moving through the area.

(2) Improving trends include:

- Steady wind direction, which indicates no change in weather systems in the area.
- Decreasing wind speeds.
- Clearing of obstructions to visibility.
- Decreasing or ending precipitation.
- Decreasing cloud coverage.
- Increasing height of cloud ceilings.
- Temperature changes slowly warmer.
- Humidity decreases.
- Increasing barometric pressure, which indicates that a higher pressure system is moving through the area.

j. **Update.** Continue to evaluate observed conditions and update the forecast.

Section III. MOUNTAIN HAZARDS

Hazards can be termed natural (caused by natural occurrence), man-made (caused by an individual, such as lack of preparation, carelessness, improper diet, equipment misuse), or as a combination (human trigger). There are two kinds of hazards while in the mountains—subjective and objective. Combinations of objective and subjective hazards are referred to as cumulative hazards.

1-22. SUBJECTIVE HAZARDS

Subjective hazards are created by humans; for example, choice of route, companions, overexertion, dehydration, climbing above one's ability, and poor judgment.

a. **Falling.** Falling can be caused by carelessness, over-fatigue, heavy equipment, bad weather, overestimating ability, a hold breaking away, or other reasons.

b. **Bivouac Site.** Bivouac sites must be protected from rockfall, wind, lightning, avalanche run-out zones, and flooding (especially in gullies). If the possibility of falling exists, rope in, the tent and all equipment may have to be tied down.

c. **Equipment.** Ropes are not total security; they can be cut on a sharp edge or break due to poor maintenance, age, or excessive use. You should always pack emergency and bivouac equipment even if the weather situation, tour, or a short climb is seemingly low of dangers.

1-23. OBJECTIVE HAZARDS

Objective hazards are caused by the mountain and weather and cannot be influenced by man; for example, storms, rockfalls, icefalls, lightning, and so on.

a. **Altitude.** At high altitudes (especially over 6,500 feet), endurance and concentration is reduced. Cut down on smoking and alcohol. Sleep well, acclimatize slowly, stay hydrated, and be aware of signs and symptoms of high-altitude illnesses. Storms can form quickly and lightning can be severe.

b. **Visibility.** Fog, rain, darkness, and or blowing snow can lead to disorientation. Take note of your exact position and plan your route to safety before visibility decreases. Cold combined with fog can cause a thin sheet of ice to form on rocks (verglas). Whiteout conditions can be extremely dangerous. If you must move under these conditions, it is best to rope up. Have the point man move to the end of the rope. The second man will use the first man as an aiming point with the compass. Use a route sketch and march table. If the tactical situation does not require it, plan route so as not to get caught by darkness.

c. **Gullies.** Rock, snow, and debris are channeled down gullies. If ice is in the gully, climbing at night may be better because the warming of the sun will loosen stones and cause rockfalls.

d. **Rockfall.** Blocks and scree at the base of a climb can indicate recurring rockfall. Light colored spots on the wall may indicate impact chips of falling rock. Spring melt or warming by the sun of the rock/ice/snow causes rockfall.

e. **Avalanches.** Avalanches are caused by the weight of the snow overloading the slope. (Refer to paragraph 1-25 for more detailed information on avalanches.)

f. **Hanging Glaciers and Seracs.** Avoid, if at all possible, hanging glaciers and seracs. They will fall without warning regardless of the time of day or time of year. One cubic meter of glacier ice weighs 910 kilograms (about 2,000 pounds). If you must cross these danger areas, do so quickly and keep an interval between each person.

g. **Crevasses.** Crevasses are formed when a glacier flows over a slope and makes a bend, or when a glacier separates from the rock walls that enclose it. A slope of only two to three degrees is enough to form a crevasse. As this slope increases from 25 to 30 degrees, hazardous icefalls can be formed. Likewise, as a glacier makes a bend, it is likely that crevasses will form at the outside of the bend. Therefore, the safest route on a glacier would be to the inside of bends, and away from steep slopes and icefalls. Extreme care must be taken when moving off of or onto the glacier because of the moat that is most likely to be present.

1-24. WEATHER HAZARDS

Weather conditions in the mountains may vary from one location to another as little as 10 kilometers apart. Approaching storms may be hard to spot if masked by local peaks. A clear, sunny day in July could turn into a snowstorm in less than an hour. Always pack some sort of emergency gear.

a. Winds are stronger and more variable in the mountains; as wind doubles in speed, the force quadruples.

b. Precipitation occurs more on the windward side than the leeward side of ranges. This causes more frequent and denser fog on the windward slope.

c. Above approximately 8,000 feet, snow can be expected any time of year in the temperate climates.

d. Air is dryer at higher altitudes, so equipment does not rust as quickly, but dehydration is of greater concern.

e. Lightning is frequent, violent, and normally attracted to high points and prominent features in mountain storms. Signs indicative of thunderstorms are tingling of the skin, hair standing on end, humming of metal objects, crackling, and a bluish light (St. Elmo's fire) on especially prominent metal objects (summit crosses and radio towers).

(1) Avoid peaks, ridges, rock walls, isolated trees, fixed wire installations, cracks that guide water, cracks filled with earth, shallow depressions, shallow overhangs, and rock needles. Seek shelter around dry, clean rock without cracks; in scree fields; or in deep indentations (depressions, caves). Keep at least half a body's length away from a cave wall and opening.

(2) Assume a one-point-of-contact body position. Squat on your haunches or sit on a rucksack or rope. Pull your knees to your chest and keep both feet together. If half way up the rock face, secure yourself with more than one point—lightning can burn through rope. If already rappelling, touch the wall with both feet together and hurry to the next anchor.

f. During and after rain, expect slippery rock and terrain in general and adjust movement accordingly. Expect flash floods in gullies or chimneys. A climber can be washed away or even drowned if caught in a gully during a rainstorm. Be especially alert for falling objects that the rain has loosened.

g. Dangers from impending high winds include frostbite (from increased wind-chill factor), windburn, being blown about (especially while rappelling), and debris being blown about. Wear protective clothing and plan the route to be finished before bad weather arrives.

h. For each 100-meter rise in altitude, the temperature drops approximately one degree Fahrenheit. This can cause hypothermia and frostbite even in summer, especially when combined with wind, rain, and snow. Always wear or pack appropriate clothing.

i. If it is snowing, gullies may contain avalanches or snow sloughs, which may bury the trail. Snowshoes or skis may be needed in autumn or even late spring. Unexpected snowstorms may occur in the summer with accumulations of 12 to 18 inches; however, the snow quickly melts.

j. Higher altitudes provide less filtering effects, which leads to greater ultraviolet (UV) radiation intensity. Cool winds at higher altitudes may mislead one into underestimating the sun's intensity, which can lead to sunburns and other heat injuries. Use sunscreen and wear hat and sunglasses, even if overcast. Drink plenty of fluids.

1-25. AVALANCHE HAZARDS

Avalanches occur when the weight of accumulated snow on a slope exceeds the cohesive forces that hold the snow in place. (Table 1-2, page 1-32, shows an avalanche hazard evaluation checklist.)

a. **Slope Stability.** Slope stability is the key factor in determining the avalanche danger.

(1) ***Slope Angle.*** Slopes as gentle as 15 degrees have avalanched. Most avalanches occur on slopes between 30 and 45 degrees. Slopes above 60 degrees often do not build up significant quantities of snow because they are too steep.

(2) ***Slope Profile.*** Dangerous slab avalanches are more likely to occur on convex slopes, but may occur on concave slopes.

(3) ***Slope Aspect.*** Snow on north facing slopes is more likely to slide in midwinter. South facing slopes are most dangerous in the spring and on sunny, warm days. Slopes on the windward side are generally more stable than leeward slopes.

(4) ***Ground Cover.*** Rough terrain is more stable than smooth terrain. On grassy slopes or scree, the snow pack has little to anchor to.

b. **Triggers.** Various factors trigger avalanches.

(1) ***Temperature.*** When the temperature is extremely low, settlement and adhesion occur slowly. Avalanches that occur during extreme cold weather usually occur during or immediately following a storm. At a temperature just below freezing, the snowpack stabilizes quickly. At temperatures above freezing, especially if temperatures rise quickly, the potential for avalanche is high. Storms with a rise in temperature can deposit dry snow early, which bonds poorly with the heavier snow deposited later. Most avalanches occur during the warmer midday.

(2) ***Precipitation.*** About 90 percent of avalanches occur during or within twenty-four hours after a snowstorm. The rate at which snow falls is important. High rates of snowfall (2.5 centimeters per hour or greater), especially when accompanied by wind, are usually responsible for major periods of avalanche activity. Rain falling on snow will increase its weight and weakens the snowpack.

(3) ***Wind.*** Sustained winds of 15 miles per hour and over transport snow and form wind slabs on the lee side of slopes.

(4) ***Weight.*** Most victims trigger the avalanches that kill them.

(5) ***Vibration.*** Passing helicopters, heavy equipment, explosions, and earth tremors have triggered avalanches.

c. **Snow Pits.** Snow pits can be used to determine slope stability.

(1) Dig the snow pit on the suspect slope or a slope with the same sun and wind conditions. Snow deposits may vary greatly within a few meters due to wind and sun variations. (On at least one occasion, a snow pit dug across the fall line triggered the suspect slope). Dig a 2-meter by 2-meter pit across the fall line, through all the snow, to the ground. Once the pit is complete, smooth the face with a shovel.

(2) Conduct a shovel shear test.

(a) A shovel shear test puts pressure on a representative sample of the snowpack. The core of this test is to isolate a column of the snowpack from three sides. The column should be of similar size to the blade of the shovel. Dig out the sides of the column without pressing against the column with the shovel (this affects the strength). To isolate the rear of the column, use a rope or string to saw from side to side to the base of the column.

(b) If the column remained standing while cutting the rear, place the shovel face down on the top of the column. Tap with varying degrees of strength on the shovel to see what force it takes to create movement on the bed of the column. The surface that eventually slides will be the layer to look at closer. This test provides a better understanding of the snowpack strength. For greater results you will need to do this test in many areas and formulate a scale for the varying methods of tapping the shovel.

(3) Conduct a Rutschblock test. To conduct the test, isolate a column slightly longer than the length of your snowshoes or skis (same method as for the shovel shear test). One person moves on their skis or snowshoes above the block without disturbing the block. Once above, the person carefully places one showshoe or ski onto the block with no body weight for the first stage of the test. The next stage is adding weight to the first leg. Next, place the other foot on the block. If the block is still holding up, squat once, then twice, and so on. The remaining stage is to jump up and land on the block.

d. **Types of Snow Avalanches.** There are two types of snow avalanches: loose snow (point) and slab.

(1) Loose snow avalanches start at one point on the snow cover and grow in the shape of an inverted "V." Although they happen most frequently during the winter snow season, they can occur at any time of the year in the mountains. They often fall as many small sluffs during or shortly after a storm. This process removes snow from steep upper slopes and either stabilizes lower slopes or loads them with additional snow.

(2) Wet loose snow avalanches occur in spring and summer in all mountain ranges. Large avalanches of this type, lubricated and weighed down by meltwater or rain can travel long distances and have tremendous destructive power. Coastal ranges that have high temperatures and frequent rain are the most common areas for this type of avalanche.

(3) Slab avalanches occur when cohesive snow begins to slide on a weak layer. The fracture line where the moving snow breaks away from the snowpack makes this type of avalanche easy to identify. Slab release is rapid. Although any avalanche can kill you, slab avalanches are generally considered more dangerous than loose snow avalanches.

(a) Most slab avalanches occur during or shortly after a storm when slopes are loaded with new snow at a critical rate. The old rule of never travel in avalanche terrain for a few days after a storm still holds true.

(b) As slabs become harder, their behavior becomes more unpredictable; they may allow several people to ski across before releasing. Many experts believe they are susceptible to rapid temperature changes. Packed snow expands and contracts with temperature changes. For normal density, settled snow, a drop in temperature of 10 degrees Celsius (18 degrees Fahrenheit) would cause a snow slope 300 meters wide to contract 2 centimeters. Early ski mountaineers in the Alps noticed that avalanches sometimes occurred when shadows struck a previously sun-warmed slope.

d. **Protective Measures.** Avoiding known or suspected avalanche areas is the easiest method of protection. Other measures include:

(1) ***Personal Safety.*** Remove your hands from ski pole wrist straps. Detach ski runaway cords. Prepare to discard equipment. Put your hood on. Close up your clothing to prepare for hypothermia. Deploy avalanche cord. Make avalanche probes and shovels accessible. Keep your pack on at all times—do not discard. Your pack can act as a flotation device, as well as protect your spine.

(2) ***Group Safety.*** Send one person across the suspect slope at a time with the rest of the group watching. All members of the group should move in the same track from safe zone to safe zone.

e. **Route Selection.** Selecting the correct route will help avoid avalanche prone areas, which is always the best choice. Always allow a wide margin of safety when making your decision.

(1) The safest routes are on ridge tops, slightly on the windward side; the next safest route is out in the valley, far from the bottom of slopes.

(2) Avoid cornices from above or below. Should you encounter a dangerous slope, either climb to the top of the slope or descend to the bottom—well out of the way of the run-out zone. If you must traverse, pick a line where you can traverse downhill as quickly as possible. When you must ascend a dangerous slope, climb to the side of the avalanche path, and not directly up the center.

(3) Take advantage of dense timber, ridges, or rocky outcrops as islands of safety. Use them for lunch and rest stops. Spend as little time as possible on open slopes.

(4) Since most avalanches occur within twenty-four hours of a storm and or at midday, avoid moving during these periods. Moving at night is tactically sound and may be safer.

f. **Stability Analysis.** Look for nature's billboards on slopes similar to the one you are on.

(1) ***Evidence of Avalanching.*** Look for recent avalanches and for signs of wind-loading and wind-slabs.

(2) ***Fracture Lines.*** Avoid any slopes showing cracks.

(3) ***Sounds.*** Beware of hollow sounds—a "whumping" noise. They may suggest a radical settling of the snowpack.

g. **Survival.** People trigger avalanches that bury people. If these people recognized the hazard and chose a different route, they would avoid the avalanche. The following steps should be followed if caught in an avalanche.

(1) Discard equipment. Equipment can injure or burden you; discarded equipment will indicate your position to rescuers.

(2) Swim or roll to stay on tope of the snow. FIGHT FOR YOUR LIFE. Work toward the edge of the avalanche. If you feel your feet touch the ground, give a hard push and try to "pop out" onto the surface.

(3) If your head goes under the snow, shut your mouth, hold your breath, and position your hands and arms to form an air pocket in front of your face. Many avalanche victims suffocate by having their mouths and noses plugged with snow.

(4) When you sense the slowing of the avalanche, you must try your hardest to reach the surface. Several victims have been found quickly because a hand or foot was sticking above the surface.

(5) When the snow comes to rest it sets up like cement and even if you are only partially buried, it may be impossible to dig yourself out. Don't shout unless you hear rescuers immediately above you; in snow, no one can hear you scream. Don't struggle to free yourself—you will only waste energy and oxygen.

(6) Try to relax. If you feel yourself about to pass out, do not fight it. The respiration of an unconscious person is more shallow, their pulse rate declines, and the body temperature is lowered, all of which reduce the amount of oxygen needed. (See Appendix C for information on search and rescue techniques.)

AVALANCHE HAZARD EVALUATION CHECKLIST

Critical Data PARAMETERS:	KEY INFORMATION	Hazard Rating G	Y	R
TERRAIN: *Is the terrain capable of producing an avalanche?*				
	-Slope angle (steep enough to slide? prime time?)	☐	☐	☐
	-Slope aspect (leeward, shadowed, or extremely sunny?)	☐	☐	☐
	-Slope configuration (anchoring? shape?)	☐	☐	☐
	Overall Terrain Rating:	☐	☐	☐
SNOWPACK: *Could the snow fail?*				
	-Slab Configuration (slab? depth and distribution?)	☐	☐	☐
	-Bonding Ability (weak layer? tender spots?)	☐	☐	☐
	-Sensitivity (how much force to fail? shear tests? clues?)	☐	☐	☐
	Overall Snowpack Rating:	☐	☐	☐
Weather: *Is the weather contributing to instability?*				
	-Precipitation (type, amount, intensity? added weight?)	☐	☐	☐
	-Wind (snow transport? amount and rate of deposition?)	☐	☐	☐
	-Temperature (storm trends? effects on snowpack?)	☐	☐	☐
	Overall Weather Rating:	☐	☐	☐
Human: *What are your alternatives and their possible consequences?*				
	-Attitude (toward life? risk? goals? assumptions?)	☐	☐	☐
	-Technical Skill Level (traveling? evaluating aval. hazard?)	☐	☐	☐
	-Strength/Equipment (strength? prepared for the worst?)	☐	☐	☐
	Overall Human Rating:	☐	☐	☐

Decision/Action:
Overall Hazard Rating/GO or NO Go? **GO ☐ or NOGO ☐**

***HAZARD LEVEL SYMBOLS:**
R = Red light (stop/dangerous)
G = Green light (go/OK)
Y = Yellow light (caution/potentially dangerous).

Table 1-2. Avalanche hazard evaluation checklist.

CHAPTER 2

MOUNTAIN LIVING

Units deploying to high elevations must receive advanced training to survive in the harsh mountain environment. Normal activities (navigation, communications, and movement) require specialized techniques. Training should be conducted as realistically as possible, preferably under severe conditions so the soldier gains confidence. Extended training exercises test support facilities and expose the soldier to the isolation common to mountain operations. Training should reflect the harsh mountain environment and should consider the following:

- *Temperature and altitude extremes.*
- *Hygiene and sanitation.*
- *Limited living space (difficulty of bivouac).*
- *Clothing requirements.*

Section I. SURVIVAL

The soldier trained to fight and survive in a mountain environment will have increased confidence in himself. Training should include: psychological preparation, locating water, shelter considerations, fire building, health hazards, and techniques for obtaining food (see FM 21-76).

2-1. WATER SUPPLY

Mountain water should never be assumed safe for consumption. Training in water discipline should be emphasized to ensure soldiers drink water only from approved sources. Fluids lost through respiration, perspiration, and urination must be replaced if the soldier is to operate efficiently.

a. Maintaining fluid balance is a major problem in mountain operations. The sense of thirst may be dulled by high elevations despite the greater threat of dehydration. Hyperventilation and the cool, dry atmosphere bring about a three- to four-fold increase in water loss by evaporation through the lungs. Hard work and overheating increase the perspiration rate. The soldier must make an effort to drink liquids even when he does not feel thirsty. One quart of water, or the equivalent, should be drunk every four hours; more should be drunk if the unit is conducting rigorous physical activity.

b. Three to six quarts of water each day should be consumed. About 75 percent of the human body is liquid. All chemical activities in the body occur in water solution, which assists in removing toxic wastes and in maintaining an even body temperature. A loss of two quarts of body fluid (2.5 percent of body weight) decreases physical efficiency by 25 percent, and a loss of 12 quarts (15 percent of body weight) is usually fatal. Salt lost by sweating should be replaced in meals to avoid a deficiency and subsequent cramping. Consuming the usual military rations (three meals a day) provides sufficient sodium replacement. Salt tablets are not necessary and may contribute to dehydration.

c. Even when water is plentiful, thirst should be satisfied in increments. Quickly drinking a large volume of water may actually slow the soldier. If he is hot and the water is cold, severe cramping may result. A basic rule is to drink small amounts often. Pure water should always be kept in reserve for first aid use. Emphasis must be placed on the three rules of water discipline:

- Drink only treated water.
- Conserve water for drinking. Potable water in the mountains may be in short supply.
- Do not contaminate or pollute water sources.

d. Snow, mountain streams, springs, rain, and lakes provide good sources of water supply. Purification must be accomplished, however, no matter how clear the snow or water appears. Fruits, juices, and powdered beverages may supplement and encourage water intake (do not add these until the water has been treated since the purification tablets may not work). Soldiers cannot adjust permanently to a decreased water intake. If the water supply is insufficient, physical activity must be reduced. Any temporary deficiency should be replaced to maintain maximum performance.

e. All water that is to be consumed must be potable. Drinking water must be taken only from approved sources or purified to avoid disease or the possible use of polluted water. Melting snow into water requires an increased amount of fuel and should be planned accordingly. Nonpotable water must not be mistaken for drinking water. Water that is unfit to drink, but otherwise not dangerous, may be used for other purposes such as bathing. Soldiers must be trained to avoid wasting water. External cooling (pouring water over the head and chest) is a waste of water and an inefficient means of cooling. Drinking water often is the best way to maintain a cool and functioning body.

f. Water is scarce above the timberline. After setting up a perimeter (patrol base, assembly area, defense), a watering party should be employed. After sundown, high mountain areas freeze, and snow and ice may be available for melting to provide water. In areas where water trickles off rocks, a shallow reservoir may be dug to collect water (after the sediment settles). Water should be treated with purification tablets (iodine tablets or calcium hypochlorite), or by boiling at least one to two minutes. Filtering with commercial water purification pumps can also be conducted. Solar stills may be erected if time and sunlight conditions permit (see FM 21-76). Water should be protected from freezing by storing it next to a soldier or by placing it in a sleeping bag at night. Water should be collected at midday when the sun thaw available.

2-2. NUTRITION

Success in mountain operations depends on proper nutrition. Because higher altitudes affect eating habits, precautions must be taken. If possible, at least one hot meal each day should be eaten, which may require personnel to heat their individual rations.

a. The following elements are characteristic of nutritional acclimatization in mountain operations:

- Weight loss during the first two to three days at high elevation.
- A loss of appetite with symptoms of mountain sickness.
- Loss of weight usually stops with acclimatization.
- At progressively higher elevations (greater than 14,000 feet), the tolerance of fatty/high-protein foods rapidly decreases. A high carbohydrate diet may lessen the symptoms of acute mountain sickness and is digested better than fat at high altitudes.

b. Increased fatigue may cause soldiers to become disinterested in eating properly. Decreased consumption may result in malnutrition because of the unpleasant taste of cold rations. Leaders should ensure that fuel tablets and squad stoves are available, or that natural flammable materials are used if possible. Although there is no physiological need for hot food, it does increase morale and a sense of well being. Loss of weight in the first few days occurs because of dehydration, metabolic changes, and loss of appetite. Carbohydrate-containing beverages, such as fruit juices and sports drinks, are an effective means of increasing carbohydrates, energy, and liquid intake when the normal appetite response is blunted at altitude.

c. Three major food components are required to maintain a well-functioning body: proteins, fats, and carbohydrates. These food components provide energy, amino acids, vitamins, fiber, and minerals. All three components must be provided in the correct proportions to maintain a healthy body.

(1) ***Protein***. Proteins consist of a large number of amino acid units that are linked together to form the protein. The amino acids, resulting from digestion of protein, are absorbed through the intestine into the blood, and are used to make or replace body proteins (muscle and body tissue). Sources of readily useable animal proteins include eggs, milk, cheese, poultry, fish, and meats. Other foods such as cereals, vegetables, and legumes also provide amino acids. These proteins are not as balanced in essential amino acid composition as meat, eggs, or milk proteins. The minimum daily protein requirement, regardless of physical activity, is 8 ounces for a 154-pound man. Since amino acids are either oxidized for energy or stored as fats, consuming excess protein is inefficient and may increase the water intake needed for urea nitrogen excretion. Protein requires water for digestion and may facilitate dehydration. Proteins provide the body about four kilocalories of energy per gram and require the most energy for the body to digest.

(2) ***Fats.*** Fats are the most concentrated form of food energy. Of the total daily caloric intake, 25 to 30 percent may be supplied as fats. Main sources of fats are meats, nuts, butter, eggs, milk, and cheese. Fats require more water and oxygen, and are harder to digest at higher altitudes. Fats are the body's natural stored source of energy. Fats provide the body around 9 kilocalories of energy per gram and require less energy for the body to digest than protein but more than carbohydrates.

(3) ***Carbohydrates.*** Carbohydrates are an important source of calories. In the form of glucose, carbohydrates are found in the most important energy-producing cycles in the body's cells. If carbohydrate intake exceeds energy needs, moderate amounts are stored in the muscles and liver. Larger amounts are converted into fats and stored in that form. Carbohydrates should compose up to 50 percent of the total daily caloric intake. Nutritionally, the most useful sources of carbohydrates are foods such as unrefined grains, vegetables, and fruit. Carbohydrates provide the body around four kilocalories of energy per gram and are the easiest to digest.

(4) ***Vitamins.*** Vitamins are classified into two groups on the basis of their ability to dissolve in fat or water. The fat-soluble vitamins include vitamins A, D, E, and K. The water-soluble vitamins include the B vitamins and vitamin C, which are found in cereals, vegetables, fruits, and meats. A well-balanced diet provides all of the required vitamins. Since most water-soluble vitamins are not stored, a proper diet is necessary to ensure adequate levels of these vitamins. If an improper and unbalanced diet is likely to occur

during a deployment, vitamin supplements should be considered, especially if this period is to exceed 10 days.

(5) ***Minerals.*** Mineral elements can be divided into two groups: those needed in the diet in amounts of 100 milligrams or more a day such as calcium, phosphorous, and magnesium; and trace elements needed in amounts of only a few milligrams a day such as iodine, iron, and zinc. Required minerals are contained in a balanced diet (meats, vegetables, fruits).

d. Eating a balanced diet provides the energy needed to conduct daily activities and to maintain the internal body processes. A balanced diet containing adequate amounts of vitamins and minerals ensures an efficient metabolism. Since climbing is a strenuous activity and demands high-energy use, a balanced diet is a necessity.

(1) The efficiency of the body to work above the basal metabolism varies from 20 to 40 percent, depending on the soldier. Over 50 percent of caloric intake is released as heat and is not available when the soldier works. (About 4,500 calories are expended for strenuous work and 3,500 calories for garrison activity.) Heat is a by-product of exertion. Exertion causes excessive bodily heat loss through perspiration and increased radiation. During inactivity in cold weather, the metabolism may not provide enough heat. The "internal thermostat" initiates and causes the muscles to shiver, thus releasing heat. Shivering also requires energy and burns up to 220 calories per hour (estimated for a 100-pound man).

(2) With an abrupt ascent to high altitudes, the soldier experiences physiological acclimatization. The circulatory system labors to provide the needed oxygen to the body. Large meals require the digestive system to work harder than usual to assimilate food. Large meals may be accompanied by indigestion, shortness of breath, cramps, and illness. Therefore, relatively light meals that are high in carbohydrates are best while acclimatizing at higher elevations. Personnel should eat moderately and rest before strenuous physical activity. Since fats and protein are harder to digest, less digestive disturbances may occur if meals are eaten before resting. A diet high in carbohydrates is not as dense in energy and may require eating more often. Carbohydrates, beginning in the morning and continuing through mid-afternoon, are important in maintaining energy levels.

(3) Extra food should be carried in case resupply operations fail. Food should be lightweight and easy to digest, and be eaten hot or cold. Meals-ready-to-eat (MREs) meet these criteria and provide all of the basic food groups. Commanders may consider supplementing MREs with breakfast bars, fruits, juices, candies, cereal bars, and chocolate. Bouillon cubes can replace water and salt as well as warming cold bodies and stimulating the appetite. Hot beverages of soup, juices, powdered milk, and cider should also be considered. Since coffee, tea, and hot chocolate are diuretics, the consumption of these beverages should not be relied upon for hydration.

(4) Warm meals should be provided when possible. When cooking, the heat source must be kept away from equipment and ammunition. At higher elevations, the cooking time may be doubled. To conserve fuel, stoves, fires, and fuel tablets should be protected from the wind. Extra fuel should be stored in tightly sealed, marked, metal containers. Use stoves and heat tabs for warming food and boiling water. Canteen cups and utensils should be cleaned after use. All food items and garbage are carried with the unit. If possible, garbage should be burned or deep buried. Caution must be taken to prevent animals from foraging through rucksacks, ahkios, and burial sites. As all missions are tactical, no trace of a unit should be detected.

(5) Certain drugs, medications, alcohol, and smoking have adverse effects on the circulation, perspiration, hydration, and judgment of soldiers. Therefore, they should be avoided when operating in extremely cold conditions or at high altitudes.

2-3. PERSONAL HYGIENE AND SANITATION

The principles of personal hygiene and sanitation that govern operations on low terrain also apply in the mountains. Commanders must conduct frequent inspections to ensure that personal habits of hygiene are not neglected. Standards must be maintained as a deterrent to disease, and as reinforcement to discipline and morale.

a. **Personal Hygiene.** This is especially important in the high mountains, mainly during periods of cold weather. In freezing weather, the soldier may neglect washing due to the cold temperatures and scarcity of water. This can result in skin infections and vermin infestation. If bathing is difficult for any extended period, the soldier should examine his skin and clean it often. Snow baths in lieu of a water bath are recommended. This helps reduce skin infections and aids the comfort of the soldier.

(1) Snow may be used instead of toilet paper. Soldiers should shave at rest periods in the shelter so that oils stripped in shaving will be replenished. A beard may mask the presence of frostbite or lice. Water-based creams and lotions should be avoided in cold environments since this will further dehydrate tissues and induce frostbite by freezing. The nonwater-based creams can be used for shaving in lieu of soap. Sunscreens and chap sticks should be used on lips, nose, and eyelids. Topical steroid ointments should be carried for rashes. The teeth must also be cleaned to avoid diseases of the teeth and gums. Underwear should be changed when possible, but this should not be considered a substitute for bathing. When operating in areas where resupply is not possible, each soldier should carry a complete change of clothing. If laundering of clothing is difficult, clothes should be shaken and air-dried. Sleeping bags must be regularly cleaned and aired.

(2) The principles of foot hygiene must be followed to protect the feet from cold injuries. The causes of such injuries are present throughout the year in high mountains. Boots should be laced tightly when climbing to provide needed support but not so tight as to constrict circulation. Socks should be worn with no wrinkles since this causes blisters on the feet. Feet should be washed daily, and kept as dry and clean as possible. If regular foot washing is impossible, socks should be changed often (at halts and rest periods or at least once a day) and feet massaged, dried, and sprinkled with foot powder. Talc or antifungal powder should be used when massaging; excess powder is brushed off to avoid clumping, which may cause blisters. Feet can be cleaned with snow, but must be quickly dried. Whenever changing socks, soldiers should closely examine their feet for wrinkles, cracks, blisters, and discoloration. Nails should be trimmed but not too short. Long nails wear out socks; short nails do not provide proper support for the ends of the toes. Medical attention should be sought for any possible problems.

(3) Feet should be sprayed two or three times a day with an aluminum chlorohydrate antiperspirant for a week and then once a day for the rest of the winter. If fissures or cracks occur in the feet, it is best to discontinue spraying until they are healed or to spray less often to control sweating. This process stops about 70 percent of the sweating in the feet.

(4) During periods of extreme cold, there is a tendency for the soldier to become constipated. This condition is brought about by the desire to avoid the inconvenience and

discomfort of defecating. Adequate water intake plus a low protein, high roughage diet can be helpful in preventing constipation.

b. **Sanitation.** In rocky or frozen ground, digging latrines is usually difficult. If latrines are constructed, they should be located downwind from the position and buried after use. In tactical situations, the soldier in a designated, downwind location away from water sources may dig "cat holes." Since waste freezes, it can be covered with snow and ice or pushed down a crevasse. In rocky areas above the timberline, waste may be covered with stones.

Section II. ACCLIMATIZATION AND CONDITIONING

Terrestrial altitude can be classified into five categories. Low altitude is sea level to 5,000 feet. Here, arterial blood is 96 percent saturated with oxygen in most people. Moderate altitude is from 5,000 to 8,000 feet. At these altitudes, arterial blood is greater than 92 percent saturated with oxygen, and effects of altitude are mild and temporary. High altitude extends from 8,000 to 14,000 feet, where arterial blood oxygen saturation ranges from 92 percent down to 80 percent. Altitude illness is common here. Very high altitude is the region from 14,000 to 18,000 feet, where altitude illness is the rule. Areas above 18,000 feet are considered extreme altitudes.

Soldiers deployed to high mountainous elevations require a period of acclimatization before undertaking extensive military operations. The expectation that freshly deployed, unacclimatized troops can go immediately into action is unrealistic, and could be disastrous if the opposing force is acclimatized. Even the physically fit soldier experiences physiological and psychological degradation when thrust into high elevations. Time must be allocated for acclimatization, conditioning, and training of soldiers. Training in mountains of low or medium elevation (5,000 to 8,000 feet) does not require special conditioning and acclimatization procedures. However, some soldiers will have some impairment of operating efficiency at these low altitudes. Above 8,000 feet (high elevation), most unacclimatized soldiers may display some altitude effects. Training should be conducted at progressively higher altitudes, starting at about 8,000 feet and ending at 14,000 feet. Attempts to acclimatize beyond 17,000 feet results in a degradation of the body greater than the benefits gained. The indigenous populations can out-perform even the most acclimatized and physically fit soldier who is brought to this altitude; therefore, employment of the local population may be advantageous.

2-4. SYMPTOMS AND ADJUSTMENTS

A person is said to be acclimatized to high elevations when he can effectively perform physically and mentally. The acclimatization process begins immediately upon arrival at the higher elevation. If the change in elevation is large and abrupt, some soldiers can suffer from acute mountain sickness (AMS), high-altitude pulmonary edema (HAPE), or high-altitude cerebral edema (HACE). Disappearance of the symptoms of acute mountain sickness (from four to seven days) does not indicate complete acclimatization. The process of adjustment continues for weeks or months. The altitude at which complete acclimatization is possible is not a set point but for most soldiers with proper ascent, nutrition and physical activity it is about 14,000 feet.

a. Immediately upon arrival at high elevations, only minimal physical work can be performed because of physiological changes. The incidence and severity of AMS symptoms vary with initial altitude, the rate of ascent, and the level of exertion and individual susceptibility. Ten to twenty percent of soldiers who ascend rapidly (in less than 24 hours) to altitudes up to 6,000 feet experience some mild symptoms. Rapid ascent to 10,000 feet causes mild symptoms in 75 percent of personnel. Rapid ascent to elevations of 12,000 to 14,000 feet will result in moderate symptoms in over 50 percent of the soldiers and 12 to 18 percent may have severe symptoms. Rapid ascent to 17,500 feet causes severe, incapacitating symptoms in almost all individuals. Vigorous activity during ascent or within the first 24 hours after ascent will increase both the incidence and severity of symptoms. Some of the behavioral effects that will be encountered in unacclimatized personnel include:

- Increased errors in performing simple mental tasks.
- Decreased ability for sustained concentration.
- Deterioration of memory.
- Decreased vigilance or lethargy.
- Increased irritability in some individuals.
- Impairment of night vision and some constriction in peripheral vision (up to 30 percent at 6,000 feet).
- Loss of appetite.
- Sleep disturbances.
- Irregular breathing.
- Slurred speech.
- Headache.

b. Judgment and self-evaluation are impaired the same as a person who is intoxicated. During the first few days at a high altitude, leaders have extreme difficulty in maintaining a coordinated, operational unit. The roughness of the terrain and the harshness and variability of the weather add to the problems of unacclimatized personnel. Although strong motivation may succeed in overcoming some of the physical handicaps imposed by the environment, the total impact still results in errors of judgment. When a soldier cannot walk a straight line and has a loss of balance, or he suffers from an incapacitating headache, he should be evacuated to a lower altitude (a descent of at least 1,000 feet for at least 24 hours).

2-5. PHYSICAL AND PSYCHOLOGICAL CONDITIONING

The commander must develop a conditioning/training program to bring his unit to a level where it can operate successfully in mountain conditions. Priorities of training must be established. As with all military operations, training is a major influence on the success of mountain operations.

a. U.S. forces do not routinely train in mountainous terrain. Therefore, extensive preparations are needed to ensure individual and unit effectiveness. Units must be physically and psychologically conditioned and adjusted before undertaking rigorous mountain operations. Units must be conditioned and trained as a team to cope with the terrain, environment, and enemy situation. Certain factors must be considered:

- What are the climatic and terrain conditions of the area of operations?
- How much time is available for conditioning and training?
- Will the unit conduct operations with other U.S. or Allied forces? Are there language barriers? What assistance will be required? Will training and conditioning be required for attached personnel?

- What additional personnel will accompany the unit? Will they be available for training and conditioning?
- What is the current level of physical fitness of the unit?
- What is the current level of individual expertise in mountaineering?
- What type of operations can be expected?
- What is the composition of the advance party? Will they be available to assist in training and acclimatization?
- What areas in the U.S. most closely resemble the area of operations?
- Are predeployment areas and ranges available?
- Does the unit have instructors qualified in mountain warfare?
- What type equipment will be required (to fit the season, mission, terrain)?
- Does the unit have enough of the required equipment? Do personnel know how to use the equipment? Will the equipment go with the advance party, with the unit, or follow after the unit's arrival?
- Does equipment require modification?
- Do weapons and equipment require special maintenance?

b. When the unit arrives in the area of operations, all personnel require a period of conditioning and acclimatization. The time schedule should allow for longer and more frequent periods of rest. The rigors of establishing an assembly area exhaust most unacclimatized personnel. Water, food, and rest must be considered as priorities, ensuring sufficient amounts while individual metabolisms and bodies become accustomed to functioning at higher elevations.

c. Since the acclimatization process cannot be shortened, and the absence of acclimatization hampers the successful execution of operations, deployment to higher elevations must consider the following:

(1) Above 8,000 feet, a unit should ascend at a rate of 1,000 to 2,000 feet per day. Units can leapfrog, taking an extended rest period.

(2) Units should not resort to the use of pharmaceutical pretreatment with carbonic anhydrase inhibitors such as acetazolamide (Diamox). These drugs have side effects that mimic the signs and symptoms of AMS. Inexperienced medics may have difficulty recognizing the differences between the side effects of the drug and a condition that could possibly be life threatening. Additionally, these drugs are diuretics, which results in higher hydration levels (at least 25 percent increase per man per day). These higher hydration levels create a larger logistical demand on the unit by requiring more water, time to acquire water, water purification supplies, and, if in a winter environment, fuels for melting snow and ice for water.

(3) Carbonic anhydrase inhibitors such as acetazolamide are effective in the treatment of mild and severe AMS. These drugs should accompany attached medical personnel because they can treat the soldier suffering the symptoms of AMS and, although rest may be required evacuation may not be needed.

(4) Do not move troops directly to high altitudes even if allowances can be made for inactivity for the first three to five days before mission commitment. Moving troops directly to high altitude can increase the probability of altitude sickness. Even if inactivity follows deployment, the incidence of altitude sickness is more likely than with a gradual ascent.

d. Training on high-altitude effects can prevent psychological preconceptions. Soldiers who have lived on flat terrain may have difficulty when learning to negotiate steep slopes or cliffs, developing a sense of insecurity and fear. They must be slowly introduced to the new terrain and encouraged to develop the confidence required to negotiate obstacles with assurance and ease. They must be taught the many climbing techniques and principles of mountain movement. They overcome their fear of heights by becoming familiar with the problem. The soldier cannot be forced to disregard this fear.

e. Regardless of previous training and the amount of flat cross-country movement practice, the untrained soldier finds mountain movement hard and tiring. A different group of muscles are used, which must be developed and hardened. A new technique of rhythmic movement must be learned. Such conditioning is attained through frequent marches and climbs, while carrying TOE and special equipment loads. This conditions the back and legs, which results in increased ability and endurance. At the same time, the men acquire confidence and ability to safely negotiate the terrain. The better the physical condition of the soldier, the better the chance of avoiding exhaustion. Proper physical conditioning ensures the soldier is an asset and not a liability. The body improves its capacity for exercise, the metabolism becomes more efficient, and blood and oxygen flow quickly and effectively.

f. A physical fitness training program that gradually increases in difficulty should include marches, climbing, and calisthenics. This increases the soldier's endurance. Through a sustained high level of muscular exertion, the soldier's capacity for exertion is increased. Physical conditioning should include long-distance running for aerobic conditioning; calisthenics and weight training to strengthen the heart, lungs, abdomen, legs, back, arms, and hands; a swimming program to increase lung efficiency; and road marches over mountainous terrain with all combat equipment. Upon deploying to high elevations, caution must be exercised by units that are in superior physical condition. The heart rate, metabolism, and lungs must become accustomed to the elevation and thinner air. A conditioning program must be set up on site and integrated in gradual stages where acclimatization, conditioning, and mountaineering skills are realized.

g. Conditioning should begin with basic climbing. It is equally important to instill the will to climb. Confidence goes hand in hand with physical conditioning and skill development. Repetitive practice, to the point of instinctive reaction, is key to learning and maintaining climbing proficiency and technical skills. There are no quick and easy methods to becoming acclimatized and conditioned. Training should gradually challenge the soldier over an extended period and reinforce learning skills.

Section III. MEDICAL CONSIDERATIONS

Improper acclimatization poses many problems for medical personnel. Facilities and supplies may be inadequate to treat all victims. After acclimatization, personnel can still become injured (sprains, strains, fractures, frostbite, hypothermia, and trench foot). Mountain sickness and other illnesses may also occur. Evacuation of the sick and wounded is compounded by the terrain and weather.

2-6. ILLNESS AND INJURY

Units operating in mountainous regions are exposed to varied types of injuries and illnesses not associated with other areas. Medical considerations are like those for other environments; however, there are some unique aspects of mountain operations to be considered if effective support is to be provided. Most injuries in the mountain environment are soft tissue injuries. These include sprains, strains, abrasions, contusions and fractures. As with any other injuries, the most life threatening are treated first with the emphasis on airway control, breathing management, and circulatory support. Skills in basic first aid are essential to the mountain leader and should be reinforced with regular sustainment training.

2-7. TREATMENT AND EVACUATION

In harsh mountain weather, the most important course of action is to provide injured soldiers with medical aid as soon as possible. Immediate first aid is given on site. Due to rough terrain, medical units can seldom reach unit aid stations by vehicle to evacuate casualties. Litter bearers are required to move casualties to the rear where they can be evacuated by ground or air to clearing stations. The victim is protected from the weather and shock during transportation. Rendezvous points are coordinated with medical units as far forward as possible. Training must be accomplished with all litter bearers on evacuation techniques and first aid. Lightly wounded personnel may need assistance to move over rough terrain.

2-8. SOLAR INJURIES

Solar injuries can happen in warm weather or in cold weather. These types of injuries can be just as incapacitating as most other injuries but usually are not fatal. The peak hours of ultraviolet (UV) radiation are between the hours of 1100 and 1500. Due to the long wavelengths of ultraviolet light, cloudy days can be more dangerous than sunny days. On sunny days the soldier takes more care due to the bright conditions. On cloudy days the soldier tends not to wear sunglasses or sunscreen.

a. **Sunburn.** Sunburn is the burning of exposed skin surfaces by ultraviolet radiation.

(1) Contributing factors include fair skin, improper use of para-amino benzoic acid (PABA)-based sunscreens, and exposure to intense ultraviolet rays for extended periods.

(2) Symptoms of sunburn are painful, burning, red or blistered skin with a slight swelling. The skin may be warm to the touch. In severe cases chills, fever, and headaches may occur.

(3) To treat sunburn, apply cool saline dressings to alleviate pain and swelling. Do not pop blisters. If blisters do break, wash thoroughly, bandage, and seek medical attention. A solution of vinegar (acetic) and water can be lightly applied with sterile gauze to alleviate burning. The tannic acid in used tea bags can also be applied to alleviate burning. Administer pain medication if needed.

(4) To prevent sunburn, skin should be covered with clothing or PABA-based sunscreens (at least sun protection factor [SPF] 15) should be applied liberally to exposed skin during the peak hours of UV exposure. The SPF means that you can stay exposed to the suns UV rays that many times longer than without it. (For example, an SPF of 15 means that skin can be exposed to UV rays 15 times longer than without sunscreen.) During sustained activity, the sunscreen should be regularly reapplied to maintain the SPF.

b. **Snowblindness.** Snowblindness is sunburn of the cornea of the eye caused by exposure to ultraviolet radiation.

(1) A contributing factor is the reflection of sunlight from all directions off the snow, ice, and water. Ultraviolet rays can cause vision problems even on cloudy days. They are less filtered at high altitudes than at low altitudes.

(2) Symptoms of snowblindness are painful, red, watery eyes; a gritty feeling; blurred vision; and a headache.

(3) To treat snowblindness, patch both eyes with cold compresses for 24 hours. Topical anesthetics such as Tetracaine Ophthalmic can be used to relieve pain. Avoid rubbing the eyes. If still painful, keep the victim's eyes patched and administer oral pain medication. Snowblindness will usually resolve in about 24 hours for mild to moderate cases. Victims are rarely in need of evacuation unless the case is unusually severe.

(4) To prevent snow blindness, use quality sunglasses even on cloudy days in snow-covered terrain. Proper sunglasses should provide 100 percent UVA and UVB protection and have hoods on the sides to prevent reflected light from entering the eye. (Currently, the U.S. Army does not have these types of "glacier" sunglasses in their inventory and they must be acquired from nonmilitary sources.) In an emergency, improvise slit glasses from materials such as cardboard or birch bark.

2-9. COLD-WEATHER INJURIES

Cold-weather injuries can occur during any season of the year. Death has resulted in temperatures as high as 10 degrees Celsius (50 degrees Fahrenheit). A loss of body heat combined with shock produces devastating results. However, most of these accidents can be prevented by proper planning to include: timely requisition and receipt of supplies and proper clothing; thorough training of personnel with respect to the hazards of cold weather; effective methods for the receipt, dissemination, and use of cold-weather data; periodic inspections of clothing, personnel, and equipment; and personnel receiving a balance of water, rest, and nutrition.

a. Soldiers must be prepared to survive, move, and fight in winter conditions. Intense cold affects the mind as well as the body. Simple tasks take longer to perform, and they take more effort than in a temperate climate. When weather conditions become extreme the problems of survival become more significant. Warmth and comfort become the top priorities. The effects of extreme cold and the probability of injury are magnified due to the lack of proper diet and sleep. The most important measure in the prevention of cold-weather injuries is the education of personnel and their leaders.

b. Cold injuries may be divided into two types: freezing and nonfreezing. The freezing type is known as frostbite. The nonfreezing type includes hypothermia, dehydration, and immersion foot. Cold injuries result from impaired circulation and the action of ice formation and cold upon the tissues of the body. Temperature alone is not a reliable guide as to whether a cold injury can occur. Low temperatures are needed for cold injuries to occur, but freezing temperatures are not. Wind speed can accelerate body heat loss under both wet and cold conditions. All commanders and subordinate leaders/instructors must be familiar with and carry GTA 5-8-12, which includes a wind chill equivalent temperature chart (Figure 2-1, page 2-12).

WIND CHILL FACTOR CHART												
COOLING POWER OF WIND EXPRESSED AS AN EQUIVALENT CHILL TEMPERATURE (UNDER CALM CONDITIONS)												
ESTIMATED WIND SPEED (IN MPH)	ACTUAL THERMOMETER READING (F)											
	50	40	30	20	10	0	-10	-20	-30	-40	-50	-60
	EQUIVALENT TEMPERATURES (F)											
Calm	50	40	30	20	10	0	-10	-20	-30	-40	-50	-60
5	48	37	27	16	6	-5	-15	-26	-36	-47	-57	-68
10	40	28	16	4	-9	-24	-33	-46	-58	-70	-83	-95
15	36	22	9	-5	-18	-32	-45	-58	-72	-85	-99	-112
20	32	18	4	-10	-25	-39	-53	-67	-82	-96	-110	-124
25	30	16	0	-15	-29	-44	-59	-74	-88	-104	-118	-133
30	28	13	-2	-18	-33	-48	-63	-79	-94	-109	-125	-140
35	27	11	-4	-21	-35	-51	-67	-82	-98	-113	-129	-145
40	26	10	-6	-21	-37	-53	-69	-85	-100	-116	-132	-148
Winds greater than 40 MPH have little additional effect.	LITTLE DANGER (For properly clothed person) Maximum danger of false sense of security.				INCREASING DANGER Danger from freezing of exposed flesh.			GREAT DANGER				
	Trench foot and immersion foot may occur at any point on this chart.											

Figure 2-1. Wind chill chart.

c. Many other factors in various combinations determine if cold injuries will occur.

(1) ***Previous Cold Injuries.*** If a soldier has had a cold injury before, he is at higher risk for subsequent cold injuries.

(2) ***Race.*** Blacks are more susceptible to cold-weather injuries than Caucasians.

(3) ***Geographic Origin.*** Personnel from warmer climates are more susceptible to cold injury than those from colder climates.

(4) ***Ambient Temperature.*** The temperature of the air (or water) surrounding the body is critical to heat regulation. For example, the body uses more heat to maintain the temperature of the skin when the temperature of the surrounding air is 37 degrees Fahrenheit than when it is 50 degrees Fahrenheit.

(5) ***Wind Chill Factor.*** The commander should know the wind chill factor. When the forecast gives a figure that falls within the increased danger zone or beyond, caution must be taken to minimize cold injury. The equivalent wind chill temperature is especially important when the ambient temperature is 0 degrees Celsius (32 degrees Fahrenheit) or less. Tissue can freeze if exposed for a prolonged period and if frequent warming is not practiced. The lower the wind chill, the faster tissue freezing can occur. Wind chill is the rate of cooling. Wind does not lower the ambient temperature. The ambient temperature alone determines freezing or nonfreezing injuries. Frostbite Wind chill may cause faster cooling due to increased convection, but not below the ambient temperature.

(6) ***Type of Mission.*** Combat action requiring prolonged immobility and long hours of exposure to low temperatures, or not having an opportunity to warm up increases the possibility of cold injuries.

(7) ***Terrain.*** Minimal cover and wet conditions increase the potential for cold injury.

(8) ***Clothing.*** Clothing for cold weather should be worn with the acronym **C.O.L.D.** in mind.

- **C**—Clothing should be clean since prolonged wear reduces its air-trapping abilities and clogs air spaces with dirt and body oils.
- **O**—Overheating. Avoid overheating. Appropriate measures should be taken when a change in weather or activity alters the amount of clothing needed to prevent overheating and, therefore, accumulation of perspiration.
- **L**—Loose and in layers (to trap air and to conserve body heat). The uniform should be worn completely and correctly to avoid injury to exposed body surfaces. The cold-weather uniform is complete when worn with gloves and inserts.
- **D**—Dry. Keep dry. Wet clothing loses insulation value.

(9) ***Moisture.*** Water conducts heat more rapidly than air (25 percent). When the skin or clothing becomes damp or wet, the risk of cold injury is greatly increased.

(10) ***Dehydration.*** The most overlooked factor causing cold injuries is dehydration. Individuals must retain their body fluids. In cold weather the human body needs special care, and the consumption of water is important to retain proper hydration.

(11) ***Age.*** Within the usual age range of combat personnel, age is not a significant factor.

(12) ***Fatigue.*** Mental weariness may cause apathy leading to neglect of duties vital to survival.

(13) ***Concomitant Injury.*** Injuries resulting in shock or blood loss reduce blood flow to extremities and may cause the injured individual to be susceptible to cold injury, which in turn can accelerate shock.

(14) ***Discipline, Training, and Experience.*** Well-trained and disciplined soldiers suffer less than others from the cold.

(15) ***Nutrition.*** Good nutrition is essential for providing the body with fuel to produce heat in cold weather. The number of calories consumed normally increases as the temperature becomes colder.

(16) ***Excess Activity.*** Excess activity (overheating) results in loss of large amounts of body heat by perspiration. This loss of body heat combined with the loss of insulation value provided by the clothing (due to perspiration dampening the clothing) can subject a soldier to cold injuries.

(17) ***Radical Changes in the Weather.*** Weather conditions in mountainous terrain are known to change considerably throughout the day. Weather can quickly change to extremely cold and wet conditions, especially in higher elevations.

d. Commanders should ensure that the following measures are taken.

(1) Soldiers' uniforms are kept as dry as possible and are protected from the elements.

(2) Soldiers are educated on proper use of clothing systems to avoid the effects of overheating and perspiration (layer dressing and ventilate).

(3) The buddy system is used to watch for early signs of cold-weather injuries.

(4) All soldiers waterproof their equipment.

(5) The rate of movement should be slow, deliberate, and careful. Soldiers should not move out at a force march pace and then be stationary after they have perspired heavily. Soldiers should not wear excessive cold-weather clothing while moving.

e. Medical procedures are needed when sickness and injuries occur. Leaders should—

- Assess the situation (tactical and environmental).
- Approach the victim safely (avoid rock or snow slide).
- Perform emergency first aid.
- Treat for shock (always assume that shock is present).
- Check for other injuries/cold injuries.
- Develop a course of action (decide on a means of evacuation).
- Execute the plan and monitor the victim's condition.

f. Body heat may be lost through radiation, conduction, convection, or evaporation.

(1) ***Radiation.*** The direct heat loss from the body to its surrounding atmosphere is called radiation heat loss. The head can radiate up to 80 percent of the total body heat output. On cold days, personnel must keep all extremities covered to retain heat. This accounts for the largest amount of heat lost from the body.

(2) ***Conduction.*** Conduction is the direct transfer of heat from one object in contact with another (being rained on or sitting in snow).

(3) ***Convection.*** Convection is the loss of heat due to moving air or water in contact with the skin. Wind chill is convection cooling. Clothing that ventilates, insulates, and protects must control the layer of warm air next to the skin.

(4) ***Evaporation.*** The evaporation of perspiration causes heat loss. Wet clothing can cause heat loss by conduction and evaporation. Dressing in layers allows soldiers to remove or add clothing as needed.

g. Some of the most common cold-weather injuries are described in the following paragraphs.

(1) ***Shock.*** Shock is the depressed state of vital organs due to the cardiovascular (heart) system not providing enough blood. Although shock is not a cold-weather injury, it is a symptom or a result of other injuries. Any illness or injury can produce shock, which increases the instance and severity of a cold-weather injury. Shock should be assumed in all injuries and treated accordingly. Even minor injuries can produce shock due to cold, pain, fear, and loss of blood.

(a) *Symptoms.* Initial symptoms of shock include apprehension, shortness of breath, sweating, cold skin, rapid and faint pulse, and excessive thirst. If the victim is not given adequate first aid immediately, his condition may digress into incoherence, slower heart beat, unconsciousness, and possibly death.

(b) *Treatment.* To treat shock, restore breathing and heart rate through artificial respiration or cardiopulmonary resuscitation. Treat the injury and control hemorrhaging. Make the victim as comfortable as possible and try to relieve the pain. Keep the victim warm but do not overheat him. Elevate the back and head, or feet. If the victim is conscious and has no abdominal injuries, administer water. The victim should receive proper medical attention as soon as possible.

(2) ***Dehydration.*** Dehydration is the loss of body fluids to the point that normal body functions are prevented or slowed. This is usually caused by overexertion and improper water intake. Dehydration precedes all cold-weather injuries and is a major symptom in acute mountain sickness. It contributes to poor performance in all physical activities—even more so than lack of food. Cold weather requirements for water are no different than in the desert. They may, in fact, exceed desert requirements because of the increased difficulty in moving with extra clothing and through the snow. At high altitudes, the air is dry. Combined

with a rapid rate of breathing, as much as two liters of liquid may be lost each day through respiration. A soldier needs about three to six quarts of water each day to prevent dehydration when living and performing physical labor in a cold or mountainous environment. Coffee and tea are diuretics and cause excessive urination and should be avoided. The adequacy of liquid intake can best be judged by the urine color and volume. Dark amber colored urine instead of light yellow or the absence of a need to urinate upon awakening from a night's sleep are indicators of dehydration. Thirst is not a good indicator of hydration.

(a) *Contributing Factors.* Factors that contribute to dehydration in cold weather are:

- The thirst mechanism does not function properly in cold weather.
- Water is often inconvenient to obtain and purify.
- The air in cold climates and at high altitudes lacks moisture.
- Cold causes frequent urination.

(b) *Symptoms.* Symptoms of dehydration include darkening urine, decreased amounts of urine being produced, dry mouth, tiredness, mental sluggishness, lack of appetite, headache, fainting, rapid heartbeat, dizziness, higher temperature, upset stomach, and unconsciousness. The symptoms of dehydration are similar to those of hypothermia. To distinguish between them, open the victim's clothes and feel the stomach. If the stomach is cold, the victim is probably hypothermic; if it is warm, he is probably dehydrated. However, this test is not conclusive since cold-weather dehydrating can also lead to total body cooling. The cold environment may act as a diuretic and impair the body's ability to conserve fluid (cold-induced diuresis and increased rate of urination).

(c) *Treatment.* Prevent dehydration by consuming three to six quarts of fluids each day (forced drinking in the absence of thirst is mandatory) and avoid caffeine and alcohol, which may chemically contribute to dehydration. Keep the victim warm and treat for shock. In advanced cases, administer fluids by mouth if the victim is conscious. Do not let him eat snow; eating snow uses body heat. Allow the victim to rest. If he fails to improve within one hour or is unconscious, evacuate him to a medical facility immediately.

(3) ***Hypothermia.*** Hypothermia is the lowering of the body core temperature at a rate faster than the body can produce heat. Hypothermia may be caused by exposure or by sudden wetting of the body such as falling into a lake or being sprayed with fuel or other liquid. Hypothermia can occur even on moderate days with temperatures of 40 to 50 degrees Fahrenheit with little precipitation if heat loss exceeds heat gain and the condition of the soldier is allowed to deteriorate. Hypothermia is classified as mild (core temperature above 90 degrees Fahrenheit or 32 degrees Celsius) or severe (core temperature below 90 degrees Fahrenheit or 32 degrees Celsius). An individual is considered to be "clinically hypothermic" when the core temperature is less than or equal to 95 degrees Fahrenheit.

(a) *Contributing Factors.* Factors that contribute to hypothermia are:

- Dehydration.
- Poor nutrition.
- Diarrhea.
- Decreased physical activity.
- Accidental immersion in water.
- Change in weather.
- High winds.
- Inadequate types or amounts of clothing.

(b) *Symptoms* The first symptom of hypothermia is when the body core (rectal) temperature falls to about 96 degrees Fahrenheit. Other symptoms include:

- Shivering, which may progress to an uncontrollable point making it hard for an individual to care for himself. Shivering begins after a drop in body temperature of one to two degrees. This is followed by clumsiness (stumbling or falling), slow reactions, mental confusion, and difficulty in speaking.
- Body temperature drop from 95 degrees Fahrenheit to 90 degrees Fahrenheit, which can cause sluggish thinking, irrational thought, apathy, and a false sense of warmth. The victim becomes cold and pale; cannot perform simple tasks; experiences amnesia and hallucinations; develops blueness of skin and decreased heart and respiratory rate with a weak pulse; pupils of the eyes dilate; speech becomes slurred; and visual disturbance occurs.
- Body temperature drop from 90 degrees Fahrenheit to 85 degrees Fahrenheit, which causes irrationality, incoherence, loss of contact with the environment, muscular rigidity, disorientation, and exhaustion. The soldier might stop shivering after his core temperature drops below 90 degrees Fahrenheit.
- Body temperature drop from 85 degrees Fahrenheit and below, which causes muscle rigidity, unconsciousness, comatose state, and faint vital signs. The pulse may be faint or impalpable, and breathing is too shallow to observe.

(c) *Prevention.* Prevent hypothermia by using the buddy system to watch each other for symptoms; consume adequate amounts of liquids daily; rest; and eat properly.

(d) *Avoidance.* Hypothermia can be avoided by dressing in layers, which permits easy additions or deletions to prevent overheating, becoming too cold, or getting wet or windblown. If the soldier is in a situation that precludes staying warm and dry, he should seek shelter. Sweets and physical activity help to produce body heat.

(e) *Treatment.* Treatment methods vary based on the severity of the hypothermia.

- Mild cases: If a soldier shows symptoms of hypothermia, prevent additional heat loss by getting the victim into a shelter; removing wet clothing and replacing it with dry, insulated clothing; insulating the victim from the ground; and sharing a sleeping bag (cover head) to transfer body heat. Make a diagnosis (rectal temperature). Rehydrate the victim with warm liquids, sweets, and food. If the tactical situation allows, build a fire. Above all else, keep the victim conscious until his vital signs are normal, and seek medical assistance. If possible, keep the victim physically active to produce body heat.
- Severe cases: If the victim is unconscious or appears dead without any obvious injury, prevent further heat loss. Rapid rewarming of an unconscious victim may create problems and should not be attempted. It is best to evacuate as soon as possible. At all times, the victim should be handled gently so as not to cause the cold blood from the extremities to rush to the heart. Do not allow the victim to perform ANY physical activity. Immediately transport the victim to the nearest medical facility. Field reheating is not effective and may be hazardous. Provide artificial respiration if breathing stops. If no pulse is detectable, be aware that in hypothermia there is often effective circulation for the victim's hypothermic state. In such a case, cardiac compression (such as CPR) may be fatal. The exception is acute hypothermia with near drowning.

- Breathing warm, moist air is the fastest way to warm the inside of the body. If breathing steam is not possible, place tubing under the rescuer's shirt so the victim will still breathe warm, moist air. This process can be done while on the move. In addition to breathing moist, warm air the victim must be gradually rewarmed using external heat sources. Padded hot water bottles or heated stones should be placed in the armpits.
- If conscious, the victim can be given warm, sweet drinks.
- The Hibler Pack is an improvised method of rewarming hypothermic victims in the field. This is used to heat the body core first so the vital organs are warmed and not the extremities. As the body warms up the warm blood will eventually warm all parts of the body. First lay out a blanket or sleeping bag and place a poncho or space blanket inside of it. The poncho or space blanket should go from the base of the skull to the base of the butt. This keeps the sleeping bag/blanket dry and acts like a vapor barrier. Lay the hypothermic patient inside the sleeping bag/blanket. Using a stove, warm water until it is hot to the touch (but not hot enough to burn the patient) and completely dampen any absorbable materials (such as T-shirt, towel, BDU top, and so on). Place the warm, wet items inside a plastic bag or directly in the armpits and chest of the patient. After the warm, wet item has been placed on the patient, wrap the patient tightly inside the poncho/space blanket and the blanket/sleeping bag. Continually check the temperature of the wet material and keep it warm.
- All bodily systems in hypothermia are brittle so treat the victim gently. As these attempts are being made, try to evacuate the victim. Severe complications may arise as the body temperature rises, which may result in cardiac arrest even though the victim seems to be doing well.

(4) ***Immersion or Trench Foot.*** This is damage to the circulatory and nervous systems of the feet that occurs from prolonged exposure to cold and wet at above freezing temperatures. This can happen wearing boots or not. A soldier may not feel uncomfortable until the injury has already begun.

(a) *Contributing Factors.* Factors that contribute to immersion or trench foot are:

- Stepping into water over the boot tops.
- Not changing socks often enough.
- Improper hygiene.
- Prolonged exposure (three to five days).

(b) *Symptoms.* Symptoms of immersion or trench foot include the sensation of tingling, numbness, and then pain. The toes are pale, and feel cold and stiff. The skin is wet and soggy with the color turning from red to bright red, progressing to pale and mottled, and then grayish blue. As symptoms progress and damage appears, the skin becomes red and then bluish or black. Swelling may occur. Because the early stages of trench foot are not painful, soldiers must be constantly aware to prevent it.

(c) *Treatment.* To prevent this condition, keep the feet dry and clean. Change socks often, drying the insides of boots, massaging the feet, and using foot powder. Drying the feet for 24 hours usually heals mild cases. Moderate cases usually heal within three to five days. The feet should be handled gently—NOT rubbed or massaged. They should be cleaned with soap and water, dried, elevated, and exposed to room temperature. The victim must stay off his feet and seek medical attention. Severe cases, when feet are not allowed to dry, are evacuated as a litter casualty.

(5) ***Blisters.*** When first noticed and before the formation of a blister, cover a hotspot with moleskin (over the area and beyond it). Use tincture benzoin to help the moleskin adhere to and toughen the skin. Once a blister has formed, cover it with a dressing large enough to fit over the blister, and then tape it. Never drain blisters unless they are surrounded by redness, or draining pus indicates infection. If this occurs, drain the blister from the side with a clean sterile needle. After cleaning with soap and water, gently press out the fluid leaving the skin intact. Make a doughnut of moleskin to go around the blister and apply to the skin. For toe blisters, wrap the entire toe with adhesive tape over the moleskin. (Toenails should be trimmed straight across the top, leaving a 90-degree angle on the sides. This provides an arch so that the corners do not irritate the skin.)

(6) ***Frostbite.*** Frostbite is the freezing or crystallization of living tissues due to heat being lost faster than it can be replaced by blood circulation, or from direct exposure to extreme cold or high winds. Exposure time can be minutes or instantaneous. The extremities are usually the first to be affected. Damp hands and feet may freeze quickly since moisture conducts heat away from the body and destroys the insulating value of clothing. Heat loss is compounded with intense cold and inactivity. With proper clothing and equipment, properly maintained and used, frostbite can be prevented. The extent of frostbite depends on temperature and duration of exposure. Frostbite is one of the major nonfatal cold-weather injuries encountered in military operations, but does not occur above an ambient temperature of 32 degrees Fahrenheit.

(a) *Categories of Frostbite.* Superficial (mild) frostbite involves only the skin (Figure 2-2). The layer immediately below usually appears white to grayish with the surface feeling hard, but the underlying tissue is soft. Deep (severe) frostbite extends beyond the first layer of skin and may include the bone (Figure 2-3). Discoloration continues from gray to black, and the texture becomes hard as the tissue freezes deeper. This condition requires immediate evacuation to a medical facility.

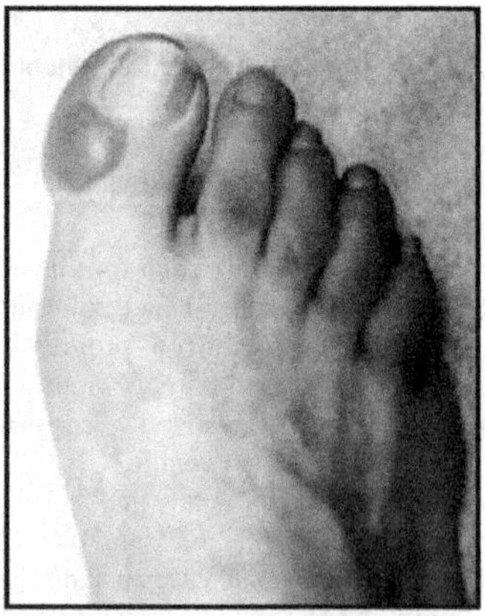

Figure 2-2. Superficial frostbite.

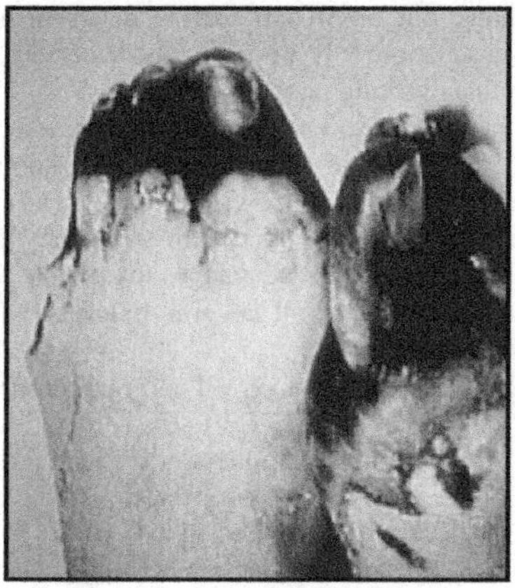

Figure 2-3. Deep frostbite.

(b) *Contributing Factors.* Factors that contribute to frostbite are:

- Dehydration.
- Below-freezing temperatures.
- Skin contact with super cooled metals or liquids.
- Use of caffeine, tobacco, or alcohol.
- Neglect.

(c) *Symptoms.* Symptoms of frostbite vary and may include a cold feeling, pain, burning, numbness, and, in the final stages, a false sense of warmth. The skin first turns red, then pale. It may be bluish in color and then may appear frosty or waxy white. The skin may feel hard, may not be movable over the joints and bony prominences, or may be frozen. Identification of deep versus superficial frostbite is difficult to determine and often requires three to seven days after rewarming for medical personnel to diagnose. Blisters, swelling, and pain may occur after thawing.

(d) *Treatment.* Using the buddy system is one of the primary ways to prevent frostbite. Buddies must watch each other for symptoms of frostbite and provide mutual aid if frostbite occurs. Frostbite should be identified early with prompt first-aid care applied to prevent further damage.

- Treat early signs of frostbite by rewarming with skin-to-skin contact or by sheltering the body part under the clothing next to the body. *Do this immediately.* If tissues have frozen, evacuate the victim before they thaw. If the feet are involved, evacuate the victim as a litter patient.
- Thawing of a frostbitten victim is a hospital procedure. If the victim has frostbite with frozen extremities, protect the frozen parts and evacuate as a litter patient.
- If frostbite is not recognized before it thaws, do not let the area refreeze since this causes more damage. The most often-affected body parts are the hands, fingers, toes, feet, ears, chin, and nose. If evacuation of the victim as a litter case

is not possible and the body part has not yet thawed, have the victim walk out on his own. Walking out on frozen feet is better than having them thaw and refreeze. Self-evacuation may be tactically necessary. Walking on frozen feet does less harm than walking on thawed feet.

- If reheating is inevitable, do not overheat the affected body parts near flame; the warming temperature should not be greater than normal body temperature. Do not rub the parts—the crystallized tissues may break internally and cause more damage. Do not pop blisters; cover them with a dry, sterile dressing. Keep the victim warm (apply loose, bulky bandages to separate toes and fingers.)
- Once a part is rewarmed it will become painful. Pain may be managed with narcotic analgesics.
- Once the foot is rewarmed it will swell and putting the boot back on will not be possible.

(7) ***Constipation.*** Constipation is the infrequent or difficult passage of stools.

(a) *Contributing Factors.* Factors that contribute to constipation are a lack of fluids, improper nutrition, and not defecating when needed.

(b) *Symptoms.* Symptoms include headache, cramping, lack of bowel movement, painful bowel movement, and loss of appetite.

(c) *Treatment.* Constipation is prevented by consuming adequate amounts and varieties of food, drinking from four to six liters of liquid each day, and defecating regularly. If allowed to progress beyond self-care stages, victims will need medical aid.

(8) ***Carbon Monoxide Poisoning.*** This is the replacement of oxygen in the blood with carbon monoxide.

(a) *Contributing Factor.* A contributing factor is inhaling fumes from burning fuel, such as fires, stoves, heaters, and running engines, without proper ventilation.

(b) *Symptoms.* Symptoms are similar to other common illnesses and include headaches, fatigue, excessive yawning, nausea, dizziness, drowsiness, confusion, and unconsciousness. Death may occur. The one visible symptom is bright red lips, mouth, and inside of the eyelids.

(c) *Treatment.* Remove the victim from the source of contamination; administer oxygen, if available; and evacuate to a medical facility. Severe complications may develop even in casualties who appear to have recovered. If the victim is unconscious, administer rescue breathing and CPR as needed.

2-10. HEAT INJURIES

Heat injuries, although associated with hot weather, can occur in cold-weather environments. Most heat injuries can be avoided by planning, periodic inspections of personnel clothing (ventilation) and equipment, a balance of water and food intake, and rest.

a. **Heat Cramps.** Heat cramps are caused by an accumulation of lactic acid in the muscles and a loss of salt through perspiration.

(1) ***Contributing Factor.*** Strenuous exertion causes the body to heat up and to produce heavy perspiration.

(2) ***Symptoms.*** Symptoms of heat cramps include pain and cramping in the arms, legs, back, and stomach. The victim sweats profusely and cannot quench his thirst.

(3) ***Treatment.*** Have the victim rest in a cool, shady area, breath deeply, and stretch the cramped muscle as soon as possible to obtain relief. Loosen the victim's clothing and have him drink cool water. Monitor his condition and seek medical attention if pain and cramps continue.

b. **Heat Exhaustion.** Heat exhaustion may occur when a soldier exerts himself in any environment and he overheats. The blood vessels in the skin become so dilated that the blood flow to the brain and other organs is reduced.

(1) ***Contributing Factors.*** Factors that contribute to heat exhaustion are strenuous activity in hot areas, unacclimatized troops, inappropriate diet, and not enough water or rest.

(2) ***Symptoms.*** Symptoms of heat exhaustion may be similar to fainting but may also include weakness; dizziness; confusion; headache; cold, clammy skin; and nausea. The victim may also have a rapid but weak pulse.

(3) ***Treatment.*** Move the victim to a cool, shady area and loosen his clothes and boots. Have the victim drink water and, if possible, immerse him in water to aid in cooling. Elevate the victim's legs to help restore proper circulation. Monitor his condition and seek medical attention if the symptoms persist.

c. **Heat Stroke.** Heat stroke is a life-threatening situation caused by overexposure to the sun. The body is so depleted of liquids that its internal cooling mechanisms fail to function.

(1) ***Contributing Factors.*** Factors that contribute to heat stroke are prolonged exposure to direct sunlight, overexertion, dehydration, and depletion of electrolytes.

(2) ***Symptoms.*** Symptoms of heat stroke include hot, dry skin; dizziness; confusion and incoherency; headache; nausea; seizures; breathing difficulty; a slow pulse; and loss of consciousness.

(3) ***Treatment.*** Cool the victim at once, and restore breathing and circulation. If the victim is conscious, administer water. If possible, submerge the victim in water to reduce his temperature, treat for shock, and prepare for immediate evacuation.

2-11. ACUTE MOUNTAIN SICKNESS

Acute mountain sickness is a temporary illness that may affect both the beginner and experienced climber. Soldiers are subject to this sickness in altitudes as low as 5,000 feet. Incidence and severity increases with altitude, and when quickly transported to high altitudes. Disability and ineffectiveness can occur in 50 to 80 percent of the troops who are rapidly brought to altitudes above 10,000 feet. At lower altitudes, or where ascent to altitudes is gradual, most personnel can complete assignments with moderate effectiveness and little discomfort.

a. Personnel arriving at moderate elevations (5,000 to 8,000 feet) usually feel well for the first few hours; a feeling of exhilaration or well-being is not unusual. There may be an initial awareness of breathlessness upon exertion and a need for frequent pauses to rest. Irregular breathing can occur, mainly during sleep; these changes may cause apprehension. Severe symptoms may begin 4 to 12 hours after arrival at higher altitudes with symptoms of nausea, sluggishness, fatigue, headache, dizziness, insomnia, depression, uncaring attitude, rapid and labored breathing, weakness, and loss of appetite.

b. A headache is the most noticeable symptom and may be severe. Even when a headache is not present, some loss of appetite and a decrease in tolerance for food occurs. Nausea, even without food intake, occurs and leads to less food intake. Vomiting may occur and contribute to dehydration. Despite fatigue, personnel are unable to sleep. The symptoms usually develop and increase to a peak by the second day. They gradually subside over the next several days so that the total course of AMS may extend from five to seven days. In some instances, the headache may become incapacitating and the soldier should be evacuated to a lower elevation.

c. Treatment for AMS includes the following:

- Oral pain medications such as ibuprofen or aspirin.
- Rest.
- Frequent consumption of liquids and light foods in small amounts.
- Movement to lower altitudes (at least 1,000 feet) to alleviate symptoms, which provides for a more gradual acclimatization.
- Realization of physical limitations and slow progression.
- Practice of deep-breathing exercises.
- Use of acetazolamide in the first 24 hours for mild to moderate cases.

d. AMS is nonfatal, although if left untreated or further ascent is attempted, development of high-altitude pulmonary edema (HAPE) and or high-altitude cerebral edema (HACE) can be seen. A severe persistence of symptoms may identify soldiers who acclimatize poorly and, thus, are more prone to other types of mountain sickness.

2-12. CHRONIC MOUNTAIN SICKNESS

Although not commonly seen in mountaineers, chronic mountain sickness (CMS) (or Monge's disease) can been seen in people who live at sufficiently high altitudes (usually at or above 10,000 feet) over a period of several years. CMS is a right-sided heart failure characterized by chronic pulmonary edema that is caused by years of strain on the right ventricle.

2-13. UNDERSTANDING HIGH-ALTITUDE ILLNESSES

As altitude increases, the overall atmospheric pressure decreases. Decreased pressure is the underlying source of altitude illnesses. Whether at sea level or 20,000 feet the surrounding atmosphere has the same percentage of oxygen. As pressure decreases the body has a much more difficult time passing oxygen from the lungs to the red blood cells and thus to the tissues of the body. This lower pressure means lower oxygen levels in the blood and increased carbon dioxide levels. Increased carbon dioxide levels in the blood cause a systemic vasodilatation, or expansion of blood vessels. This increased vascular size stretches the vessel walls causing leakage of the fluid portions of the blood into the interstitial spaces, which leads to cerebral edema or HACE. Unless treated, HACE will continue to progress due to the decreased atmospheric pressure of oxygen. Further ascent will hasten the progression of HACE and could possibly cause death.

While the body has an overall systemic vasodilatation, the lungs initially experience pulmonary vasoconstriction. This constricting of the vessels in the lungs causes increased workload on the right ventricle, the chamber of the heart that receives de-oxygenated blood from the right atrium and pushes it to the lungs to be re-oxygenated. As the right ventricle works harder to force blood to the lungs, its overall output is decreased thus decreasing the overall pulmonary perfusion. Decreased pulmonary perfusion causes decreased cellular respiration—the transfer of oxygen from the alveoli to the red blood cells. The body is now experiencing increased carbon dioxide levels due to the decreased oxygen levels, which now causes pulmonary vasodilatation. Just as in HACE, this expanding of the vascular structure causes leakage into interstitial space resulting in pulmonary edema or HAPE. As the edema or fluid in the lungs increases, the capability to pass oxygen to the red blood cells decreases thus creating a vicious cycle, which can quickly become fatal if left untreated.

2-14. HIGH-ALTITUDE PULMONARY EDEMA

HAPE is a swelling and filling of the lungs with fluid, caused by rapid ascent. It occurs at high altitudes and limits the oxygen supply to the body.

a. HAPE occurs under conditions of low oxygen pressure, is encountered at high elevations (over 8,000 feet), and can occur in healthy soldiers. HAPE may be considered a form of, or manifestation of, AMS since it occurs during the period of susceptibility to this disorder.

b. HAPE can cause death. Incidence and severity increase with altitude. Except for acclimatization to altitude, no known factors indicate resistance or immunity. Few cases have been reported after 10 days at high altitudes. When remaining at the same altitude, the incidence of HAPE is less frequent than that of AMS. No common indicator dictates how a soldier will react from one exposure to another. Contributing factors are:

- A history of HAPE.
- A rapid or abrupt transition to high altitudes.
- Strenuous physical exertion.
- Exposure to cold.
- Anxiety.

c. Symptoms of AMS can mask early pulmonary difficulties. Symptoms of HAPE include:

- Progressive dry coughing with frothy white or pink sputum (this is usually a later sign) and then coughing up of blood.
- Cyanosis—a blue color to the face, hands, and feet.
- An increased ill feeling, labored breathing, dizziness, fainting, repeated clearing of the throat, and development of a cough.
- Respiratory difficulty, which may be sudden, accompanied by choking and rapid deterioration.
- Progressive shortness of breath, rapid heartbeat (pulse 120 to 160), and coughing (out of contrast to others who arrived at the same time to that altitude).
- Crackling, cellophane-like noises (rales) in the lungs caused by fluid buildup (a stethoscope is usually needed to hear them).
- Unconsciousness, if left untreated. Bubbles form in the nose and mouth, and death results.

d. HAPE is prevented by good nutrition, hydration, and gradual ascent to altitude (no more than 1,000 to 2,000 feet per day to an area of sleep). A rest day, with no gain in altitude or heavy physical exertion, is planned for every 3,000 feet of altitude gained. If a soldier develops symptoms despite precautions, immediate descent is mandatory where he receives prompt treatment, rest, warmth, and oxygen. He is quickly evacuated to lower altitudes as a litter patient. A descent of 300 meters may help; manual descent is not delayed to await air evacuation. If untreated, HAPE may become irreversible and cause death. Cases that are recognized early and treated promptly may expect to recover with no aftereffects. Soldiers who have had previous attacks of HAPE are prone to second attacks.

e. Treatment of HAPE includes:

- Immediate descent (2,000 to 3,000 feet minimum) if possible; if not, then treatment in a monoplace hyperbaric chamber.
- Rest (litter evacuation)
- Supplemental oxygen if available.
- Morphine for the systemic vasodilatation and reduction of preload. This should be carefully considered due to the respiratory depressive properties of the drug.
- Furosemide (Lasix), which is a diuretic, given orally can also be effective.
- The use of mannitol should not be considered due to the fact that it crystallizes at low temperatures. Since almost all high-altitude environments are cold, using mannitol could be fatal.
- Nifidipine (Procardia), which inhibits calcium ion flux across cardiac and smooth muscle cells, decreasing contractility and oxygen demand. It may also dilate coronary arteries and arterioles.
- Diphenhydramine (Benadryl), which can help alleviate the histamine response that increases mucosal secretions.

2-15. HIGH-ALTITUDE CEREBRAL EDEMA

HACE is the accumulation of fluid in the brain, which results in swelling and a depression of brain function that may result in death. It is caused by a rapid ascent to altitude without progressive acclimatization. Prevention of HACE is the same as for HAPE. HAPE and HACE may occur in experienced, well-acclimated mountaineers without warning or obvious predisposing conditions. They can be fatal; when the first symptoms occur, immediate descent is mandatory.

a. Contributing factors include rapid ascent to heights over 8,000 feet and aggravation by overexertion.

b. Symptoms of HACE include mild personality changes, paralysis, stupor, convulsions, coma, inability to concentrate, headaches, vomiting, decrease in urination, and lack of coordination. The main symptom of HACE is a severe headache. A headache combined with any other physical or psychological disturbances should be assumed to be manifestations of HACE. Headaches may be accompanied by a loss of coordination, confusion, hallucinations, and unconsciousness. These may be combined with symptoms of HAPE. The victim is often mistakenly left alone since others may think he is only irritable or temperamental; no one should ever be ignored. The symptoms may rapidly progress to death. Prompt descent to a lower altitude is vital.

c. Preventive measures include good eating habits, maintaining hydration, and using a gradual ascent to altitude. Rest, warmth, and oxygen at lower elevations enhance recovery. Left untreated, HACE can cause death.

d. Treatment for HACE includes:

- Dexamethasone injection immediately followed by oral dexamethasone.
- Supplemental oxygen.
- Rapid descent and medical attention.
- Use of a hyberbaric chamber if descent is delayed.

2-16. HYDRATION IN HAPE AND HACE

HAPE and HACE cause increased proteins in the plasma, or the fluid portion of the blood, which in turn increases blood viscosity. Increased viscosity increases vascular pressure. Vascular leakage caused by stretching of the vessel walls is made worse because of this increased vascular pressure. From this, edema, both cerebral and pulmonary, occurs. Hydration simply decreases viscosity.

CHAPTER 3
MOUNTAINEERING EQUIPMENT

Commanders at every level must understand the complexity of operations in a mountainous environment where every aspect of combat operations becomes more difficult. Leaders must understand that each individual has a different metabolism and, therefore, cools down and heats up differently, which requires soldiers to dress-up and dress-down at different intervals. Provided all tactical concerns are met, the concept of uniformity is outdated and only reduces the unit's ability to fight and function at an optimum level. The extreme cold weather clothing system (ECWCS) is specifically designed to allow for rapid moisture transfer and optimum heat retention while protecting the individual from the elements. Every leader is responsible for ensuring that the ECWCS is worn in accordance with the manufacturers' recommendations. Commanders at all levels must also understand that skills learned at an Army mountaineering school are perishable and soldiers need constant practice to remain proficient. The properly trained mountain soldier of today can live better, move faster, and fight harder in an environment that is every bit as hostile as the enemy.

Section I. EQUIPMENT DESCRIPTION AND MAINTENANCE

With mountainous terrain encompassing a large portion of the world's land mass, the proper use of mountaineering equipment will enhance a unit's combat capability and provide a combat multiplier. The equipment described in this chapter is produced by many different manufacturers; however, each item is produced and tested to extremely high standards to ensure safety when being used correctly. The weak link in the safety chain is the user. Great care in performing preventative maintenance checks and services and proper training in the use of the equipment is paramount to ensuring safe operations. The manufacturers of each and every piece of equipment provide recommendations on how to use and care for its product. It is imperative to follow these instructions explicitly.

3-1. FOOTWEAR

Currently, CTA 50-900 provides adequate footwear for most operations in mountainous terrain. In temperate climates a combination of footwear is most appropriate to accomplish all tasks.

a. The hot weather boot provides an excellent all-round platform for movement and climbing techniques and should be the boot of choice when the weather permits. The intermediate cold weather boot provides an acceptable platform for operations when the weather is less than ideal. These two types of boots issued together will provide the unit with the footwear necessary to accomplish the majority of basic mountain missions.

b. Mountain operations are encumbered by extreme cold, and the extreme cold weather boot (with vapor barrier) provides an adequate platform for many basic mountain missions. However, plastic mountaineering boots should be incorporated into training as soon as possible. These boots provide a more versatile platform for any condition that would be encountered in the mountains, while keeping the foot dryer and warmer.

c. Level 2 and level 3 mountaineers will need mission-specific footwear that is not currently available in the military supply system. The two types of footwear they will need are climbing shoes and plastic mountaineering boots.

(1) Climbing shoes are made specifically for climbing vertical or near vertical rock faces. These shoes are made with a soft leather upper, a lace-up configuration, and a smooth "sticky rubber" sole (Figure 3-1). The smooth "sticky rubber" sole is the key to the climbing shoe, providing greater friction on the surface of the rock, allowing the climber access to more difficult terrain.

(2) The plastic mountaineering boot is a double boot system (Figure 3-1). The inner boot provides support, as well as insulation against the cold. The inner boot may or may not come with a breathable membrane. The outer boot is a molded plastic (usually with a lace-up configuration) with a lug sole. The welt of the boot is molded in such a way that crampons, ski bindings, and snowshoes are easily attached and detached.

Note: Maintenance of all types of footwear must closely follow the manufacturers' recommendations.

Figure 3-1. Climbing shoes and plastic mountaineering boots.

3-2. CLOTHING

Clothing is perhaps the most underestimated and misunderstood equipment in the military inventory. The clothing system refers to every piece of clothing placed against the skin, the insulation layers, and the outer most garments, which protect the soldier from the elements. When clothing is worn properly, the soldier is better able to accomplish his tasks. When worn improperly, he is, at best, uncomfortable and, at worst, develops hypothermia or frostbite.

a. **Socks.** Socks are one of the most under-appreciated part of the entire clothing system. Socks are extremely valuable in many respects, if worn correctly. As a system, socks provide cushioning for the foot, remove excess moisture, and provide insulation from cold temperatures. Improper wear and excess moisture are the biggest causes of hot spots and blisters. Regardless of climatic conditions, socks should always be worn in layers.

(1) The first layer should be a hydrophobic material that moves moisture from the foot surface to the outer sock.

(2) The outer sock should also be made of hydrophobic materials, but should be complimented with materials that provide cushioning and abrasion resistance.

(3) A third layer can be added depending upon the climatic conditions.

(a) In severe wet conditions, a waterproof type sock can be added to reduce the amount of water that would saturate the foot. This layer would be worn over the first two layers if conditions were extremely wet.

(b) In extremely cold conditions a vapor barrier sock can be worn either over both of the original pairs of socks or between the hydrophobic layer and the insulating layer. If the user is wearing VB boots, the vapor barrier sock is not recommended.

b. **Underwear.** Underwear should also be made of materials that move moisture from the body. Many civilian companies manufacture this type of underwear. The primary material in this product is polyester, which moves moisture from the body to the outer layers keeping the user drier and more comfortable in all climatic conditions. In colder environments, several pairs of long underwear of different thickness should be made available. A lightweight set coupled with a heavyweight set will provide a multitude of layering combinations.

c. **Insulating Layers.** Insulating layers are those layers that are worn over the underwear and under the outer layers of clothing. Insulating layers provide additional warmth when the weather turns bad. For the most part, today's insulating layers will provide for easy moisture movement as well as trap air to increase the insulating factor. The insulating layers that are presently available are referred to as pile or fleece. The ECWCS (Figure 3-2, page 3-4) also incorporates the field jacket and field pants liner as additional insulating layers. However, these two components do not move moisture as effectively as the pile or fleece.

d. **Outer Layers.** The ECWCS provides a jacket and pants made of a durable waterproof fabric. Both are constructed with a nylon shell with a laminated breathable membrane attached. This membrane allows the garment to release moisture to the environment while the nylon shell provides a degree of water resistance during rain and snow. The nylon also acts as a barrier to wind, which helps the garment retain the warm air trapped by the insulating layers. Leaders at all levels must understand the importance of wearing the ECWCS correctly.

Note: Cotton layers must not be included in any layer during operations in a cold environment.

Figure 3-2. Extreme cold weather clothing system.

e. **Gaiters.** Gaiters are used to protect the lower leg from snow and ice, as well as mud, twigs, and stones. The use of waterproof fabrics or other breathable materials laminated to the nylon makes the gaiter an integral component of the cold weather clothing system. Gaiters are not presently fielded in the standard ECWCS and, in most cases, will need to be locally purchased. Gaiters are available in three styles (Figure 3-3).

(1) The most common style of gaiter is the open-toed variety, which is a nylon shell that may or may not have a breathable material laminated to it. The open front allows the boot to slip easily into it and is closed with a combination of zipper, hook-pile tape, and snaps. It will have an adjustable neoprene strap that goes under the boot to keep it snug to the boot. The length should reach to just below the knee and will be kept snug with a drawstring and cord lock.

(2) The second type of gaiter is referred to as a full or randed gaiter. This gaiter completely covers the boot down to the welt. It can be laminated with a breathable material and can also be insulated if necessary. This gaiter is used with plastic mountaineering boots and should be glued in place and not removed.

(3) The third type of gaiter is specific to high-altitude mountaineering or extremely cold temperatures and is referred to as an overboot. It is worn completely over the boot and must be worn with crampons because it has no traction sole.

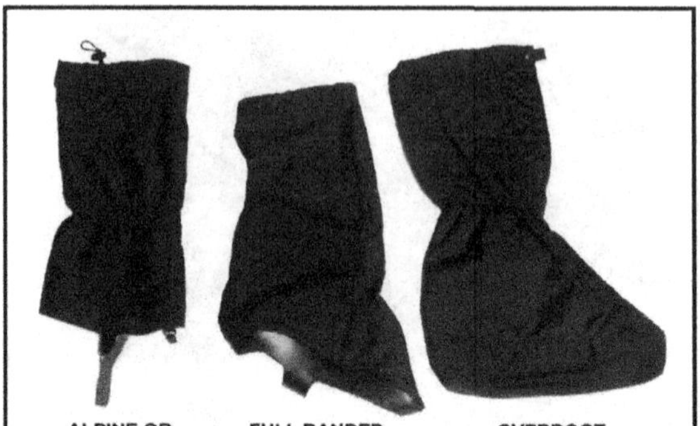

Figure 3-3. Three types of gaiters.

f. **Hand Wear.** During operations in mountainous terrain the use of hand wear is extremely important. Even during the best climatic conditions, temperatures in the mountains will dip below the freezing point. While mittens are always warmer than gloves, the finger dexterity needed to do most tasks makes gloves the primary cold weather hand wear (Figure 3-4, page 3-6).

(1) The principals that apply to clothing also apply to gloves and mittens. They should provide moisture transfer from the skin to the outer layers—the insulating layer must insulate the hand from the cold and move moisture to the outer layer. The outer layer must be weather resistant and breathable. Both gloves and mittens should be required for all soldiers during mountain operations, as well as replacement liners for both. This will provide enough flexibility to accomplish all tasks and keep the users' hands warm and dry.

(2) Just as the clothing system is worn in layers, gloves and mittens work best using the same principle. Retention cords that loop over the wrist work extremely well when the wearer needs to remove the outer layer to accomplish a task that requires fine finger dexterity. Leaving the glove or mitten dangling from the wrist ensures the wearer knows where it is at all times.

Figure 3-4. Hand wear.

g. **Headwear.** A large majority of heat loss (25 percent) occurs through the head and neck area. The most effective way to counter heat loss is to wear a hat. The best hat available to the individual soldier through the military supply system is the black watch cap. Natural fibers, predominately wool, are acceptable but can be bulky and difficult to fit under a helmet. As with clothes and hand wear, man-made fibers are preferred. For colder climates a neck gaiter can be added. The neck gaiter is a tube of man-made material that fits around the neck and can reach up over the ears and nose (Figure 3-5). For extreme cold, a balaclava can be added. This covers the head, neck, and face leaving only a slot for the eyes (Figure 3-5). Worn together the combination is warm and provides for moisture movement, keeping the wearer drier and warmer.

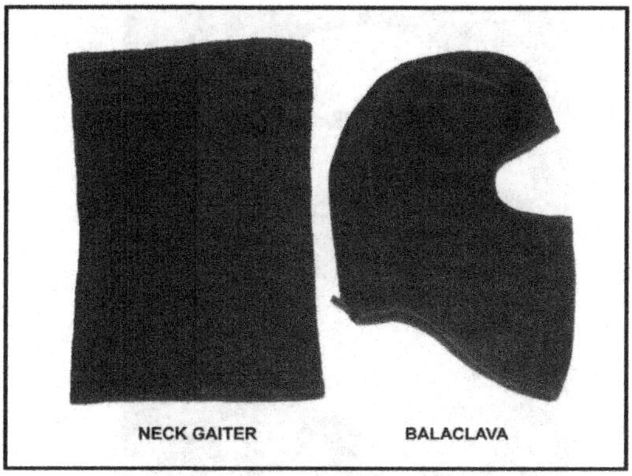

Figure 3-5. Neck gaiter and balaclava.

h. **Helmets**. The Kevlar ballistic helmet can be used for most basic mountaineering tasks. It must be fitted with parachute retention straps and the foam impact pad (Figure 3-6). The level 2 and 3 mountaineer will need a lighter weight helmet for specific climbing scenerios. Several civilian manufacturers produce an effective helmet. Whichever helmet is selected, it should be designed specifically for mountaineering and adjustable so the user can add a hat under it when needed.

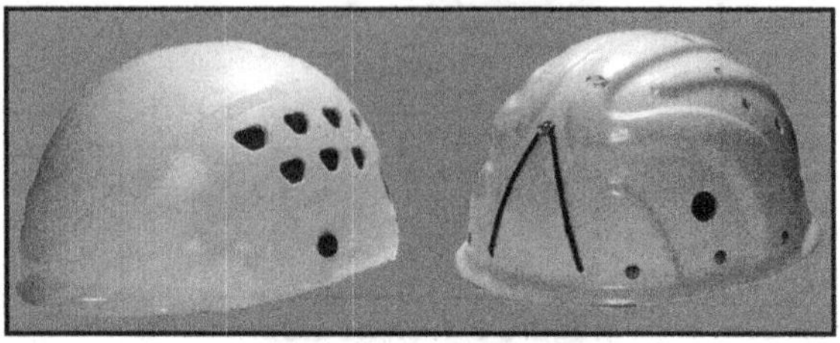

Figure 3-6. Helmets.

i. **Eyewear.** The military supply system does not currently provide adequate eyewear for mountaineering. Eyewear is divided into two catagories: glacier glasses and goggles (Figure 3-7). Glacier glasses are sunglasses that cover the entire eye socket. Many operations in the mountains occur above the tree line or on ice and snow surfaces where the harmful UV rays of the sun can bombard the eyes from every angle increasing the likelihood of snowblindness. Goggles for mountain operations should be antifogging. Double or triple lenses work best. UV rays penetrate clouds so the goggles should be UV protected. Both glacier glasses and goggles are required equipment in the mountains. The lack of either one can lead to severe eye injury or blindness.

Figure 3-7. Glacier glasses and goggles.

j. **Maintenance of Clothing.** Clothing and equipment manufacturers provide specific instructions for proper care. Following these instructions is necessary to ensure the equipment works as intended.

3-3. CLIMBING SOFTWARE

Climbing software refers to rope, cord, webbing, and harnesses. All mountaineering specific equipment, to include hardware (see paragraph 3-4), should only be used if it has the UIAA certificate of safety. UIAA is the organization that oversees the testing of mountaineering equipment. It is based in Paris, France, and comprises several commissions. The safety commission has established standards for mountaineering and climbing equipment that have become well recognized throughout the world. Their work continues as new equipment develops and is brought into common use. Community Europe (CE) recognizes UIAA testing standards and, as the broader-based testing facility for the combined European economy, meets or exceeds the UIAA standards for all climbing and mountaineering equipment produced in Europe. European norm (EN) and CE have been combined to make combined European norm (CEN). While the United States has no specific standards, American manufacturers have their equipment tested by UIAA to ensure safe operating tolerances.

a. **Ropes and Cord.** Ropes and cords are the most important pieces of mountaineering equipment and proper selection deserves careful thought. These items are your lifeline in the mountains, so selecting the right type and size is of the utmost importance. All ropes and cord used in mountaineering and climbing today are constructed with the same basic configuration. The construction technique is referred to as Kernmantle, which is, essentially, a core of nylon fibers protected by a woven sheath, similar to parachute or 550 cord (Figure 3-8).

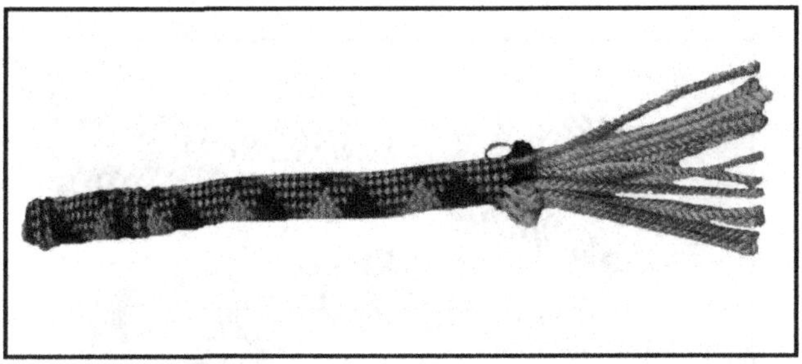

Figure 3-8. Kernmantle construction.

(1) Ropes come in two types: static and dynamic. This refers to their ability to stretch under tension. A static rope has very little stretch, perhaps as little as one to two percent, and is best used in rope installations. A dynamic rope is most useful for climbing and general mountaineering. Its ability to stretch up to 1/3 of its overall length makes it the right choice any time the user might take a fall. Dynamic and static ropes come in various

diameters and lengths. For most military applications, a standard 10.5- or 11-millimeter by 50-meter dynamic rope and 11-millimeter by 45-meter static rope will be sufficient.

(2) When choosing dynamic rope, factors affecting rope selection include intended use, impact force, abrasion resistance, and elongation. Regardless of the rope chosen, it should be UIAA certified.

(3) Cord or small diameter rope is indispensable to the mountaineer. Its many uses make it a valuable piece of equipment. All cord is static and constructed in the same manner as larger rope. If used for Prusik knots, the cord's diameter should be 5 to 7 millimeters when used on an 11-mm rope.

b. **Webbing and Slings.** Loops of tubular webbing or cord, called slings or runners, are the simplest pieces of equipment and some of the most useful. The uses for these simple pieces are endless, and they are a critical link between the climber, the rope, carabiners, and anchors. Runners are predominately made from either 9/16-inch or 1-inch tubular webbing and are either tied or sewn by a manufacturer (Figure 3-9). Runners can also be made from a high-performance fiber known as spectra, which is stronger, more durable, and less susceptible to ultraviolet deterioration. Runners should be retired regularly following the same considerations used to retire a rope. For most military applications, a combination of different lengths of runners is adequate.

(1) Tied runners have certain advantages over sewn runners—they are inexpensive to make, can be untied and threaded around natural anchors, and can be untied and retied to other pieces of webbing to create extra long runners.

(2) Sewn runners have their own advantages—they tend to be stronger, are usually lighter, and have less bulk than the tied version. They also eliminate a major concern with the homemade knotted runner—the possibility of the knot untying. Sewn runners come in four standard lengths: 2 inches, 4 inches, 12 inches, and 24 inches. They also come in three standard widths: 9/16 inch, 11/16 inch, and 1 inch.

Figure 3-9. Tied or sewn runners.

c. **Harnesses.** Years ago climbers secured themselves to the rope by wrapping the rope around their bodies and tying a bowline-on-a-coil. While this technique is still a viable way of attaching to a rope, the practice is no longer encouraged because of the increased possibility of injury from a fall. The bowline-on-a-coil is best left for low-angle climbing or an emergency situation where harness material is unavailable. Climbers today can select from a wide range of manufactured harnesses. Fitted properly, the harness should ride high on the hips and have snug leg loops to better distribute the force of a fall to the entire pelvis. This type of harness, referred to as a seat harness, provides a comfortable seat for rappelling (Figure 3-10).

(1) Any harness selected should have one very important feature—a double-passed buckle. This is a safety standard that requires the waist belt to be passed over and back through the main buckle a second time. At least 2 inches of the strap should remain after double-passing the buckle.

(2) Another desirable feature on a harness is adjustable leg loops, which allows a snug fit regardless of the number of layers of clothing worn. Adjustable leg loops allow the soldier to make a latrine call without removing the harness or untying the rope.

(3) Equipment loops are desirable for carrying pieces of climbing equipment. For safety purposes always follow the manufacturer's directions for tying-in.

(4) A field-expedient version of the seat harness can be constructed by using 22 feet of either 1-inch or 2-inch (preferred) tubular webbing (Figure 3-10). Two double-overhand knots form the leg loops, leaving 4 to 5 feet of webbing coming from one of the leg loops. The leg loops should just fit over the clothing. Wrap the remaining webbing around the waist ensuring the first wrap is routed through the 6- to 10-inch long strap between the double-overhand knots. Finish the waist wrap with a water knot tied as tightly as possible. With the remaining webbing, tie a square knot without safeties over the water knot ensuring a minimum of 4 inches remains from each strand of webbing.

(5) The full body harness incorporates a chest harness with a seat harness (Figure 3-10). This type of harness has a higher tie-in point and greatly reduces the chance of flipping backward during a fall. This is the only type of harness that is approved by the UIAA. While these harnesses are safer, they do present several disadvantages—they are more expensive, are more restrictive, and increase the difficulty of adding or removing clothing. Most mountaineers prefer to incorporate a separate chest harness with their seat harness when warranted.

(6) A separate chest harness can be purchased from a manufacturer, or a field-expedient version can be made from either two runners or a long piece of webbing. Either chest harness is then attached to the seat harness with a carabiner and a length of webbing or cord.

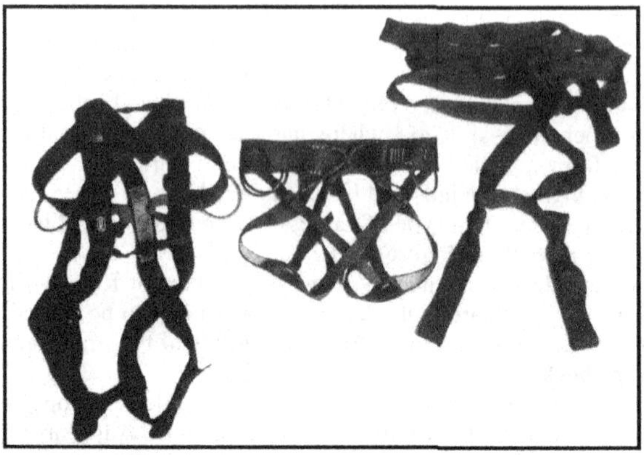

Figure 3-10. Seat harness, field-expedient harness, and full body harness.

3-4. CLIMBING HARDWARE

Climbing hardware refers to all the parts and pieces that allow the trained mountain soldier to accomplish many tasks in the mountains. The importance of this gear to the mountaineer is no less than that of the rifle to the infantryman.

a. **Carabiners.** One of the most versatile pieces of equipment available to the mountaineer is the carabiner. This simple piece of gear is the critical connection between the climber, his rope, and the protection attaching him to the mountain. Carabiners must be strong enough to hold hard falls, yet light enough for the climber to easily carry a quantity of them. Today's high tech metal alloys allow carabiners to meet both of these requirements. Steel is still widely used, but is not preferred for general mountaineering, given other options. Basic carabiner construction affords the user several different shapes. The oval, the D-shaped, and the pear-shaped carabiner are just some of the types currently available. Most models can be made with or without a locking mechanism for the gate opening (Figure 3-11, page 3-12). If the carabiner does have a locking mechanism, it is usually referred to as a locking carabiner. When using a carabiner, great care should be taken to avoid loading the carabiner on its minor axis and to avoid three-way loading (Figure 3-12, page 3-12).

Note: Great care should be used to ensure all carabiner gates are closed and locked during use.

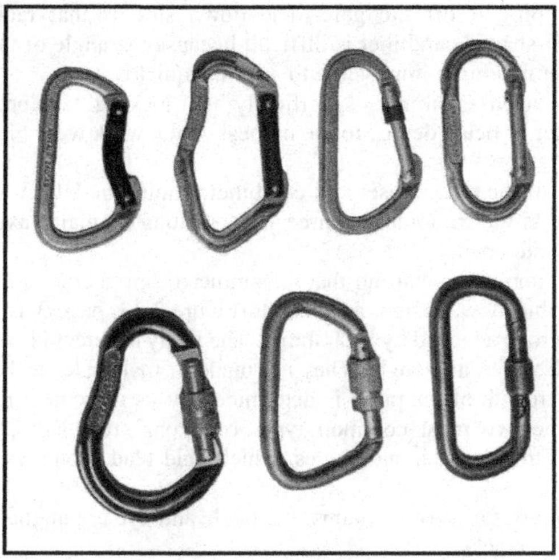

Figure 3-11. Nonlocking and locking carabiners.

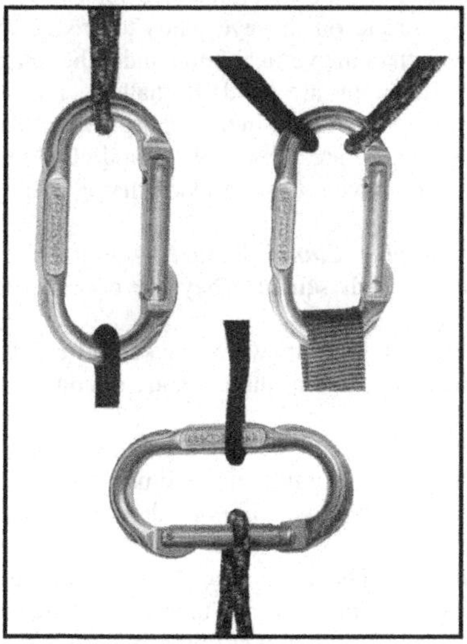

Figure 3-12. Major and minor axes and three-way loading.

(1) The major difference between the oval and the D-shaped carabiner is strength. Because of the design of the D-shaped carabiner, the load is angled onto the spine of the

carabiner thus keeping it off the gate. The down side is that racking any gear or protection on the D-shaped carabiner is difficult because the angle of the carabiner forces all the gear together making it impossible to separate quickly.

(2) The pear-shaped carabiner, specifically the locking version, is excellent for clipping a descender or belay device to the harness. They work well with the munter hitch belaying knot.

(3) Regardless of the type chosen, all carabiners should be UIAA tested. This testing is extensive and tests the carabiner in three ways—along its major axis, along its minor axis, and with the gate open.

b. **Pitons.** A piton is a metal pin that is hammered into a crack in the rock. They are described by their thickness, design, and length (Figure 3-13, page 3-14). Pitons provide a secure anchor for a rope attached by a carabiner. The many different kinds of pitons include: vertical, horizontal, wafer, and angle. They are made of malleable steel, hardened steel, or other alloys. The strength of the piton is determined by its placement rather than its rated tensile strength. The two most common types of pitons are: blades, which hold when wedged into tight-fitting cracks, and angles, which hold blade compression when wedged into a crack.

(1) ***Vertical Pitons***. On vertical pitons, the blade and eye are aligned. These pitons are used in flush, vertical cracks.

(2) ***Horizontal Pitons***. On horizontal pitons, the eye of the piton is at right angles to the blade. These pitons are used in flush, horizontal cracks and in offset or open-book type vertical or horizontal cracks. They are recommended for use in vertical cracks instead of vertical pitons because the torque on the eye tends to wedge the piton into place. This provides more holding power than the vertical piton under the same circumstances.

(3) ***Wafer Pitons***. These pitons are used in shallow, flush cracks. They have little holding power and their weakest points are in the rings provided for the carabiner.

(4) ***Knife Blade Pitons***. These are used in direct-aid climbing. They are small and fit into thin, shallow cracks. They have a tapered blade that is optimum for both strength and holding power.

(5) ***Realized Ultimate Reality Pitons***. Realized ultimate reality pitons (RURPs) are hatchet-shaped pitons about 1-inch square. They are designed to bite into thin, shallow cracks.

(6) ***Angle Pitons***. These are used in wide cracks that are flush or offset. Maximum strength is attained only when the legs of the piton are in contact with the opposite sides of the crack.

(7) ***Bong Pitons***. These are angle pitons that are more than 3.8 centimeters wide. Bongs are commonly made of steel or aluminum alloy and usually contain holes to reduce weight and accommodate carabiners. They have a high holding power and require less hammering than other pitons.

(8) ***Skyhook (Cliffhangers).*** These are small hooks that cling to tiny rock protrusions, ledges, or flakes. Skyhooks require constant tension and are used in a downward pull direction. The curved end will not straighten under body weight. The base is designed to prevent rotation and aid stability.

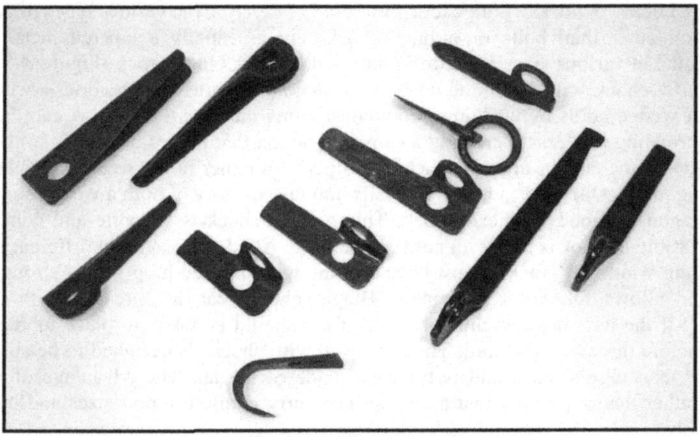

Figure 3-13. Various pitons.

c. **Piton Hammers.** A piton hammer has a flat metal head; a handle made of wood, metal, or fiberglass; and a blunt pick on the opposite side of the hammer (Figure 3-14). A safety lanyard of nylon cord, webbing, or leather is used to attach it to the climber The lanyard should be long enough to allow for full range of motion. Most hammers are approximately 25.5 centimeters long and weigh 12 to 25 ounces. The primary use for a piton hammer is to drive pitons, to be used as anchors, into the rock. The piton hammer can also be used to assist in removing pitons, and in cleaning cracks and rock surfaces to prepare for inserting the piton. The type selected should suit individual preference and the intended use.

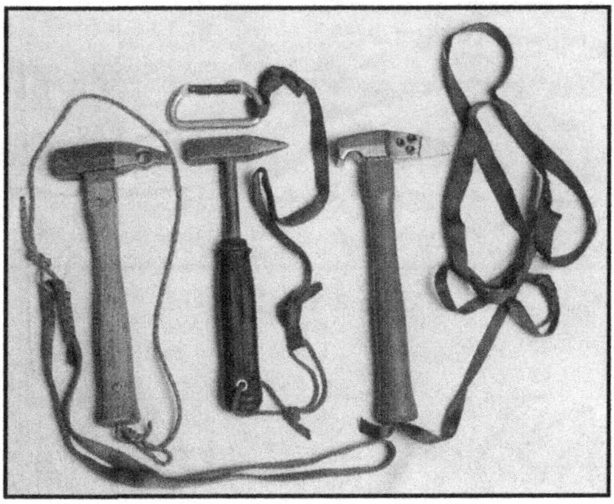

Figure 3-14. Piton hammer.

d. **Chocks.** "Chocks" is a generic term used to describe the various types of artificial protection other than bolts or pitons. Chocks are essentially a tapered metal wedge constructed in various sizes to fit different sized openings in the rock (Figure 3-15). The design of a chock will determine whether it fits into one of two categories—wedges or cams. A wedge holds by wedging into a constricting crack in the rock. A cam holds by slightly rotating in a crack, creating a camming action that lodges the chock in the crack or pocket. Some chocks are manufactured to perform either in the wedging mode or the camming mode. One of the chocks that falls into the category of both a wedge and cam is the hexagonal-shaped or "hex" chock. This type of chock is versatile and comes with either a cable loop or is tied with cord or webbing. All chocks come in different sizes to fit varying widths of cracks. Most chocks come with a wired loop that is stronger than cord and allows for easier placement. Bigger chocks can be threaded with cord or webbing if the user ties the chock himself. Care should be taken to place tubing in the chock before threading the cord. The cord used with chocks is designed to be stiffer and stronger than regular cord and is typically made of Kevlar. The advantage of using a chock rather than a piton is that a climber can carry many different sizes and use them repeatedly.

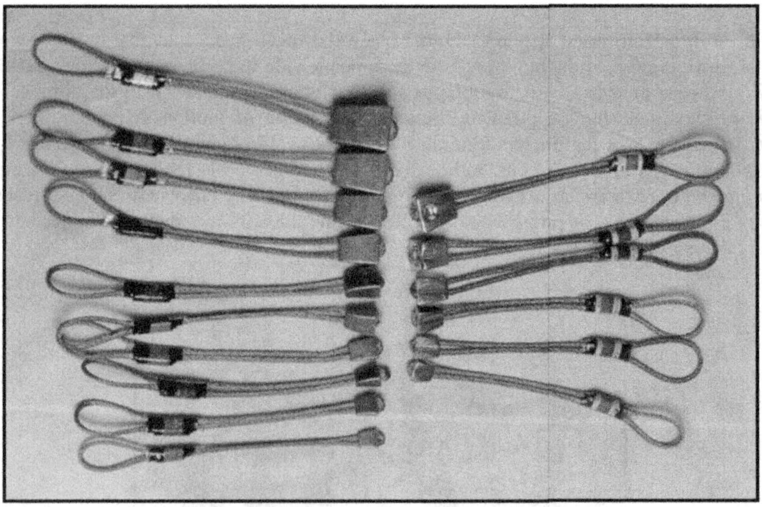

Figure 3-15. Chocks.

e. **Three-Point Camming Device.** The three-point camming device's unique design allows it to be used both as a camming piece and a wedging piece (Figure 3-16). Because of this design it is extremely versatile and, when used in the camming mode, will fit a wide range of cracks. The three-point camming device comes in several different sizes with the smaller sizes working in pockets that no other piece of gear would fit in.

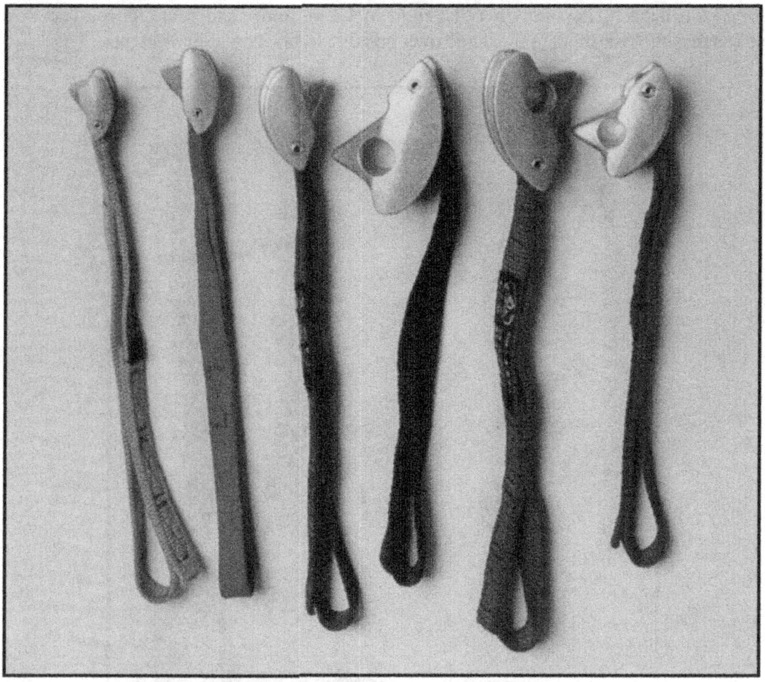

Figure 3-16. Three-point camming device.

f. **Spring-Loaded Camming Devices.** Spring-loaded camming devices (SLCDs) (Figure 3-17) provide convenient, reliable placement in cracks where standard chocks are not practical (parallel or flaring cracks or cracks under roofs). SLCDs have three or four cams rotating around a single or double axis with a rigid or semi-rigid point of attachment. These are placed quickly and easily, saving time and effort. SLCDs are available in many sizes to accommodate different size cracks. Each fits a wide range of crack widths due to the rotating cam heads. The shafts may be rigid metal or semi-rigid cable loops. The flexible cable reduces the risk of stem breakage over an edge in horizontal placements.

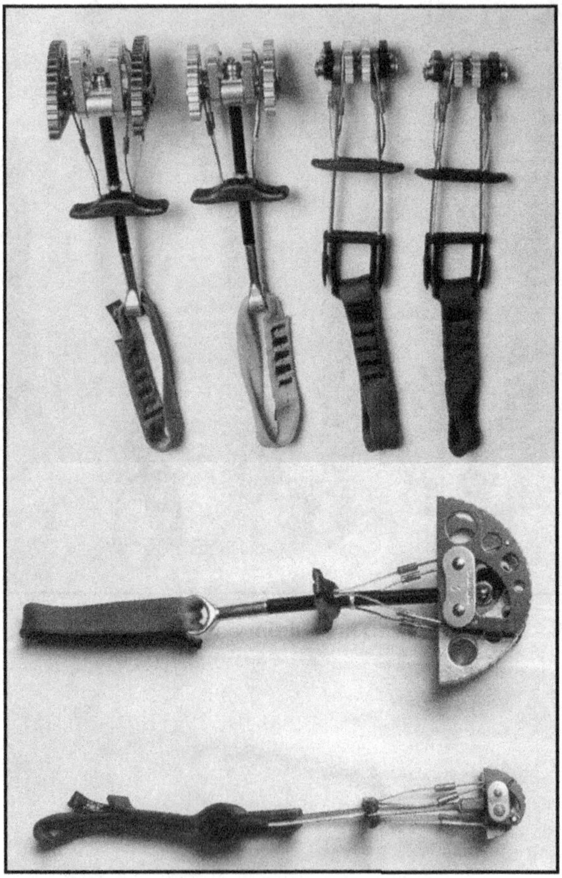

Figure 3-17. Spring-loaded camming devices.

g. **Chock Picks.** Chock picks are primarily used to extract chocks from rock when the they become severely wedged (Figure 3-18). They are also handy to clean cracks with. Made from thin metal, they can be purchased or homemade. When using a chock pick to extract a chock be sure no force is applied directly to the cable juncture. One end of the chock pick should have a hook to use on jammed SLCDs.

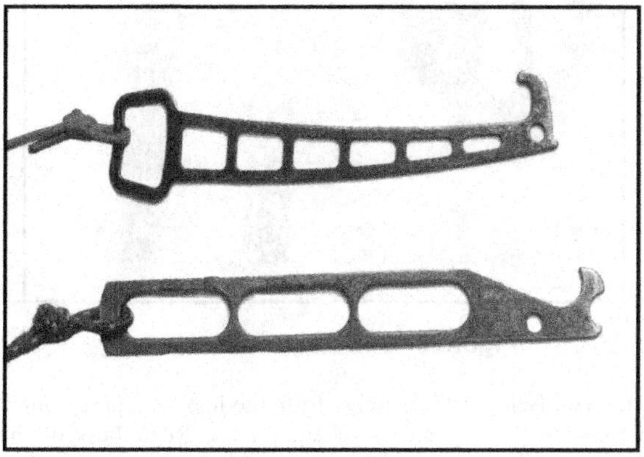

Figure18. Chock picks.

h. **Bolts.** Bolts are screw-like shafts made from metal that are drilled into rock to provide protection (Figure 3-19). The two types are contraction bolts and expansion bolts. Contraction bolts are squeezed together when driven into a rock. Expansion bolts press around a surrounding sleeve to form a snug fit into a rock. Bolts require drilling a hole into a rock, which is time-consuming, exhausting, and extremely noisy. Once emplaced, bolts are the most secure protection for a multidirectional pull. Bolts should be used only when chocks and pitons cannot be emplaced. A bolt is hammered only when it is the nail or self-driving type.

(1) A hanger (for carabiner attachment) and nut are placed on the bolt. The bolt is then inserted and driven into the hole. Because of this requirement, a hand drill must be carried in addition to a piton hammer. Hand drills (also called star drills) are available in different sizes, brands, and weights. A hand drill should have a lanyard to prevent loss.

(2) Self-driving bolts are quicker and easier to emplace. These require a hammer, bolt driver, and drilling anchor, which is driven into the rock. A bolt and carrier are then secured to the emplaced drilling anchor. All metal surfaces should be smooth and free of rust, corrosion, dirt, and moisture. Burrs, chips, and rough spots should be filed smooth and wire-brushed or rubbed clean with steel wool. Items that are cracked or warped indicate excessive wear and should be discarded.

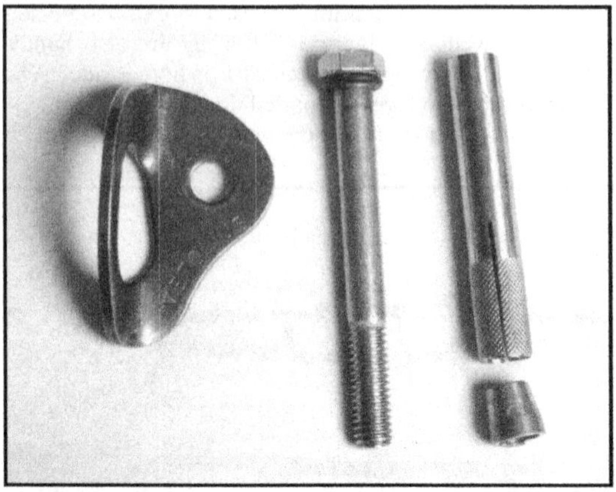

Figure 3-19. Bolts and hangers.

i. **Belay Devices.** Belay devices range from the least equipment intensive (the body belay) to high-tech metal alloy pieces of equipment. Regardless of the belay device choosen, the basic principal remains the same—friction around or through the belay device controls the ropes' movement. Belay devices are divided into three categories: the slot, the tuber, and the mechanical camming device (Figure 3-20).

(1) The slot is a piece of equipment that attaches to a locking carabiner in the harness; a bight of rope slides through the slot and into the carabiner for the belay. The most common slot type belay device is the Sticht plate.

(2) The tuber is used exactly like the slot but its shape is more like a cone or tube.

(3) The mechanical camming device is a manufactured piece of equipment that attaches to the harness with a locking carabiner. The rope is routed through this device so that when force is applied the rope is cammed into a highly frictioned position.

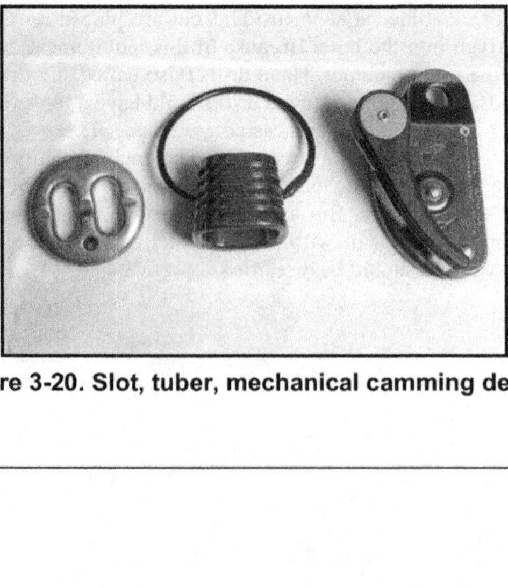

Figure 3-20. Slot, tuber, mechanical camming device.

j. **Descenders.** One piece of equipment used for generations as a descender is the carabiner. A figure-eight is another useful piece of equipment and can be used in conjunction with the carabiner for descending (Figure 3-21).

Note: All belay devices can also be used as descending devices.

Figure 3-21. Figure-eights.

k. **Ascenders.** Ascenders may be used in other applications such as a personal safety or hauling line cam. All modern ascenders work on the principle of using a cam-like device to allow movement in one direction. Ascenders are primarily made of metal alloys and come in a variety of sizes (Figure 3-22). For difficult vertical terrain, two ascenders work best. For lower angle movement, one ascender is sufficient. Most manufacturers make ascenders as a right and left-handed pair.

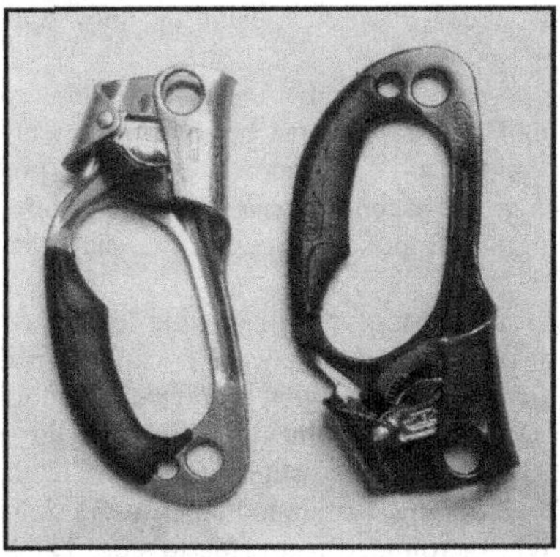

Figure 3-22. Ascenders.

l. **Pulleys.** Pulleys are used to change direction in rope systems and to create mechanical advantage in hauling systems. A pulley should be small, lightweight, and strong. They should accommodate the largest diameter of rope being used. Pulleys are made with several bearings, different-sized sheaves (wheel), and metal alloy sideplates (Figure 3-23). Plastic pulleys should always be avoided. The sideplate should rotate on the pulley axle to allow the pulley to be attached at any point along the rope. For best results, the sheave diameter must be at least four times larger than the rope's diameter to maintain high rope strength.

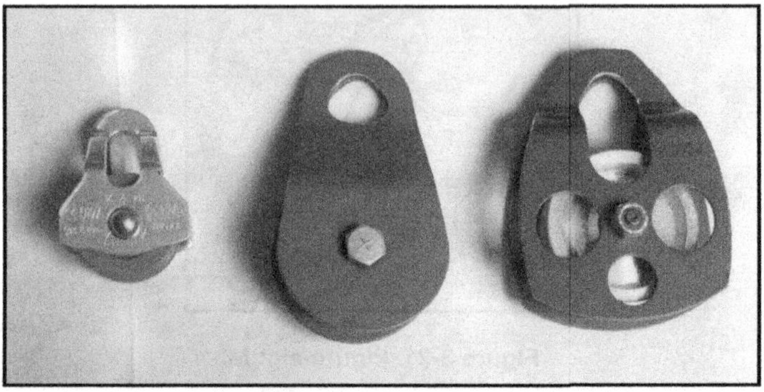

Figure 3-23. Pulley.

3-5. SNOW AND ICE CLIMBING HARDWARE

Snow and ice climbing hardware is the equipment that is particular to operations in some mountainous terrain. Specific training on this type of equipment is essential for safe use. Terrain that would otherwise be inaccessible—snowfields, glaciers, frozen waterfalls—can now be considered avenues of approach using the snow and ice climbing gear listed in this paragraph.

a. **Ice Ax.** The ice ax is one of the most important tools for the mountaineer operating on snow or ice. The climber must become proficient in its use and handling. The versatility of the ax lends itself to balance, step cutting, probing, self-arrest, belays, anchors, direct-aid climbing, and ascending and descending snow and ice covered routes.

(1) Several specific parts comprise an ice ax: the shaft, head (pick and adze), and spike (Figure 3-24, page 3-22).

(a) The shaft (handle) of the ax comes in varying lengths (the primary length of the standard mountaineering ax is 70 centimeters). It can be made of fiberglass, hollow aluminum, or wood; the first two are stronger, therefore safer for mountaineering.

(b) The head of the ax, which combines the pick and the adze, can have different configurations. The pick should be curved slightly and have teeth at least one-fourth of its length. The adze, used for chopping, is perpendicular to the shaft. It can be flat or curved along its length and straight or rounded from side to side. The head can be of one-piece construction or have replaceable picks and adzes. The head should have a hole directly above the shaft to allow for a leash to be attached.

(c) The spike at the bottom of the ax is made of the same material as the head and comes in a variety of shapes.

(2) As climbing becomes more technical, a shorter ax is much more appropriate, and adding a second tool is a must when the terrain becomes vertical. The shorter ax has all the attributes of the longer ax, but it is anywhere from 40 to 55 centimeters long and can have a straight or bent shaft depending on the preference of the user.

b. **Ice Hammer.** The ice hammer is as short or shorter than the technical ax (Figure 3-24). It is used for pounding protection into the ice or pitons into the rock. The only difference between the ice ax and the ice hammer is the ice hammer has a hammerhead instead of an adze. Most of the shorter ice tools have a hole in the shaft to which a leash is secured, which provides a more secure purchase in the ice.

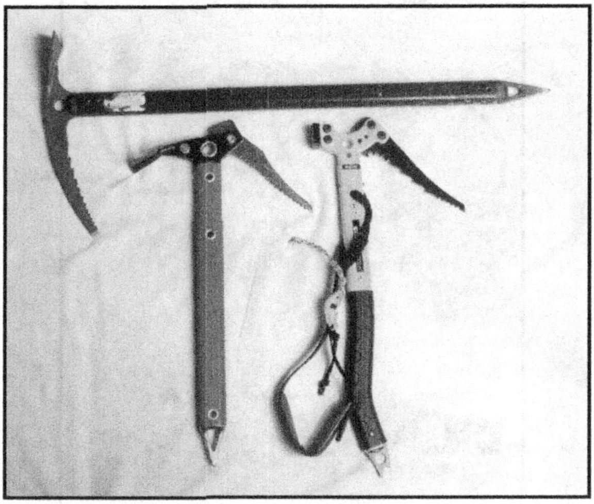

Figure 3-24. Ice ax and ice hammers.

c. **Crampons.** Crampons are used when the footing becomes treacherous. They have multiple spikes on the bottom and spikes protruding from the front (Figure 3-25). Two types of crampons are available: flexible and rigid. Regardless of the type of crampon chosen, fit is the most important factor associated with crampon wear. The crampon should fit snugly on the boot with a minimum of 1 inch of front point protruding. Straps should fit snugly around the foot and any long, loose ends should be trimmed. Both flexible and rigid crampons come in pairs, and any tools needed for adjustment will be provided by the manufacturer.

(1) The hinged or flexible crampon is best used when no technical ice climbing will be done. It is designed to be used with soft, flexible boots, but can be attached to plastic

mountaineering boots. The flexible crampon gets its name from the flexible hinge on the crampon itself. All flexible crampons are adjustable for length while some allow for width adjustment. Most flexible crampons will attach to the boot by means of a strap system. The flexible crampon can be worn with a variety of boot types.

(2) The rigid crampon, as its name implies, is rigid and does not flex. This type of crampon is designed for technical ice climbing, but can be used on less vertical terrain. The rigid crampon can only be worn with plastic mountaineering boots. Rigid crampons will have a toe and heel bail attachment with a strap that wraps around the ankle.

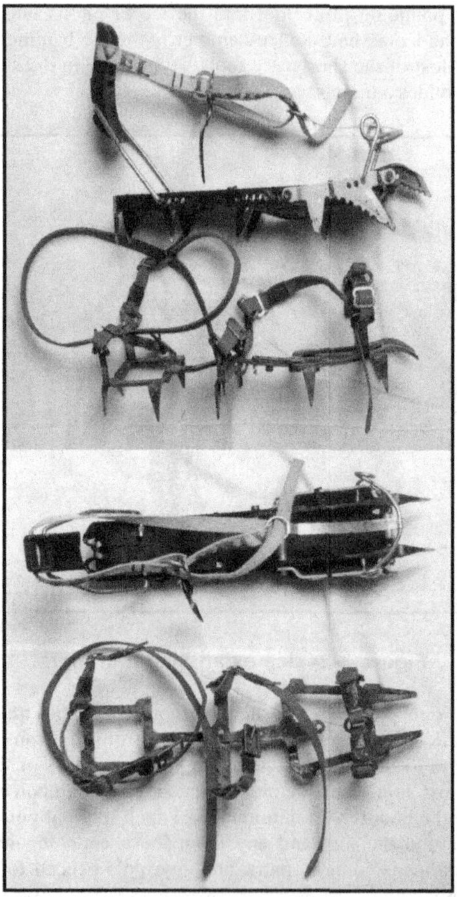

Figure 3-25. Crampons.

d. **Ice Screws.** Ice screws provide artificial protection for climbers and equipment for operations in icy terrain. They are screwed into ice formations. Ice screws are made of chrome-molybdenum steel and vary in lengths from 11 centimeters to 40 centimeters (Figure 3-26). The eye is permanently affixed to the top of the ice screw. The tip consists of milled or hand-ground teeth, which create sharp points to grab the ice when being emplaced. The ice screw has right-hand threads to penetrate the ice when turned clockwise.

(1) When selecting ice screws, choose a screw with a large thread count and large hollow opening. The close threads will allow for ease in turning and better strength. The large hollow opening will allow snow and ice to slide through when turning.

- Type I is 17 centimeters in length with a hollow inner tube.
- Type II is 22 centimeters in length with a hollow inner tube.
- Other variations are hollow alloy screws that have a tapered shank with external threads, which are driven into ice and removed by rotation.

(2) Ice screws should be inspected for cracks, bends, and other deformities that may impair strength or function. If any cracks or bends are noticed, the screw should be turned in. A file may be used to sharpen the ice screw points. Steel wool should be rubbed on rusted surfaces and a thin coat of oil applied when storing steel ice screws.

Note: Ice screws should always be kept clean and dry. The threads and teeth should be protected and kept sharp for ease of application.

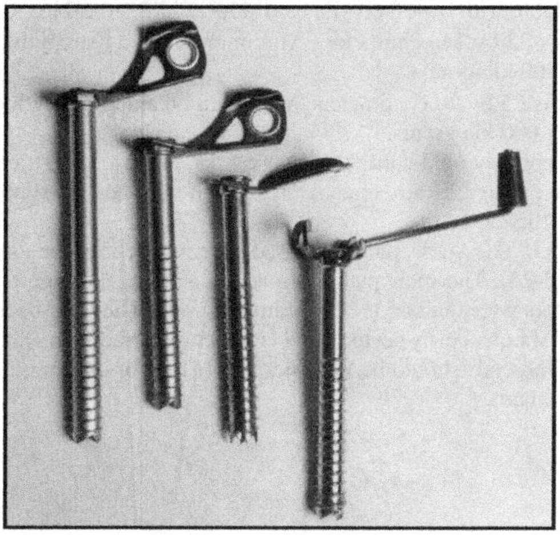

Figure 3-26. Ice screws.

e. **Ice Pitons.** Ice pitons are used to establish anchor points for climbers and equipment when conducting operations on ice. They are made of steel or steel alloys (chrome-molybdenum), and are available in various lengths and diameters (Figure 3-27). They are tubular with a hollow core and are hammered into ice with an ice hammer. The eye is permanently fixed to the top of the ice piton. The tip may be beveled to help grab the ice to facilitate insertion. Ice pitons are extremely strong when placed properly in hard ice. They can, however, pull out easily on warm days and require a considerable amount of effort to extract in cold temperatures.

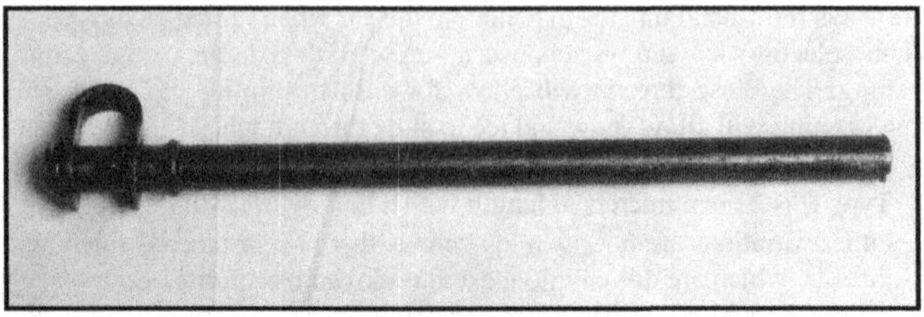

Figure 3-27. Ice piton.

f. **Wired Snow Anchors.** The wired snow anchor (or fluke) provides security for climbers and equipment in operations involving steep ascents by burying the snow anchor into deep snow (Figure 3-28, page 3-26). The fluted anchor portion of the snow anchor is made of aluminum. The wired portion is made of either galvanized steel or stainless steel. Fluke anchors are available in various sizes—their holding ability generally increases with size. They are available with bent faces, flanged sides, and fixed cables. Common types are:

- Type I is 22 by 14 centimeters. Minimum breaking strength of the swaged wire loop is 600 kilograms.
- Type II is 25 by 20 centimeters. Minimum breaking strength of the swaged wire loop is 1,000 kilograms.

The wired snow anchor should be inspected for cracks, broken wire strands, and slippage of the wire through the swage. If any cracks, broken wire strands, or slippage is noticed, the snow anchor should be turned in.

g. **Snow Picket.** The snow picket is used in constructing anchors in snow and ice (Figure 3-28, page 3-26). The snow picket is made of a strong aluminum alloy 3 millimeters thick by 4 centimeters wide, and 45 to 90 centimeters long. They can be angled or T–section stakes. The picket should be inspected for bends, chips, cracks, mushrooming ends, and other deformities. The ends should be filed smooth. If bent or cracked, the picket should be turned in for replacement.

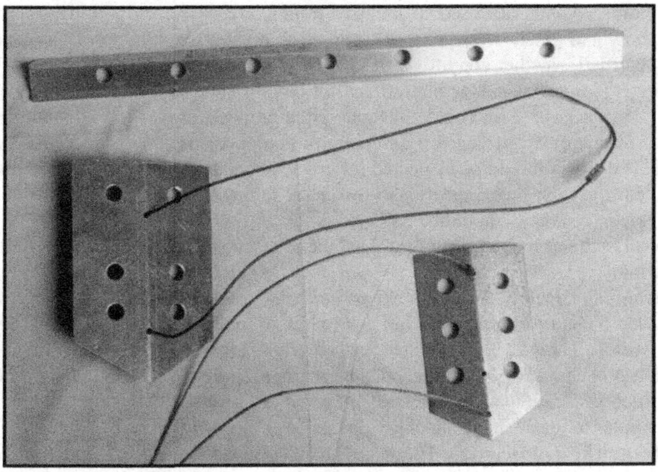

Figure 3-28. Snow anchors, flukes, and pickets.

3-6. SUSTAINABILITY EQUIPMENT

This paragraph describes all additional equipment not directly involved with climbing. This equipment is used for safety (avalanche equipment, wands), bivouacs, movement, and carrying gear. While not all of it will need to be carried on all missions, having the equipment available and knowing how to use it correctly will enhance the unit's capability in mountainous terrain.

a. **Snow Saw**. The snow saw is used to cut into ice and snow. It can be used in step cutting, in shelter construction, for removing frozen obstacles, and for cutting snow stability test pits. The special tooth design of the snow saw easily cuts into frozen snow and ice. The blade is a rigid aluminum alloy of high strength about 3 millimeters thick and 38 centimeters long with a pointed end to facilitate entry on the forward stroke. The handle is either wooden or plastic and is riveted to the blade for a length of about 50 centimeters. The blade should be inspected for rust, cracks, warping, burrs, and missing or dull teeth. A file can repair most defects, and steel wool can be rubbed on rusted areas. The handle should be inspected for cracks, bends, and stability. On folding models, the hinge and nuts should be secure. If the saw is beyond repair, it should not be used.

b. **Snow Shovel.** The snow shovel is used to cut and remove ice and snow. It can be used for avalanche rescue, shelter construction, step cutting, and removing obstacles. The snow shovel is made of a special, lightweight aluminum alloy. The handle should be telescopic, folding, or removable to be compact when not in use. The shovel should have a flat or rounded bottom and be of strong construction. The shovel should be inspected for cracks, bends, rust, and burrs. A file and steel wool can remove rust and put an edge on the blade of the shovel. The handle should be inspected for cracks, bends, and stability. If the shovel is beyond repair, it should be turned in.

c. **Wands.** Wands are used to identify routes, crevasses, snow-bridges, caches, and turns on snow and glaciers. Spacing of wands depends on the number of turns, number of hazards identified, weather conditions (and visibility), and number of teams in the climbing party. Carry too many wands is better than not having enough if they become lost. Wands are 1 to 1.25 meters long and made of lightweight bamboo or plastic shafts pointed on one end with a plastic or nylon flag (bright enough in color to see at a distance) attached to the other end. The shafts should be inspected for cracks, bends, and deformities. The flag should be inspected for tears, frays, security to the shaft, fading, and discoloration. If any defects are discovered, the wands should be replaced.

d. **Avalanche rescue equipment.** Avalanche rescue equipment (Figure 3-29) includes the following:

(1) ***Avalanche Probe.*** Although ski poles may be used as an emergency probe when searching for a victim in an avalanche, commercially manufactured probes are better for a thorough search. They are 9-millimeter thick shafts made of an aluminum alloy, which can be joined to probe up to 360 centimeters. The shafts must be strong enough to probe through avalanche debris. Some manufacturers of ski poles design poles that are telescopic and mate with other poles to create an avalanche probe.

(2) ***Avalanche Transceivers.*** These are small, compact radios used to identify avalanche burial sites. They transmit electromagnetic signals that are picked up by another transceiver on the receive mode.

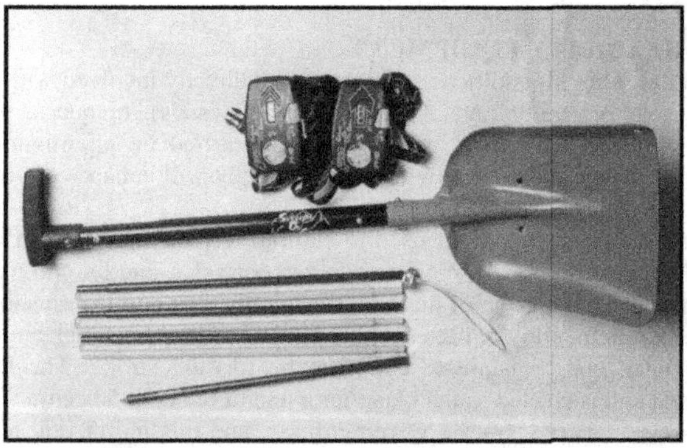

Figure 3-29. Avalanche rescue equipment.

e. **Packs.** Many types and brands of packs are used for mountaineering. The two most common types are internal and external framed packs.

(1) Internal framed packs have a rigid frame within the pack that help it maintain its shape and hug the back. This assists the climber in keeping their balance as they climb or ski. The weight in an internal framed pack is carried low on the body assisting with balance. The body-hugging nature of this type pack also makes it uncomfortable in warm weather.

(2) External framed packs suspend the load away from the back with a ladder-like frame. The frame helps transfer the weight to the hips and shoulders easier, but can be cumbersome when balance is needed for climbing and skiing.

(3) Packs come in many sizes and should be sized appropriately for the individual according to manufacturer's specifications. Packs often come with many unneeded features. A good rule of thumb is: The simpler the pack, the better it will be.

f. **Stoves.** When selecting a stove one must define its purpose—will the stove be used for heating, cooking or both? Stoves or heaters for large elements can be large and cumbersome. Stoves for smaller elements might just be used for cooking and making water, and are simple and lightweight. Stoves are a necessity in mountaineering for cooking and making water from snow and ice. When choosing a stove, factors that should be considered are weight, altitude and temperature where it will be used, fuel availability, and its reliability.

(1) There are many choices in stove design and in fuel types. White gas, kerosene, and butane are the common fuels used. All stoves require a means of pressurization to force the fuel to the burner. Stoves that burn white gas or kerosene have a hand pump to generate the pressurization and butane stoves have pressurized cartridges. All stoves need to vaporize the liquid fuel before it is burned. This can be accomplished by burning a small amount of fuel in the burner cup assembly, which will vaporize the fuel in the fuel line.

(2) Stoves should be tested and maintained prior to a mountaineering mission. They should be easy to clean and repair during an operation. The reliability of the stove has a huge impact on the success of the mission and the morale of personnel.

g. **Tents.** When selecting a tent, the mission must be defined to determine the number of people the tent will accommodate. The climate the tents will be used in is also of concern. A tent used for warmer temperatures will greatly differ from tents used in a colder, more harsh environment. Manufacturers of tents offer many designs of different sizes, weights, and materials.

(1) Mountaineering tents are made out of a breathable or weatherproof material. A single-wall tent allows for moisture inside the tent to escape through the tent's material. A double-wall tent has a second layer of material (referred to as a fly) that covers the tent. The fly protects against rain and snow and the space between the fly and tent helps moisture to escape from inside. Before using a new tent, the seams should be treated with seam sealer to prevent moisture from entering through the stitching.

(2) The frame of a tent is usually made of an aluminum or carbon fiber pole. The poles are connected with an elastic cord that allows them to extend, connect, and become long and rigid. When the tent poles are secured into the tent body, they create the shape of the tent.

(3) Tents are rated by a "relative strength factor," the speed of wind a tent can withstand before the frame deforms. Temperature and expected weather for the mission should be determined before choosing the tent.

h. **Skis.** Mountaineering skis are wide and short. They have a binding that pivots at the toe and allows for the heel to be free for uphill travel or locked for downhill. Synthetic skins with fibers on the bottom can be attached to the bottom of the ski and allow the ski to travel forward and prevent slipping backward. The skins aid in traveling uphill and slow down the rate of descents. Wax can be applied to the ski to aid in ascents instead of skins. Skis can decrease the time needed to reach an objective depending on the ability of the user. Skis can make crossing crevasses easier because of the load distribution, and they can become a makeshift stretcher for casualties. Ski techniques can be complicated and require thorough training for adequate proficiency.

i. **Snowshoes.** Snowshoes are the traditional aid to snow travel that attach to most footwear and have been updated into small, lightweight designs that are more efficient than older models. Snowshoes offer a large displacement area on top of soft snow preventing tiresome post-holing. Some snowshoes come equipped with a crampon like binding that helps in ascending steep snow and ice. Snowshoes are slower than skis, but are better suited for mixed terrain, especially if personnel are not experienced with the art of skiing. When carrying heavy packs, snowshoes can be easier to use than skis.

j. **Ski poles**. Ski poles were traditionally designed to assist in balance during skiing. They have become an important tool in mountaineering for aid in balance while hiking, snowshoeing, and carrying heavy packs. They can take some of the weight off of the lower body when carrying a heavy pack. Some ski poles are collapsible for ease of packing when not needed (Figure 3-30). The basket at the bottom prevents the pole from plunging deep into the snow and, on some models, can be detached so the pole becomes an avalanche or crevasse probe. Some ski poles come with a self-arrest grip, but should not be the only means of protection on technical terrain.

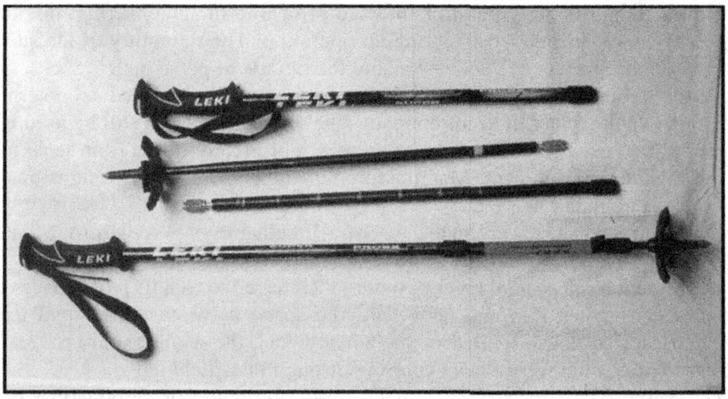

Figure 3-30. Collapsible ski poles.

k. **Sleds**. Sleds vary greatly in size, from the squad-size Ahkio, a component of the 10-man arctic tent system, to the one-person skow. Regardless of the size, sleds are an invaluable asset during mountainous operations when snow and ice is the primary surface on which to travel. Whichever sled is chosen, it must be attachable to the person or people that will be pulling it. Most sleds are constructed using fiberglass bottoms with or without exterior runners. Runners will aid the sleds ability to maintain a true track in the snow. The sled should also come with a cover of some sort—whether nylon or canvas, a cover is essential for keeping the components in the sled dry. Great care should be taken when packing the sled, especially when hauling fuel. Heavier items should be carried towards the rear of the sled and lighter items towards the front.

l. **Headlamps**. A headlamp is a small item that is not appreciated until it is needed. It is common to need a light source and the use of both hands during limited light conditions in mountaineering operations. A flashlight can provide light, but can be cumbersome when both hands are needed. Most headlamps attach to helmets by means of elastic bands.

(1) When choosing a headlamp, ensure it is waterproof and the battery apparatus is small. All components should be reliable in extreme weather conditions. When the light is being packed, care should be taken that the switch doesn't accidentally activate and use precious battery life.

(2) The battery source should compliment the resupply available. Most lights will accept alkaline, nickel-cadmium, or lithium batteries. Alkaline battery life diminishes quickly in cold temperatures, nickel-cadmium batteries last longer in cold but require a recharging unit, and lithium batteries have twice the voltage so modifications are required.

Section II. EQUIPMENT PACKING

Equipment brought on a mission is carried in the pack, worn on the body, or hauled in a sled (in winter). Obviously, the rucksack and sled (or Ahkio) can hold much more than a climber can carry. They would be used for major bivouac gear, food, water, first aid kits, climbing equipment, foul weather shells, stoves, fuel, ropes, and extra ammunition and demolition materials, if needed.

3-7. CHOICE OF EQUIPMENT

Mission requirements and unit SOP will influence the choice of gear carried but the following lists provide a sample of what should be considered during mission planning.

a. **Personal Gear.** Personal gear includes emergency survival kit containing signaling material, fire starting material, food procurement material, and water procurement material. Pocket items should include a knife, whistle, pressure bandage, notebook with pen or pencil, sunglasses, sunblock and lip protection, map, compass and or altimeter.

b. **Standard Gear.** Standard gear that can be individually worn or carried includes cushion sole socks; combat boots or mountain boots, if available; BDU and cap; LCE with canteens, magazine pouches, and first aid kit; individual weapon; a large rucksack containing waterproof coat and trousers, polypropylene top, sweater, or fleece top; helmet; poncho; and sleeping bag.

CAUTION

Cotton clothing, due to its poor insulating and moisture-wicking characteristics, is virtually useless in most mountain climates, the exception being hot, desert, or jungle mountain environments. Cotton clothing should be replaced with synthetic fabric clothing.

c. **Mountaineering Equipment and Specialized Gear.** This gear includes:

- Sling rope or climbing harness.
- Utility cord(s).
- Nonlocking carabiners.
- Locking carabiner(s).
- Rappelling gloves.
- Rappel/belay device.
- Ice ax.
- Crampons.
- Climbing rope, one per climbing team.
- Climbing rack, one per climbing team.

d. **Day Pack.** When the soldier plans to be away from the bivouac site for the day on a patrol or mountaineering mission, he carries a light day pack. This pack should contain the following items:

- Extra insulating layer: polypropylene, pile top, or sweater.
- Protective layer: waterproof jacket and pants, rain suit, or poncho.
- First aid kit.
- Flashlight or headlamp.
- Canteen.
- Cold weather hat or scarf.
- Rations for the time period away from the base camp.
- Survival kit.
- Sling rope or climbing harness.
- Carabiners.
- Gloves.
- Climbing rope, one per climbing team.
- Climbing rack, one per climbing team.

e. **Squad or Team Safety Pack.** When a squad-sized element leaves the bivouac site, squad safety gear should be carried in addition to individual day packs. This can either be loaded into one rucksack or cross-loaded among the squad members. In the event of an injury, casualty evacuation, or unplanned bivouac, these items may make the difference between success and failure of the mission.

- Sleeping bag.
- Sleeping mat.

- Squad stove.
- Fuel bottle.

f. **The Ten Essentials.** Regardless of what equipment is carried, the individual military mountaineer should always carry the "ten essentials" when moving through the mountains.

(1) ***Map.***

(2) ***Compass, Altimeter, and or GPS.***

(3) ***Sunglasses and Sunscreen.***

(a) In alpine or snow-covered sub-alpine terrain, sunglasses are a vital piece of equipment for preventing snow blindness. They should filter 95 to 100 percent of ultraviolet light. Side shields, which minimize the light entering from the side, should permit ventilation to help prevent lens fogging. At least one extra pair of sunglasses should be carried by each independent climbing team.

(b) Sunscreens should have an SPF factor of 15 or higher. For lip protection, a total UV blocking lip balm that resists sweating, washing, and licking is best. This lip protection should be carried in the chest pocket or around the neck to allow frequent reapplication.

(4) ***Extra Food.*** One day's worth extra of food should be carried in case of delay caused by bad weather, injury, or navigational error.

(5) ***Extra Clothing.*** The clothing used during the active part of a climb, and considered to be the basic climbing outfit, includes socks, boots, underwear, pants, blouse, sweater or fleece jacket, hat, gloves or mittens, and foul weather gear (waterproof, breathable outerwear or waterproof rain suit).

(a) Extra clothing includes additional layers needed to make it through the long, inactive hours of an unplanned bivouac. Keep in mind the season when selecting this gear.

- Extra underwear to switch out with sweat-soaked underwear.
- Extra hats or balaclavas.
- Extra pair of heavy socks.
- Extra pair of insulated mittens or gloves.
- In winter or severe mountain conditions, extra insulation for the upper body and the legs.

(b) To back up foul weather gear, bring a poncho or extra-large plastic trash bag. A reflective emergency space blanket can be used for hypothermia first aid and emergency shelter. Insulated foam pads prevent heat loss while sitting or lying on snow. Finally, a bivouac sack can help by protecting insulating layers from the weather, cutting the wind, and trapping essential body heat inside the sack.

(6) ***Headlamp and or Flashlight.*** Headlamps provide the climber a hands-free capability, which is important while climbing, working around the camp, and employing weapons systems. Miniature flashlights can be used, but commercially available headlamps are best. Red lens covers can be fabricated for tactical conditions. Spare batteries and spare bulbs should also be carried.

(7) ***First-aid Kit.*** Decentralized operations, the mountain environment—steep, slick terrain and loose rock combined with heavy packs, sharp tools, and fatigue—requires each climber to carry his own first-aid kit. Common mountaineering injuries that can be expected are punctures and abrasions with severe bleeding, a broken bone, serious sprain,

and blisters. Therefore, the kit should contain at least enough material to stabilize these conditions. Pressure dressings, gauze pads, elastic compression wrap, small adhesive bandages, butterfly bandages, moleskin, adhesive tape, scissors, cleanser, latex gloves and splint material (if above tree line) should all be part of the kit.

(8) ***Fire Starter.*** Fire starting material is key to igniting wet wood for emergency campfires. Candles, heat tabs, and canned heat all work. These can also be used for quick warming of water or soup in a canteen cup. In alpine zones above tree line with no available firewood, a stove works as an emergency heat source.

(9) ***Matches and Lighter.*** Lighters are handy for starting fires, but they should be backed up by matches stored in a waterproof container with a strip of sandpaper.

(10) ***Knife.*** A multipurpose pocket tool should be secured with cord to the belt, harness, or pack.

g. **Other Essential Gear.** Other essential gear may be carried depending on mission and environmental considerations.

(1) ***Water and Water Containers.*** These include wide-mouth water bottles for water collection; camel-back type water holders for hands-free hydration; and a small length of plastic tubing for water procurement at snow-melt seeps and rainwater puddles on bare rock.

(2) ***Ice Ax.*** The ice ax is essential for travel on snowfields and glaciers as well as snow-covered terrain in spring and early summer. It helps for movement on steep scree and on brush and heather covered slopes, as well as for stream crossings.

(3) ***Repair Kit.*** A repair kit should include:

- Stove tools and spare parts.
- Duct tape.
- Patches.
- Safety pins.
- Heavy-duty thread.
- Awl and or needles.
- Cord and or wire.
- Small pliers (if not carrying a multipurpose tool).
- Other repair items as needed.

(4) ***Insect Repellent.***

(5) ***Signaling Devices.***

(6) ***Snow Shovel.***

3-8. TIPS ON PACKING

When loading the internal frame pack the following points should be considered.

a. In most cases, speed and endurance are enhanced if the load is carried more by the hips (using the waist belt) and less by the shoulders and back. This is preferred for movement over trails or less difficult terrain. By packing the lighter, more compressible items (sleeping bag, clothing) in the bottom of the rucksack and the heavier gear (stove, food, water, rope, climbing hardware, extra ammunition) on top, nearer the shoulder blades, the load is held high and close to the back, thus placing the most weight on the hips.

b. In rougher terrain it pays to modify the pack plan. Heavy articles of gear are placed lower in the pack and close to the back, placing more weight on the shoulders and back. This lowers the climber's center of gravity and helps him to better keep his balance.

c. Equipment that may be needed during movement should be arranged for quick access using either external pockets or placing immediately underneath the top flap of the pack. As much as possible, this placement should be standardized across the team so that necessary items can be quickly reached without unnecessary unpacking of the pack in emergencies.

d. The pack and its contents should be soundly waterproofed. Clothing and sleeping bag are separately sealed and then placed in the larger wet weather bag that lines the rucksack. Zip-lock plastic bags can be used for small items, which are then organized into color-coded stuffsacks. A few extra-large plastic garbage bags should be carried for a variety of uses—spare waterproofing, emergency bivouac shelter, and water procurement, among others.

e. The ice ax, if not carried in hand, should be stowed on the outside of the pack with the spike up and the adze facing forward or to the outside, and be securely fastened. Mountaineering packs have ice ax loops and buckle fastening systems for this. If not, the ice ax is placed behind one of the side pockets, as stated above, and then tied in place.

f. Crampons should be secured to the outside rear of the pack with the points covered.

CHAPTER 4

ROPE MANAGEMENT AND KNOTS

The rope is a vital piece of equipment to the mountaineer. When climbing, rappelling, or building various installations, the mountaineer must know how to properly utilize and maintain this piece of equipment. If the rope is not managed or maintained properly, serious injury may occur. This chapter discusses common rope terminology, management techniques, care and maintenance procedures, and knots.

SECTION I. PREPARATION, CARE AND MAINTENANCE, INSPECTION, TERMINOLOGY

The service life of a rope depends on the frequency of use, applications (rappelling, climbing, rope installations), speed of descent, surface abrasion, terrain, climate, and quality of maintenance. Any rope may fail under extreme conditions (shock load, sharp edges, misuse).

4-1. PREPARATION

The mountaineer must select the proper rope for the task to be accomplished according to type, diameter, length, and tensile strength. It is important to prepare all ropes before departing on a mission. Avoid rope preparation in the field.

a. **Packaging.** New rope comes from the manufacturer in different configurations—boxed on a spool in various lengths, or coiled and bound in some manner. Precut ropes are usually packaged in a protective cover such as plastic or burlap. Do not remove the protective cover until the rope is ready for use.

b. **Securing the Ends of the Rope:** If still on a spool, the rope must be cut to the desired length. All ropes will fray at the ends unless they are bound or seared. Both static and dynamic rope ends are secured in the same manner. The ends must be heated to the melting point so as to attach the inner core strands to the outer sheath. By fusing the two together, the sheath cannot slide backward or forward. Ensure that this is only done to the ends of the rope. If the rope is exposed to extreme temperatures, the sheath could be weakened, along with the inner core, reducing overall tensile strength. The ends may also be dipped in enamel or lacquer for further protection.

4-2. CARE AND MAINTENANCE

The rope is a climber's lifeline. It must be cared for and used properly. These general guidelines should be used when handling ropes.

a. Do not step on or drag ropes on the ground unnecessarily. Small particles of dirt will be ground between the inner strands and will slowly cut them.

b. While in use, do not allow the rope to come into contact with sharp edges. Nylon rope is easily cut, particularly when under tension. If the rope must be used over a sharp edge, pad the edge for protection.

c. Always keep the rope as dry as possible. Should the rope become wet, hang it in large loops off the ground and allow it to dry. Never dry a rope with high heat or in direct sunlight.

d. Never leave a rope knotted or tightly stretched for longer than necessary. Over time it will reduce the strength and life of the rope.

e. Never allow one rope to continuously rub over or against another. Allowing rope-on-rope contact with nylon rope is extremely dangerous because the heat produced by the friction will cause the nylon to melt.

f. Inspect the rope before each use for frayed or cut spots, mildew or rot, or defects in construction (new rope).

g. The ends of the rope should be whipped or melted to prevent unraveling.

h. Do not splice ropes for use in mountaineering.

i. Do not mark ropes with paints or allow them to come in contact with oils or petroleum products. Some of these will weaken or deteriorate nylon.

j. Never use a mountaineering rope for any purpose except mountaineering.

k. Each rope should have a corresponding rope log (DA Form 5752-R, Rope History and Usage), which is also a safety record. It should annotate use, terrain, weather, application, number of falls, dates, and so on, and should be annotated each time the rope is used (Figure 4-1). DA Form 5752-R is authorized for local reproduction on 8 1/2- by 11-inch paper.

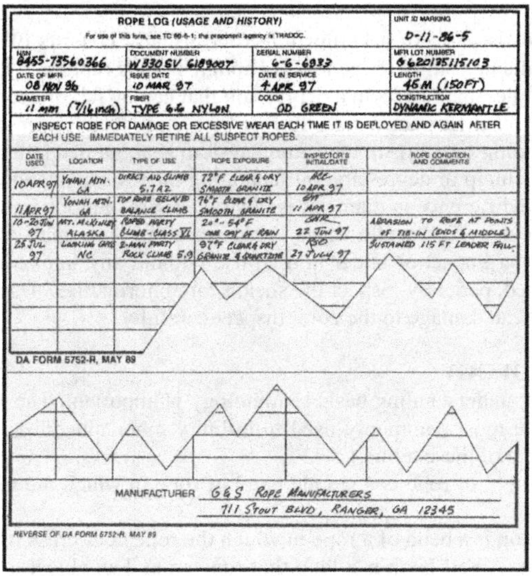

ROPE LOG (USAGE AND HISTORY)
For use of this form, see TC 90-6-1; the proponent agency is TRADOC.

UNIT ID MARKING: D-11-86-5

NSN	DOCUMENT NUMBER	SERIAL NUMBER	MFR LOT NUMBER
8455-73560366	W330SV 6189007	6-6-6933	66201351/5103
DATE OF MFR: 08 NOV 96	ISSUE DATE: 10 MAR 97	DATE IN SERVICE: 4 APR 97	LENGTH: 46 M (150 FT)
DIAMETER: 11 mm (7/16 inch)	FIBER: TYPE 6.6 NYLON	COLOR: OD GREEN	CONSTRUCTION: DYNAMIC KERMANTLE

INSPECT ROBE FOR DAMAGE OR EXCESSIVE WEAR EACH TIME IT IS DEPLOYED AND AGAIN AFTER EACH USE. IMMEDIATELY RETIRE ALL SUSPECT ROPES.

DATE USED	LOCATION	TYPE OF USE	ROPE EXPOSURE	INSPECTOR'S INITIAL/DATE	ROPE CONDITION AND COMMENTS
10 APR 97	YONAH MTN. GA	DIRECT AID CLIMB 5.7 A2	72°F CLEAR & DRY SMOOTH GRANITE	ACE 10 APR 97	
11 APR 97	YONAH MTN. GA	TOP ROPE BELAYED BALANCE CLIMB	76°F CLEAR & DRY SMOOTH GRANITE	GH 11 APR 97	
10-20 JUN 97	MT. McKINLEY ALASKA	ROPED PARTY CLIMB-CLASS VI	20°-54°F ONE DAY OF RAIN	GNR 22 JUN 97	ABRASION TO ROPE AT POINTS OF TIE-IN (ENDS & MIDDLE)
25 JUL 97	LOOKING GLASS NC	2-MAN PARTY ROCK CLIMB 5.9	97°F CLEAR & DRY GRANITE & QUARTZITE	RSO 27 JULY 97	SUSTAINED 115 FT LEADER FALL.

DA FORM 5752-R, MAY 89

MANUFACTURER G&S ROPE MANUFACTURERS
711 STOUT BLVD, RANGER, GA 12345

REVERSE OF DA FORM 5752-R, MAY 89

Figure 4-1. Example of completed DA Form 5752-R.

l. Never subject the rope to high heat or flame. This will significantly weaken it.

m. All ropes should be washed periodically to remove dirt and grit, and rinsed thoroughly. Commercial rope washers are made from short pieces of modified pipe that connect to any faucet. Pinholes within the pipe force water to circulate around and scrub the rope as you slowly feed it through the washer. Another method is to machine wash, on a gentle cycle, in cold water with a nylon safe soap, never bleach or harsh cleansers. Ensure that only front loading washing machine are used to wash ropes.

n. Ultraviolet radiation (sunlight) tends to deteriorate nylon over long periods of time. This becomes important if rope installations are left in place over a number of months.

o. When not in use, ropes should be loosely coiled and hung on wooden pegs rather than nails or other metal objects. Storage areas should be relatively cool with low humidity levels to prevent mildew or rotting. Rope may also be loosely stacked and placed in a rope bag and stored on a shelf. Avoid storage in direct sunlight, as the ultraviolet radiation will deteriorate the nylon over long periods

4-3. INSPECTION

Ropes should be inspected before and after each use, especially when working around loose rock or sharp edges.

a. Although the core of the kernmantle rope cannot be seen, it is possible to damage the core without damaging the sheath. Check a kernmantle rope by carefully inspecting the sheath before and after use while the rope is being coiled. When coiling, be aware of how the rope feels as it runs through the hands. Immediately note and tie off any lumps or depressions felt.

b. Damage to the core of a kernmantle rope usually consists of filaments or yarn breakage that results in a slight retraction. If enough strands rupture, a localized reduction in the diameter of the rope results in a depression that can be felt or even seen.

c. Check any other suspected areas further by putting them under tension (the weight of one person standing on a Prusik tensioning system is about maximum). This procedure will emphasize the lump or depression by separating the broken strands and enlarging the dip. If a noticeable difference in diameter is obvious, retire the rope immediately.

d. Many dynamic kernmantle ropes are quite soft. They may retain an indention occasionally after an impact or under normal use without any trauma to the core. When damage is suspected, patiently inspect the sheath for abnormalities. Damage to the sheath does not always mean damage to the core. Inspect carefully.

4-4. TERMINOLOGY

When using ropes, understanding basic terminology is important. The terms explained in this section are the most commonly used in military mountaineering. (Figure 4-2, page 4-4, illustrates some of these terms.)

a. **Bight**. A bight of rope is a simple bend of rope in which the rope does not cross itself.

b. **Loop.** A loop is a bend of a rope in which the rope does cross itself.

c. **Half Hitch**. A half hitch is a loop that runs around an object in such a manner as to lock or secure itself.

d. **Turn.** A turn wraps around an object, providing 360-degree contact.

e. **Round Turn**. A round turn wraps around an object one and one-half times. A round turn is used to distribute the load over a small diameter anchor (3 inches or less). It may also be used around larger diameter anchors to reduce the tension on the knot, or provide added friction.

f. **Running End**. A running end is the loose or working end of the rope.

g. **Standing Part.** The standing part is the static, stationary, or nonworking end of the rope.

h. **Lay.** The lay is the direction of twist used in construction of the rope.

i. **Pigtail.** The pigtail (tail) is the portion of the running end of the rope between the safety knot and the end of the rope.

j. **Dress.** Dress is the proper arrangement of all the knot parts, removing unnecessary kinks, twists, and slack so that all rope parts of the knot make contact.

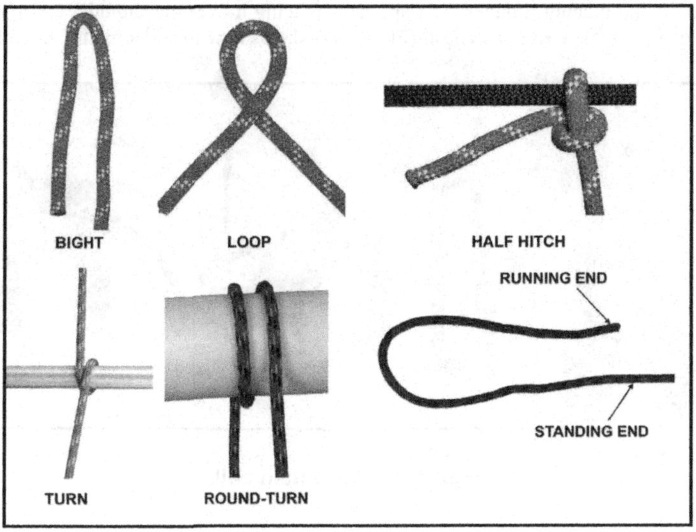

Figure 4-2. Examples of roping terminology.

Section II. COILING, CARRYING, THROWING

The ease and speed of rope deployment and recovery greatly depends upon technique and practice.

4-5. COILING AND CARRYING THE ROPE

Use the butterfly or mountain coil to coil and carry the rope. Each is easy to accomplish and results in a minimum amount of kinks, twists, and knots later during deployment.

a. **Mountain Coil.** To start a mountain coil, grasp the rope approximately 1 meter from the end with one hand. Run the other hand along the rope until both arms are outstretched. Grasping the rope firmly, bring the hands together forming a loop, which is laid in the hand closest to the end of the rope. This is repeated, forming uniform loops that run in a clockwise direction, until the rope is completely coiled. The rope may be given a 1/4 twist as each loop is formed to overcome any tendency for the rope to twist or form figure-eights.

(1) In finishing the mountain coil, form a bight approximately 30 centimeters long with the starting end of the rope and lay it along the top of the coil. Uncoil the last loop and, using this length of the rope, begin making wraps around the coil and the bight, wrapping toward the closed end of the bight and making the first wrap bind across itself so as to lock it into place. Make six to eight wraps to adequately secure the coil, and then route the end of the rope through the closed end of the bight. Pull the running end of the bight tight, securing the coil.

(2) The mountain coil may be carried either in the pack (by forming a figure eight), doubling it and placing it under the flap, or by placing it over the shoulder and under the opposite arm, slung across the chest. (Figure 4-3 shows how to coil a mountain coil.)

Figure 4-3. Mountain coil.

b. **Butterfly Coil.** The butterfly coil is the quickest and easiest technique for coiling (Figure 4-4).

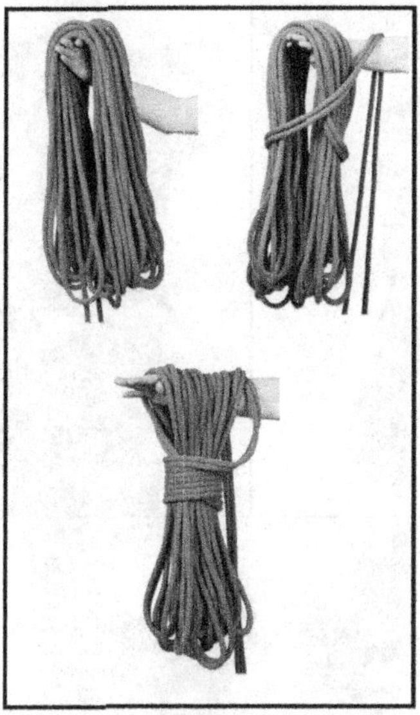

Figure 4-4. Butterfly coil.

(1) ***Coiling.*** To start the double butterfly, grasp both ends of the rope and begin back feeding. Find the center of the rope forming a bight. With the bight in the left hand, grasp both ropes and slide the right hand out until there is approximately one arms length of rope. Place the doubled rope over the head, draping it around the neck and on top of the shoulders. Ensure that it hangs no lower than the waist. With the rest of the doubled rope in front of you, make doubled bights placing them over the head in the same manner as the first bight. Coil alternating from side to side (left to right, right to left) while maintaining equal-length bights. Continue coiling until approximately two arm-lengths of rope remain. Remove the coils from the neck and shoulders carefully, and hold the center in one hand. Wrap the two ends around the coils a minimum of three doubled wraps, ensuring that the first wrap locks back on itself.

(2) ***Tie-off and Carrying.*** Take a doubled bight from the loose ends of rope and pass it through the apex of the coils. Pull the loose ends through the doubled bight and dress it down. Place an overhand knot in the loose ends, dressing it down to the apex of the bight securing coils. Ensure that the loose ends do not exceed the length of the coils. In this configuration the coiled rope is secure enough for hand carrying or carrying in a rucksack, or for storage. (Figure 4-5 shows a butterfly coil tie-off.)

Figure 4-5. Butterfly coil tie-off.

c. **Coiling Smaller Diameter Rope.** Ropes of smaller diameters may be coiled using the butterfly or mountain coil depending on the length of the rope. Pieces 25 feet and shorter (also known as cordage, sling rope, utility cord) may be coiled so that they can be hung from the harness. Bring the two ends of the rope together, ensuring no kinks are in the rope. Place the ends of the rope in the left hand with the two ends facing the body. Coil the doubled rope in a clockwise direction forming 6- to 8-inch coils (coils may be larger depending on the length of rope) until an approximate 12-inch bight is left. Wrap that bight around the coil, ensuring that the first wrap locks on itself. Make three or more wraps. Feed the bight up through the bights formed at the top of the coil. Dress it down tightly. Now the piece of rope may be hung from a carabiner on the harness.

e. **Uncoiling, Back-feeding, and Stacking.** When the rope is needed for use, it must be uncoiled and stacked on the ground properly to avoid kinks and snarls.

(1) Untie the tie-off and lay the coil on the ground. Back-feed the rope to minimize kinks and snarls. (This is also useful when the rope is to be moved a short distance and

coiling is not desired.) Take one end of the rope in the left hand and run the right hand along the rope until both arms are outstretched. Next, lay the end of the rope in the left hand on the ground. With the left hand, re-grasp the rope next to the right hand and continue laying the rope on the ground.

(2) The rope should be laid or stacked in a neat pile on the ground to prevent it from becoming tangled and knotted when throwing the rope, feeding it to a lead climber, and so on. This technique can also be started using the right hand.

4-6. THROWING THE ROPE

Before throwing the rope, it must be properly managed to prevent it from tangling during deployment. The rope should first be anchored to prevent complete loss of the rope over the edge when it is thrown. Several techniques can be used when throwing a rope. Personal preference and situational and environmental conditions should be taken into consideration when determining which technique is best.

a. Back feed and neatly stack the rope into coils beginning with the anchored end of the rope working toward the running end. Once stacked, make six to eight smaller coils in the left hand. Pick up the rest of the larger coils in the right hand. The arm should be generally straight when throwing. The rope may be thrown underhanded or overhanded depending on obstacles around the edge of the site. Make a few preliminary swings to ensure a smooth throw. Throw the large coils in the right hand first. Throw up and out. A slight twist of the wrist, so that the palm of the hand faces up as the rope is thrown, allows the coils to separate easily without tangling. A smooth follow through is essential. When a slight tug on the left hand is felt, toss the six to eight smaller coils out. This will prevent the ends of the rope from becoming entangled with the rest of the coils as they deploy. As soon as the rope leaves the hand, the thrower should sound off with a warning of "ROPE" to alert anyone below the site.

b. Another technique may also be used when throwing rope. Anchor, back feed, and stack the rope properly as described above. Take the end of the rope and make six to eight helmet-size coils in the right hand (more may be needed depending on the length of the rope). Assume a "quarterback" simulated stance. Aiming just above the horizon, vigorously throw the rope overhanded, up and out toward the horizon. The rope must be stacked properly to ensure smooth deployment.

c. When windy weather conditions prevail, adjustments must be made. In a strong cross wind, the rope should be thrown angled into the wind so that it will land on the desired target. The stronger the wind, the harder the rope must be thrown to compensate.

SECTION III. KNOTS

All knots used by a mountaineer are divided into four classes: Class I—joining knots, Class II—anchor knots, Class III—middle rope knots, and Class IV—special knots. The variety of knots, bends, bights, and hitches is almost endless. These classes of knots are intended only as a general guide since some of the knots discussed may be appropriate in more than one class. The skill of knot tying can perish if not used and practiced. With experience and practice, knot tying becomes instinctive and helps the mountaineer in many situations.

4-7. SQUARE KNOT

The square knot is used to tie the ends of two ropes of equal diameter (Figure 4-6). It is a joining knot.

a. **Tying the Knot.**

STEP 1. Holding one working end in each hand, place the working end in the right hand over the one in the left hand.

STEP 2. Pull it under and back over the top of the rope in the left hand.

STEP 3. Place the working end in the left hand over the one in the right hand and repeat STEP 2.

STEP 4. Dress the knot down and secure it with an overhand knot on each side of the square knot.

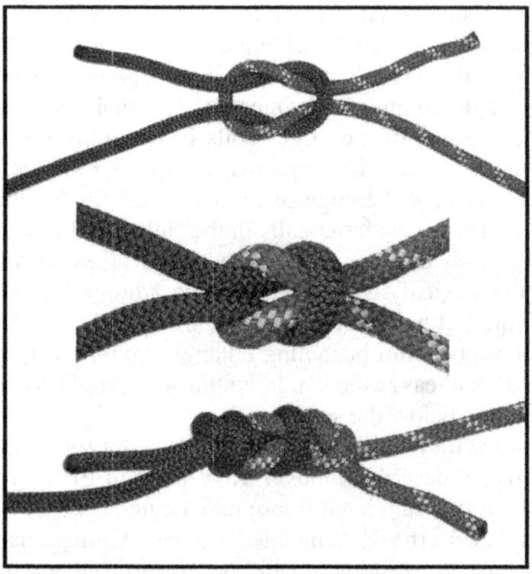

Figure 4-6. Square knot.

b. **Checkpoints.**

(1) There are two interlocking bights.

(2) The running end and standing part are on the same side of the bight formed by the other rope.

(3) The running ends are parallel to and on the same side of the standing ends with 4-inch minimum pig tails after the overhand safeties are tied.

4-8. FISHERMAN'S KNOT

The fisherman's knot is used to tie two ropes of the same or approximately the same diameter (Figure 4-7, page 4-10). It is a joining knot.

a. **Tying the Knot.**

STEP 1. Tie an overhand knot in one end of the rope.

STEP 2. Pass the working end of the other rope through the first overhand knot. Tie an overhand knot around the standing part of the first rope with the working end of the second rope.

STEP 3. Tightly dress down each overhand knot and tightly draw the knots together.

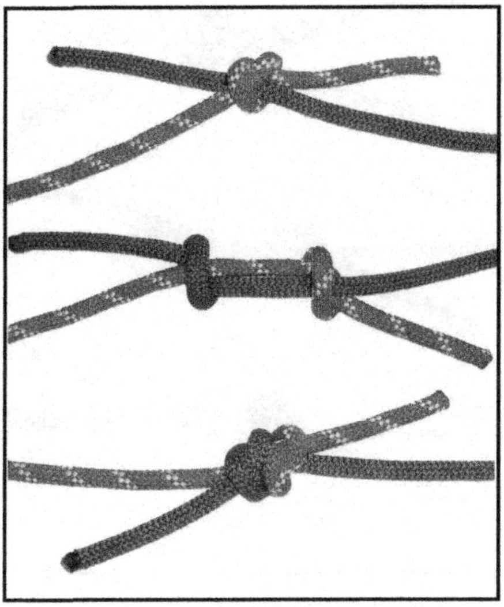

Figure 4-7. Fisherman's knot.

b. **Checkpoints.**

(1) The two separate overhand knots are tied tightly around the long, standing part of the opposing rope.

(2) The two overhand knots are drawn snug.

(3) Ends of rope exit knot opposite each other with 4-inch pigtails.

4-9. DOUBLE FISHERMAN'S KNOT

The double fisherman's knot (also called double English or grapevine) is used to tie two ropes of the same or approximately the same diameter (Figure 4-8). It is a joining knot.

a. **Tying the Knot.**

STEP 1. With the working end of one rope, tie two wraps around the standing part of another rope.

STEP 2. Insert the working end (STEP 1) back through the two wraps and draw it tight.

STEP 3. With the working end of the other rope, which contains the standing part (STEPS 1 and 2), tie two wraps around the standing part of the other rope (the working end in STEP 1). Insert the working end back through the two wraps and draw tight.

STEP 4. Pull on the opposing ends to bring the two knots together.

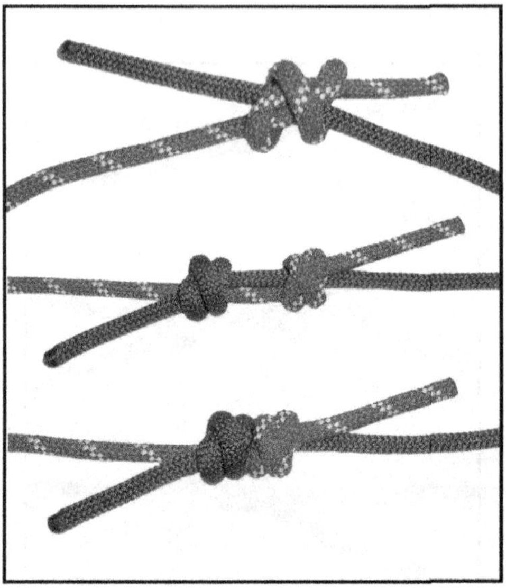

Figure 4-8. Double fisherman's knot.

b. **Checkpoints.**

(1) Two double overhand knots securing each other as the standing parts of the rope are pulled apart.

(2) Four rope parts on one side of the knot form two "x" patterns, four rope parts on the other side of the knot are parallel.

(3) Ends of rope exit knot opposite each other with 4-inch pigtails.

4-10. FIGURE-EIGHT BEND

The figure-eight bend is used to join the ends of two ropes of equal or unequal diameter within 5-mm difference (Figure 4-9, page 4-12).

a. **Tying the Knot.**

STEP 1. Grasp the top of a 2-foot bight.

STEP 2. With the other hand, grasp the running end (short end) and make a 360-degree turn around the standing end.

STEP 3. Place the running end through the loop just formed creating an in-line figure eight.

STEP 4. Route the running end of the other ripe back through the figure eight starting from the original rope's running end. Trace the original knot to the standing end.

STEP 5. Remove all unnecessary twists and crossovers. Dress the knot down.

Figure 4-9. Figure-eight bend.

b. **Checkpoints.**

(1) There is a figure eight with two ropes running side by side.

(2) The running ends are on opposite sides of the knot.

(3) There is a minimum 4-inch pigtail.

4-11. WATER KNOT

The water knot is used to attach two webbing ends (Figure 4-10). It is also called a ring bend, overhand retrace, or tape knot. It is used in runners and harnesses and is a joining knot.

a. **Tying the Knot.**

STEP 1. Tie an overhand knot in one of the ends.

STEP 2. Feed the other end back through the knot, following the path of the first rope in reverse.

STEP 3. Draw tight and pull all of the slack out of the knot. The remaining tails must extend at least 4 inches beyond the knot in both directions.

Figure 4-10. Water knot.

b. **Checkpoints.**

(1) There are two overhand knots, one retracing the other.

(2) There is no slack in the knot, and the working ends come out of the knot in opposite directions.

(3) There is a minimum 4-inch pigtail.

4-12. BOWLINE

The bowline is used to tie the end of a rope around an anchor. It may also be used to tie a single fixed loop in the end of a rope (Figure 4-11, page 4-14). It is an anchor knot.

a. **Tying the Knot.**

STEP 1. Bring the working end of the rope around the anchor, from right to left (as the climber faces the anchor).

STEP 2. Form an overhand loop in the standing part of the rope (on the climber's right) toward the anchor.

STEP 3. Reach through the loop and pull up a bight.

STEP 4. Place the working end of the rope (on the climber's left) through the bight, and bring it back onto itself. Now dress the knot down.

STEP 5. Form an overhand knot with the tail from the bight.

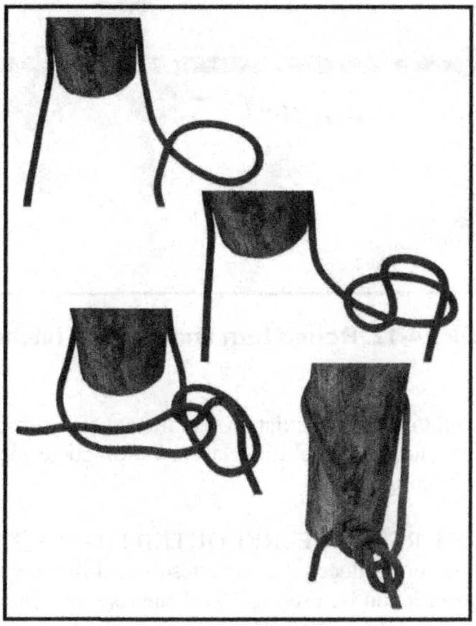

Figure 4-11. Bowline knot.

b. **Checkpoints.**

(1) The bight is locked into place by a loop.

(2) The short portion of the bight is on the inside and on the loop around the anchor (or inside the fixed loop).

(3) There is a minimum 4-inch pigtail after tying the overhand safety.

4-13. ROUND TURN AND TWO HALF HITCHES

This knot is used to tie the end of a rope to an anchor, and it must have constant tension (Figure 4-12). It is an anchor knot.

a. **Tying the Knot.**

STEP 1. Route the rope around the anchor from right to left and wrap down (must have two wraps in the rear of the anchor, and one in the front). Run the loop around the object to provide 360-degree contact, distributing the load over the anchor.

STEP 2. Bring the working end of the rope left to right and over the standing part, forming a half hitch (first half hitch).

STEP 3. Repeat STEP 2 (last half hitch has a 4 inch pigtail).

STEP 4. Dress the knot down.

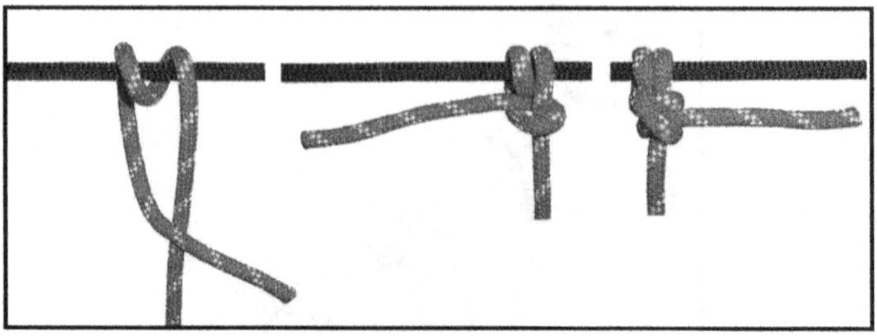

Figure 4-12. Round turn and two half hitches.

b. **Checkpoints.**

(1) A complete round turn should exist around the anchor with no crosses.

(2) Two half hitches should be held in place by a diagonal locking bar with no less than a 4-inch pigtail remaining.

4-14. FIGURE-EIGHT RETRACE (REROUTED FIGURE-EIGHT)

The figure-eight retrace knot produces the same result as a figure-eight loop. However, by tying the knot in a retrace, it can be used to fasten the rope to trees or to places where the loop cannot be used (Figure 4-13, page 4-16). It is also called a rerouted figure-eight and is an anchor knot.

a. **Tying the Knot.**

STEP 1. Use a length of rope long enough to go around the anchor, leaving enough rope to work with.

STEP 2. Tie a figure-eight knot in the standing part of the rope, leaving enough rope to go around the anchor. To tie a figure-eight knot form a loop in the rope, wrap the working end around the standing part, and route the working end through the loop. The finished knot is dressed loosely.

STEP 3. Take the working end around the anchor point.

STEP 4. With the working end, insert the rope back through the loop of the knot in reverse.

STEP 5. Keep the original figure eight as the outside rope and retrace the knot around the wrap and back to the long-standing part.

STEP 6. Remove all unnecessary twists and crossovers; dress the knot down.

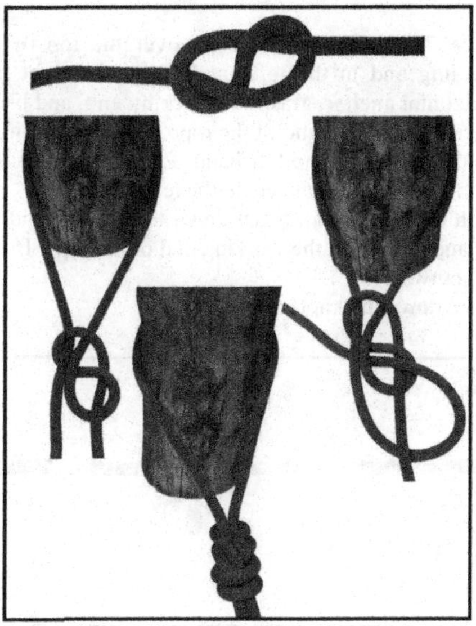

Figure 4-13. Figure-eight retrace.

b. **Checkpoints**

(1) A figure eight with a doubled rope running side by side, forming a fixed loop around a fixed object or harness.

(2) There is a minimum 4-inch pigtail.

4-15. CLOVE HITCH

The clove hitch is an anchor knot that can be used in the middle of the rope as well as at the end (Figure 4-14). The knot must have constant tension on it once tied to prevent slipping. It can be used as either an anchor or middle of the rope knot, depending on how it is tied.

a. **Tying the Knot.**

(1) ***Middle of the Rope.***

STEP 1. Hold rope in both hands, palms down with hands together. Slide the left hand to the left from 20 to 25 centimeters.

STEP 2. Form a loop away from and back toward the right.

STEP 3. Slide the right hand from 20 to 25 centimeters to the right. Form a loop inward and back to the left hand.

STEP 4. Place the left loop on top of the right loop. Place both loops over the anchor and pull both ends of the rope in opposite directions. The knot is tied.

(2) ***End of the Rope.***

Note: For instructional purposes, assume that the anchor is horizontal.

STEP 1. Place 76 centimeters of rope over the top of the anchor. Hold the standing end in the left hand. With the right hand, reach under the horizontal anchor, grasp the working end, and bring it inward.

STEP 2. Place the working end of the rope over the standing end (to form a loop). Hold the loop in the left hand. Place the working end over the anchor from 20 to 25 centimeters to the left of the loop.

STEP 3. With the right hand, reach down to the left hand side of the loop under the anchor. Grasp the working end of the rope. Bring the working end up and outward.

STEP 4. Dress down the knot.

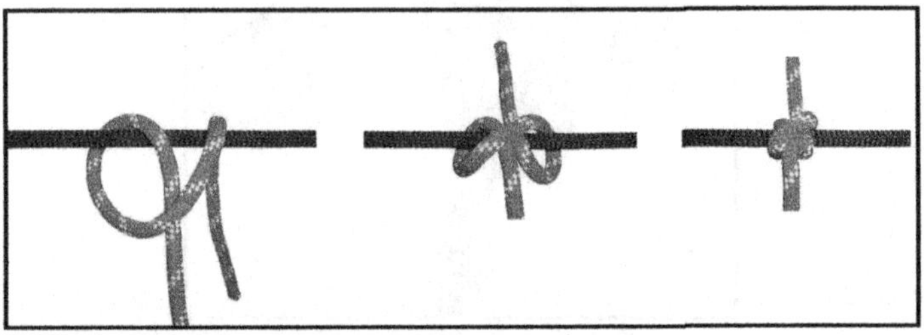

Figure 4-14. Clove hitch.

b. **Checkpoints.**

(1) The knot has two round turns around the anchor with a diagonal locking bar.

(2) The locking bar is facing 90 degrees from the direction of pull.

(3) The ends exit 180 degrees from each other.

(4) The knot has more than a 4-inch pigtail remaining.

4-16. WIREMAN'S KNOT

The wireman's knot forms a single, fixed loop in the middle of the rope (Figure 4-15, page 4-18). It is a middle rope knot.

a. **Tying the Knot.**

STEP 1. When tying this knot, face the anchor that the tie-off system will be tied to. Take up the slack from the anchor, and wrap two turns around the left hand (palm up) from left to right.

STEP 2. A loop of 30 centimeters is taken up in the second round turn to create the fixed loop of the knot.

STEP 3. Name the wraps from the palm to the fingertips: heel, palm, and fingertip.

STEP 4. Secure the palm wrap with the right thumb and forefinger, and place it over the heel wrap.

STEP 5. Secure the heel wrap and place it over the fingertip wrap.

STEP 6. Secure the fingertip wrap and place it over the palm wrap.

STEP 7. Secure the palm wrap and pull up to form a fixed loop.

STEP 8. Dress the knot down by pulling on the fixed loop and the two working ends.

STEP 9. Pull the working ends apart to finish the knot.

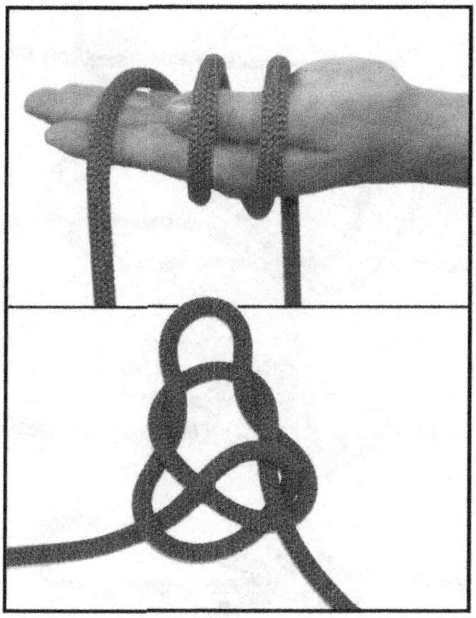

Figure 4-15. Wireman's knot.

b. **Checkpoints.**

(1) The completed knot should have four separate bights locking down on themselves with the fixed loop exiting from the top of the knot and laying toward the near side anchor point.

(2) Both ends should exit opposite each other without any bends.

4-17. DIRECTIONAL FIGURE-EIGHT

The directional figure-eight knot forms a single, fixed loop in the middle of the rope that lays back along the standing part of the rope (Figure 4-16). It is a middle rope knot.

a. **Tying the Knot.**

STEP 1. Face the far side anchor so that when the knot is tied, it lays inward.

STEP 2. Lay the rope from the far side anchor over the left palm. Make one wrap around the palm.

STEP 3. With the wrap thus formed, tie a figure-eight knot around the standing part that leads to the far side anchor.

STEP 4. When dressing the knot down, the tail and the bight must be together.

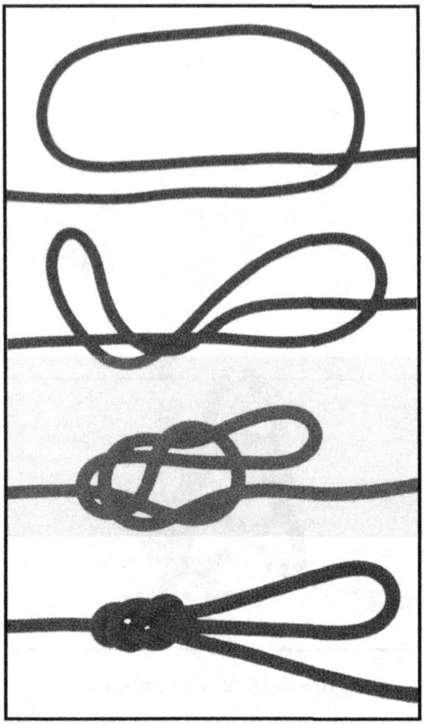

Figure 4-16. Directional figure-eight.

b. **Checkpoints.**

(1) The loop should be large enough to accept a carabiner but no larger than a helmet-size loop.

(2) The tail and bight must be together.

(3) The figure eight is tied tightly.

(4) The bight in the knot faces back toward the near side.

4-18. BOWLINE-ON-A-BIGHT (TWO-LOOP BOWLINE)

The bowline-on-a-bight is used to form two fixed loops in the middle of a rope (Figure 4-17, page 4-20). It is a middle rope knot.

a. **Tying the Knot.**

STEP 1. Form a bight in the rope about twice as long as the finished loops will be.

STEP 2. Tie an overhand knot on a bight.

STEP 3. Hold the overhand knot in the left hand so that the bight is running down and outward.

STEP 4. Grasp the bight with the right hand; fold it back over the overhand knot so that the overhand knot goes through the bight.

STEP 5. From the end (apex) of the bight, follow the bight back to where it forms the cross in the overhand knot. Grasp the two ropes that run down and outward and pull up, forming two loops.

STEP 6. Pull the two ropes out of the overhand knot and dress the knot down.

STEP 7. A final dress is required: grasp the ends of the two fixed loops and pull, spreading them apart to ensure the loops do not slip.

Figure 4-17. Bowline-on-a-bight.

b. **Checkpoints.**

(1) There are two fixed loops that will not slip.

(2) There are no twists in the knot.

(3) A double loop is held in place by a bight.

4-19. TWO-LOOP FIGURE-EIGHT

The two-loop figure-eight is used to form two fixed loops in the middle of a rope (Figure 4-18.) It is a middle rope knot.

a. **Tying the Knot.**

STEP 1. Using a doubled rope, form an 18-inch bight in the left hand with the running end facing to the left.

STEP 2. Grasp the bight with the right hand and make a 360-degree turn around the standing end in a counterclockwise direction.

STEP 3. With the working end, form another bight and place that bight through the loop just formed in the left hand.

STEP 4. Hold the bight with the left hand, and place the original bight (moving toward the left hand) over the knot.

STEP 5. Dress the knot down.

Figure 4-18. Two-loop figure-eight.

b. **Checkpoints.**

(1) There is a double figure-eight knot with two loops that share a common locking bar.

(2) The two loops must be adjustable by means of a common locking bar.

(3) The common locking bar is on the bottom of the double figure-eight knot.

4-20. FIGURE-EIGHT LOOP (FIGURE-EIGHT-ON-A-BIGHT)

The figure-eight loop, also called the figure-eight-on-a-bight, is used to form a fixed loop in a rope (Figure 4-19). It is a middle of the rope knot.

a. **Tying the Knot.**

STEP 1. Form a bight in the rope about as large as the diameter of the desired loop.

STEP 2. With the bight as the working end, form a loop in rope (standing part).

STEP 3. Wrap the working end around the standing part 360 degrees and feed the working end through the loop. Dress the knot tightly.

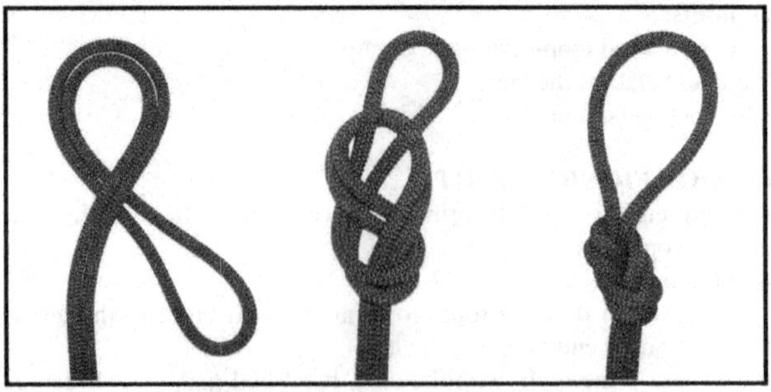

Figure 4-19. Figure-eight loop.

b. **Checkpoints.**

(1) The loop is the desired size.

(2) The ropes in the loop are parallel and do not cross over each other.

(3) The knot is tightly dressed.

4-21. PRUSIK KNOT

The Prusik knot is used to put a moveable rope on a fixed rope such as a Prusik ascent or a tightening system. This knot can be tied as a middle or end of the rope Prusik. It is a specialty knot.

a. **Tying the Knot.**

(1) ***Middle-of-the-Rope Prusik.*** The middle-of-the-rope Prusik knot can be tied with a short rope to a long rope as follows (Figure 4-20.):

STEP 1. Double the short rope, forming a bight, with the working ends even. Lay it over the long rope so that the closed end of the bight is 12 inches below the long rope and the remaining part of the rope (working ends) is the closest to the climber; spread the working end apart.

STEP 2. Reach down through the 12-inch bight. Pull up both of the working ends and lay them over the long rope. Repeat this process making sure that the working ends pass in the middle of the first two wraps. Now there are four wraps and a locking bar working across them on the long rope.

STEP 3. Dress the wraps and locking bar down to ensure they are tight and not twisted. Tying an overhand knot with both ropes will prevent the knot from slipping during periods of variable tension.

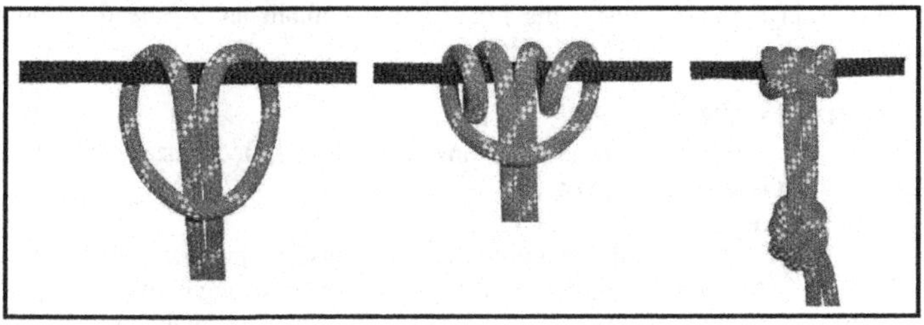

Figure 4-20. Middle-of-the-rope Prusik.

(2) ***End-of-the-Rope Prusik*** (Figure 4-21).

STEP 1. Using an arm's length of rope, and place it over the long rope.

STEP 2. Form a complete round turn in the rope.

STEP 3. Cross over the standing part of the short rope with the working end of the short rope.

STEP 4. Lay the working end under the long rope.

STEP 5. Form a complete round turn in the rope, working back toward the middle of the knot.

STEP 6. There are four wraps and a locking bar running across them on the long rope. Dress the wraps and locking bar down. Ensure they are tight, parallel, and not twisted.

STEP 7. Finish the knot with a bowline to ensure that the Prusik knot will not slip out during periods of varying tension.

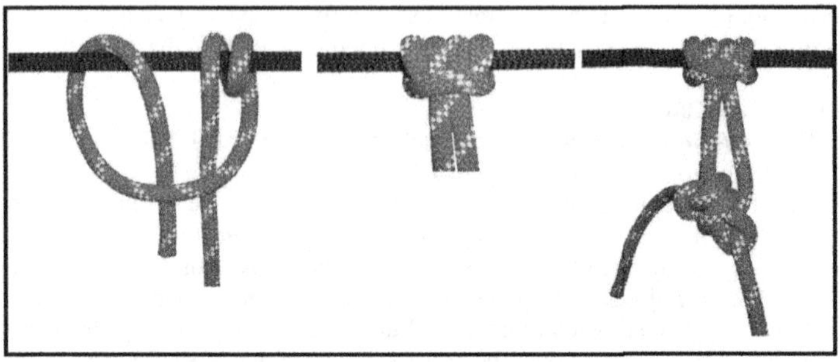

Figure 4-21. End-of-the-rope Prusik knot.

b. **Checkpoints.**

(1) Four wraps with a locking bar.

(2) The locking bar faces the climber.

(3) The knot is tight and dressed down with no ropes twisted or crossed.

(4) Other than a finger Prusik, the knot should contain an overhand or bowline to prevent slipping.

4-22. BACHMAN KNOT

The Bachman knot provides a means of using a makeshift mechanized ascender (Figure 4-22, page 4-24). It is a specialty knot.

a. **Tying the Knot.**

STEP 1. Find the middle of a utility rope and insert it into a carabiner.

STEP 2. Place the carabiner and utility rope next to a long climbing rope.

STEP 3. With the two ropes parallel from the carabiner, make two or more wraps around the climbing rope and through the inside portion of the carabiner.

Note: The rope can be tied into an etrier (stirrup) and used as a Prusik-friction principle ascender.

b. **Checkpoints.**

(1) The bight of the climbing rope is at the top of the carabiner.

(2) The two ropes run parallel without twisting or crossing.

(3) Two or more wraps are made around the long climbing rope and through the inside portion of the carabiner.

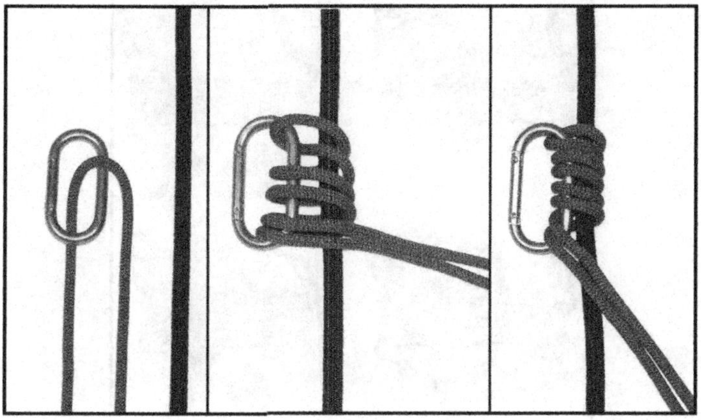

Figure 4-22. Bachman knot.

4-23. BOWLINE-ON-A-COIL

The bowline-on-a-coil is an expedient tie-in used by climbers when a climbing harness is not available (Figure 4-23). It is a specialty knot.

a. **Tying the Knot.**

STEP 1. With the running end, place 3 feet of rope over your right shoulder. The running end is to the back of the body.

STEP 2. Starting at the bottom of your rib cage, wrap the standing part of the rope around your body and down in a clockwise direction four to eight times.

STEP 3. With the standing portion of the rope in your left hand, make a clockwise loop toward the body. The standing portion is on the bottom.

STEP 4. Ensuring the loop does not come uncrossed, bring it up and under the coils between the rope and your body.

STEP 5. Using the standing part, bring a bight up through the loop. Grasp the running end of the rope with the right hand. Pass it through the bight from right to left and back on itself.

STEP 6. Holding the bight loosely, dress the knot down by pulling on the standing end.

STEP 7. Safety the bowline with an overhand around the top, single coil. Then, tie an overhand around all coils, leaving a minimum 4-inch pigtail.

b. **Checkpoints.**

(1) A minimum of four wraps, not crossed, with a bight held in place by a loop.

(2) The loop must be underneath all wraps.

(3) A minimum 4-inch pigtail after the second overhand safety is tied.

(4) Must be centered on the mid-line of the body.

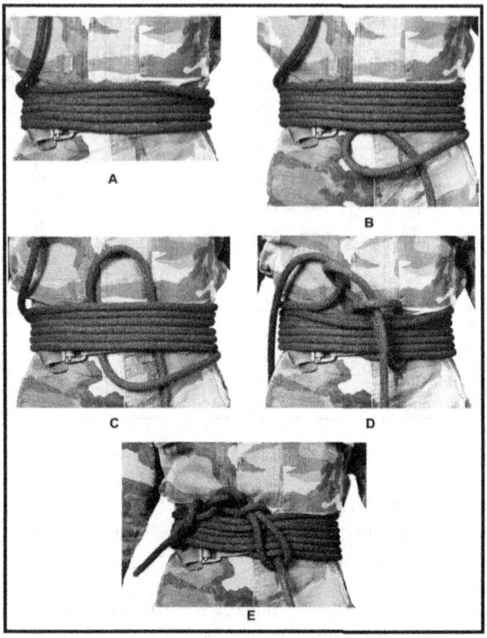

Figure 4-23. Bowline-on-a-coil.

4-24. THREE-LOOP BOWLINE

The three-loop bowline is used to form three fixed loops in the middle of a rope (Figure 4-24, page 4-26). It is used in a self-equalizing anchor system. It is a specialty knot.

a. **Tying the Knot.**

STEP 1. Form an approximate 24-inch bight.

STEP 2. With the right thumb facing toward the body, form a doubled loop in the standing part by turning the wrist clockwise. Lay the loops to the right.

STEP 3. With the right hand, reach down through the loops and pull up a doubled bight from the standing part of the rope.

STEP 4. Place the running end (bight) of the rope (on the left) through the doubled bight from left to right and bring it back on itself. Hold the running end loosely and dress the knot down by pulling on the standing parts.

STEP 5. Safety it off with a doubled overhand knot.

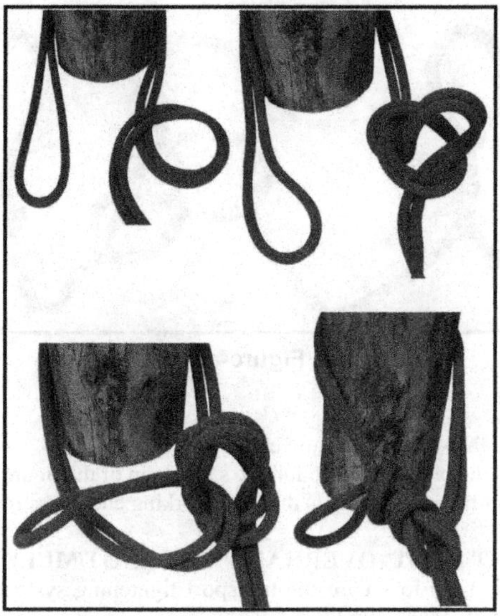

Figure 4-24. Three-loop bowline.

b. **Checkpoints.**

(1) There are two bights held in place by two loops.

(2) The bights form locking bars around the standing parts.

(3) The running end (bight) must be on the inside of the fixed loops.

(4) There is a minimum 4-inch pigtail after the double overhand safety knot is tied.

4-25. FIGURE-EIGHT SLIP KNOT

The figure eight slip knot forms an adjustable bight in a rope (Figure 4-25). It is a specialty knot.

a. **Tying the Knot.**

STEP 1. Form a 12-inch bight in the end of the rope.

STEP 2. Hold the center of the bight in the right hand. Hold the two parallel ropes from the bight in the left hand about 12 inches up the rope.

STEP 3. With the center of the bight in the right hand, twist two complete turns clockwise.

STEP 4. Reach through the bight and grasp the long, standing end of the rope. Pull another bight (from the long standing end) back through the original bight.

STEP 5. Pull down on the short working end of the rope and dress the knot down.

STEP 6. If the knot is to be used in a transport tightening system, take the working end of the rope and form a half hitch around the loop of the figure eight knot.

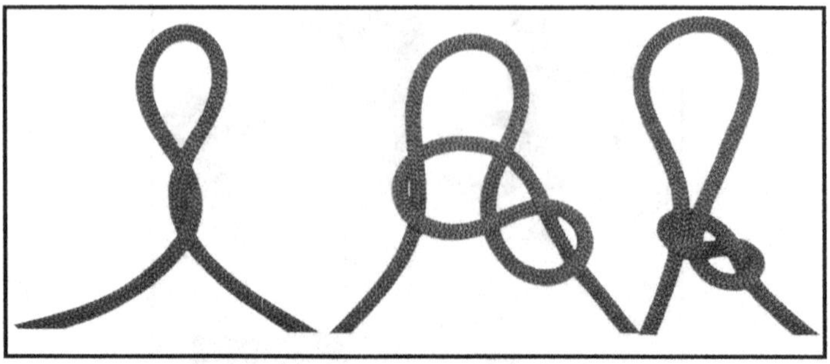

Figure 4-25. Figure-eight slip knot.

b. **Checkpoints.**
(1) The knot is in the shape of a figure eight.
(2) Both ropes of the bight pass through the same loop of the figure eight.
(3) The sliding portion of the rope is the long working end of the rope.

4-26. TRANSPORT KNOT (OVERHAND SLIP KNOT/MULE KNOT)
The transport knot is used to secure the transport tightening system (Figure 4-26, page 4-28). It is simply an overhand slip knot.

a. **Tying the Knot.**

STEP 1. Pass the running end of the rope around the anchor point passing it back under the standing portion (leading to the far side anchor) forming a loop.

STEP 2. Form a bight with the running end of the rope. Pass over the standing portion and down through the loop and dress it down toward the anchor point.

STEP 3. Secure the knot by tying a half hitch around the standing portion with the bight.

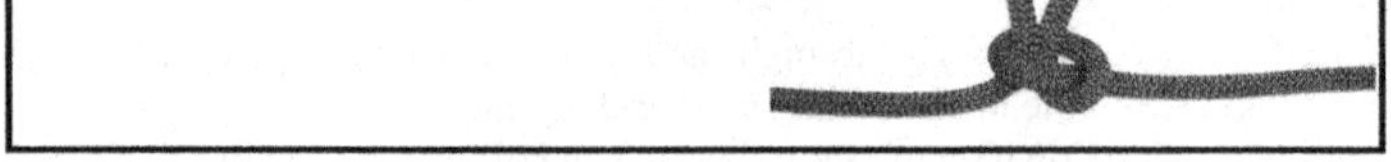

Figure 4-26. Transport knot.

b. **Check Points.**
(1) There is a single overhand slip knot.

(2) The knot is secured using a half hitch on a bight.
(3) The bight is a minimum of 12 inches long.

4-27. KLEIMHIEST KNOT

The Kleimhiest knot provides a moveable, easily adjustable, high-tension knot capable of holding extremely heavy loads while being pulled tight (Figure 4-27). It is a special-purpose knot.

a. **Tying the Knot.**

STEP 1. Using a utility rope or webbing offset the ends by 12 inches. With the ends offset, find the center of the rope and form a bight. Lay the bight over a horizontal rope.

STEP 2. Wrap the tails of the utility rope around the horizontal rope back toward the direction of pull. Wrap at least four complete turns.

STEP 3. With the remaining tails of the utility rope, pass them through the bight (see STEP 1).

STEP 4. Join the two ends of the tail with a joining knot.

STEP 5. Dress the knot down tightly so that all wraps are touching.

Note: Spectra should not be used for the Kleimhiest knot. It has a low melting point and tends to slip .

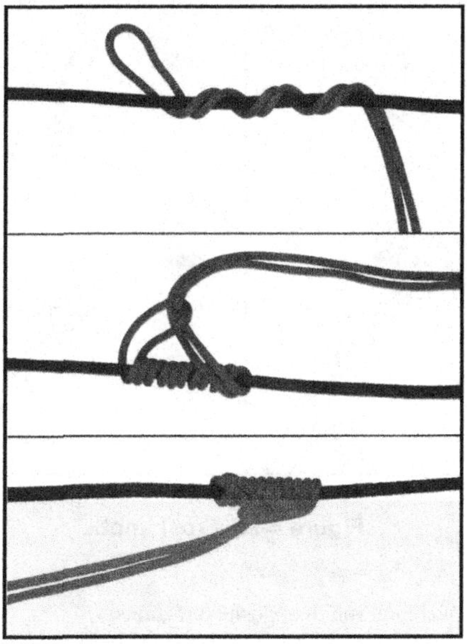

Figure 4-27. Kleimhiest knot.

b. **Checkpoints.**

(1) The bight is opposite the direction of pull.

(2) All wraps are tight and touching.

(3) The ends of the utility rope are properly secured with a joining knot.

4-28. FROST KNOT

The frost knot is used when working with webbing (Figure 4-28, page 4-30). It is used to create the top loop of an etrier. It is a special-purpose knot.

a. **Tying the Knot.**

STEP 1. Lap one end (a bight) of webbing over the other about 10 to 12 inches.

STEP 2. Tie an overhand knot with the newly formed triple-strand webbing; dress tightly.

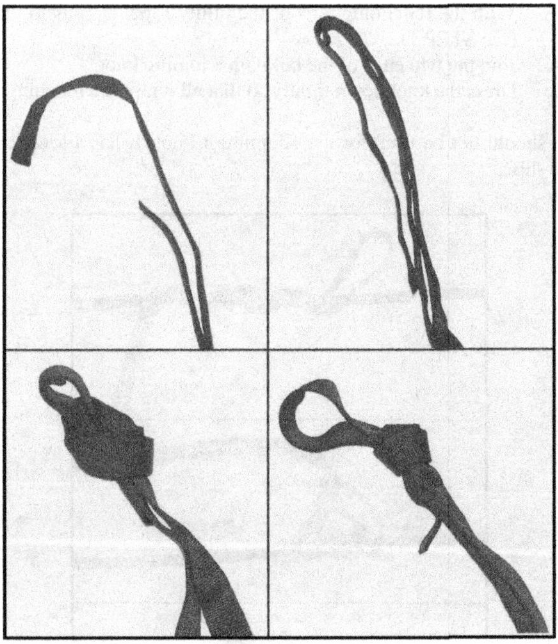

Figure 4-28. Frost knot.

b. **Checkpoints.**

(1) The tails of the webbing run in opposite directions.

(2) Three strands of webbing are formed into a tight overhand knot.

(3) There is a bight and tail exiting the top of the overhand knot.

4-29. GIRTH HITCH

The girth hitch is used to attach a runner to an anchor or piece of equipment (Figure 4-29). It is a special-purpose knot.

a. **Tying the Knot.**

STEP 1: Form a bight.
STEP 2: Bring the runner back through the bight.
STEP 3: Cinch the knot tightly.

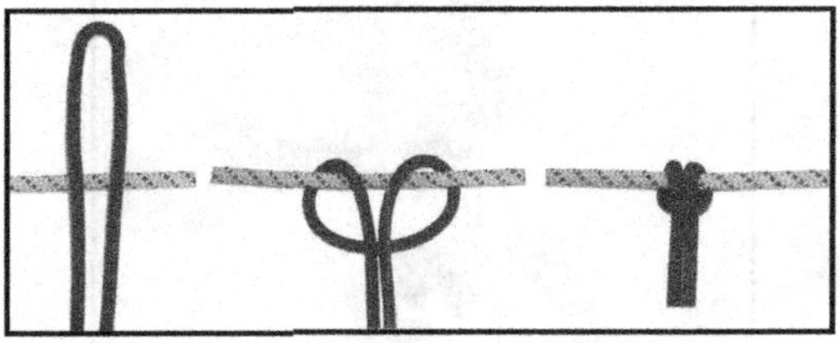

Figure 4-29. Girth hitch.

b. **Checkpoint.**

(1) Two wraps exist with a locking bar running across the wraps.
(2) The knot is dressed tightly.

4-30. MUNTER HITCH

The munter hitch, when used in conjunction with a pear-shaped locking carabiner, is used to form a mechanical belay (Figure 4-30).

a. **Tying the Knot.**

STEP 1. Hold the rope in both hands, palms down about 12 inches apart.
STEP 2. With the right hand, form a loop away from the body toward the left hand. Hold the loop with the left hand.
STEP 3. With the right hand, place the rope that comes from the bottom of the loop over the top of the loop.
STEP 4. Place the bight that has just been formed around the rope into the pear shaped carabiner. Lock the locking mechanism.

b. **Check Points.**

(1) A bight passes through the carabiner, with the closed end around the standing or running part of the rope.
(2) The carabiner is locked.

Figure 4-30. Munter hitch.

4-31. RAPPEL SEAT

The rappel seat is an improvised seat rappel harness made of rope (Figure 4-31, page 4-32). It usually requires a sling rope 14 feet or longer.

a. **Tying the Knot.**

STEP 1. Find the middle of the sling rope and make a bight.

STEP 2. Decide which hand will be used as the brake hand and place the bight on the opposite hip.

STEP 3. Reach around behind and grab a single strand of rope. Bring it around the waist to the front and tie two overhands on the other strand of rope, thus creating a loop around the waist.

STEP 4. Pass the two ends between the legs, ensuring they do not cross.

STEP 5. Pass the two ends up under the loop around the waist, bisecting the pocket flaps on the trousers. Pull up on the ropes, tightening the seat.

STEP 6. From rear to front, pass the two ends through the leg loops creating a half hitch on both hips.

STEP 7. Bring the longer of the two ends across the front to the nonbrake hand hip and secure the two ends with a square knot safetied with overhand knots. Tuck any excess rope in the pocket below the square knot.

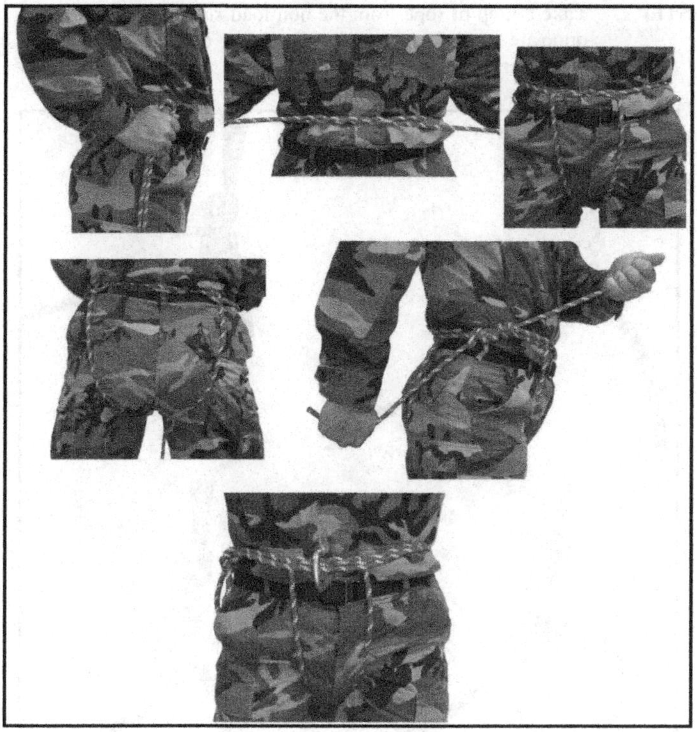

Figure 4-31. Rappel seat.

b. **Check Points.**

(1) There are two overhand knots in the front.

(2) The ropes are not crossed between the legs.

(3) A half hitch is formed on each hip.

(4) Seat is secured with a square knot with overhand safeties on the non-brake hand side.

(5) There is a minimum 4-inch pigtail after the overhand safeties are tied.

4-32. GUARDE KNOT

The guarde knot (ratchet knot, alpine clutch) is a special purpose knot primarily used for hauling systems or rescue (Figure 4-32). The knot works in only one direction and cannot be reversed while under load.

a. **Tying the Knot.**

STEP 1. Place a bight of rope into the two anchored carabiners (works best with two like carabiners, preferably ovals).

STEP 2. Take a loop of rope from the non-load side and place it down into the opposite cararabiner so that the rope comes out between the two carabiners.

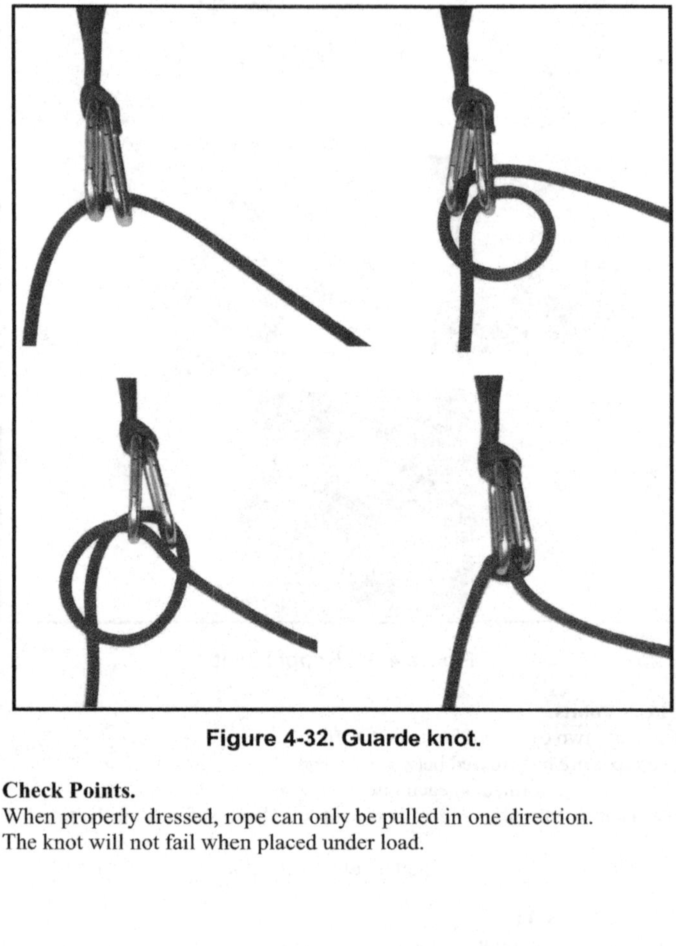

Figure 4-32. Guarde knot.

b. **Check Points.**

(1) When properly dressed, rope can only be pulled in one direction.

(2) The knot will not fail when placed under load.

CHAPTER 5

ANCHORS

This chapter discusses different types of anchors and their application in rope systems and climbing. Proper selection and placement of anchors is a critical skill that requires a great deal of practice. Failure of any system is most likely to occur at the anchor point. If the anchor is not strong enough to support the intended load, it will fail. Failure is usually the result of poor terrain features selected for the anchor point, or the equipment used in rigging the anchor was placed improperly or in insufficient amounts.

When selecting or constructing anchors, always try to make sure the anchor is "bombproof." A bombproof anchor is stronger than any possible load that could be placed on it. An anchor that has more strength than the climbing rope is considered bombproof.

Section I. NATURAL ANCHORS

Natural anchors should be considered for use first. They are usually strong and often simple to construct with minimal use of equipment. Trees, boulders, and other terrain irregularities are already in place and simply require a method of attaching the rope. However, natural anchors should be carefully studied and evaluated for stability and strength before use. Sometimes the climbing rope is tied directly to the anchor, but under most circumstances a sling is attached to the anchor and then the climbing rope is attached to the sling with a carabiner(s). (See paragraph 5-7 for slinging techniques.)

5-1. TREES

Trees are probably the most widely used of all natural anchors depending on the terrain and geographical region (Figure 5-1). However, trees must be carefully checked for suitability.

a. In rocky terrain, trees usually have a shallow root system. This can be checked by pushing or tugging on the tree to see how well it is rooted. Anchoring as low as possible to prevent excess leverage on the tree may be necessary.

b. Use padding on soft, sap producing trees to keep sap off ropes and slings.

Figure 5-1. Trees used as anchors.

5-2. BOULDERS

Boulders and rock nubbins make ideal anchors (Figure 5-2). The rock can be firmly tapped with a piton hammer to ensure it is solid. Sedimentary and other loose rock formations are not stable. Talus and scree fields are an indicator that the rock in the area is not solid. All areas around the rock formation that could cut the rope or sling should be padded.

Figure 5-2. Boulders used as anchors.

5-3. CHOCKSTONES

A chockstone is a rock that is wedged in a crack because the crack narrows downward (Figure 5-3). Chockstones should be checked for strength, security, and crumbling and should always be tested before use. All chockstones must be solid and strong enough to support the load. They must have maximum surface contact and be well tapered with the surrounding rock to remain in position.

a. Chockstones are often directional—they are secure when pulled in one direction but may pop out if pulled in another direction.

b. A creative climber can often make his own chockstone by wedging a rock into position, tying a rope to it, and clipping on a carabiner.

c. Slings should not be wedged between the chockstone and the rock wall since a fall could cut the webbing runner.

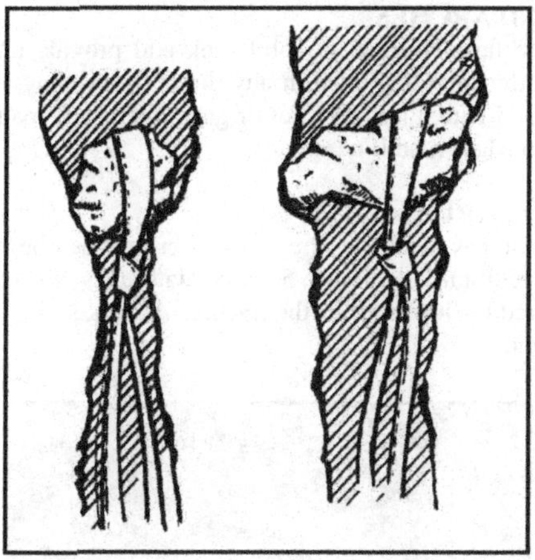

Figure 5-3. Chockstones.

5-4. ROCK PROJECTIONS

Rock projections (sometimes called nubbins) often provide suitable protection (Figure 5-4). These include blocks, flakes, horns, and spikes. If rock projections are used, their firmness is important. They should be checked for cracks or weathering that may impair their firmness. If any of these signs exist, the projection should be avoided.

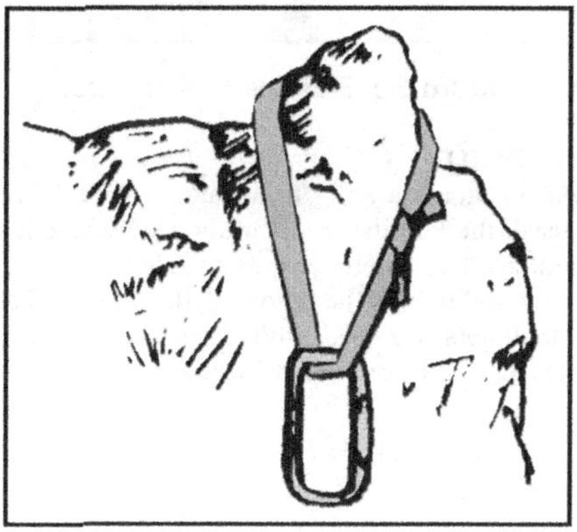

Figure 5-4. Rock projections.

5-5. TUNNELS AND ARCHES

Tunnels and arches are holes formed in solid rock and provide one of the more secure anchor points because they can be pulled in any direction. A sling is threaded through the opening hole and secured with a joining knot or girth hitch. The load-bearing hole must be strong and free of sharp edges (pad if necessary).

5-6. BUSHES AND SHRUBS

If no other suitable anchor is available, the roots of bushes can be used by routing a rope around the bases of several bushes (Figure 5-5). As with trees, the anchoring rope is placed as low as possible to reduce leverage on the anchor. All vegetation should be healthy and well rooted to the ground.

Figure 5-5. Bushes and shrubs.

5-7. SLINGING TECHNIQUES

Three methods are used to attach a sling to a natural anchor—drape, wrap, and girth. Whichever method is used, the knot is set off to the side where it will not interfere with normal carabiner movement. The carabiner gate should face away from the ground and open away from the anchor for easy insertion of the rope. When a locking carabiner cannot be used, two carabiners are used with gates opposed. Correctly opposed gates should open on opposite sides and form an "X" when opened (Figure 5-6).

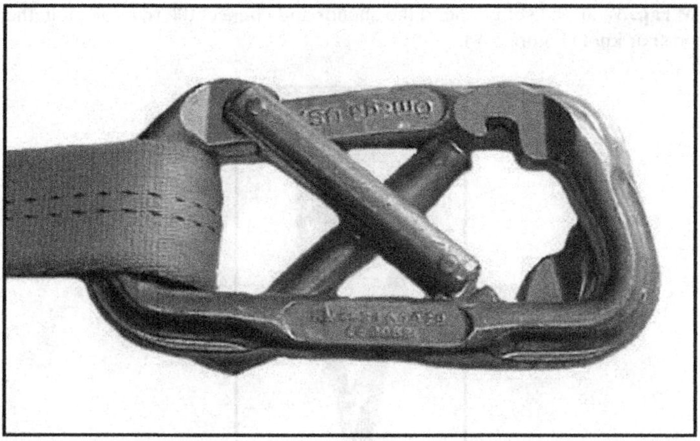

Figure 5-6. Correctly opposed carabiners.

a. **Drape.** Drape the sling over the anchor (Figure 5-7). Untying the sling and routing it around the anchor and then retying is still considered a drape.

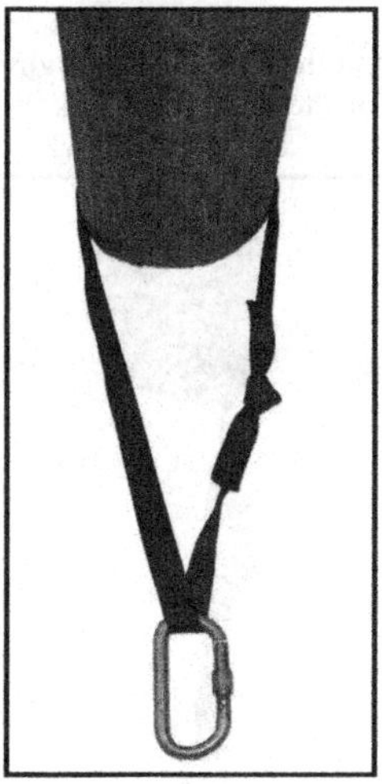

Figure 5-7. Drape.

b. **Wrap.** Wrap the sling around the anchor and connect the two ends together with a carabiner(s) or knot (Figure 5-8).

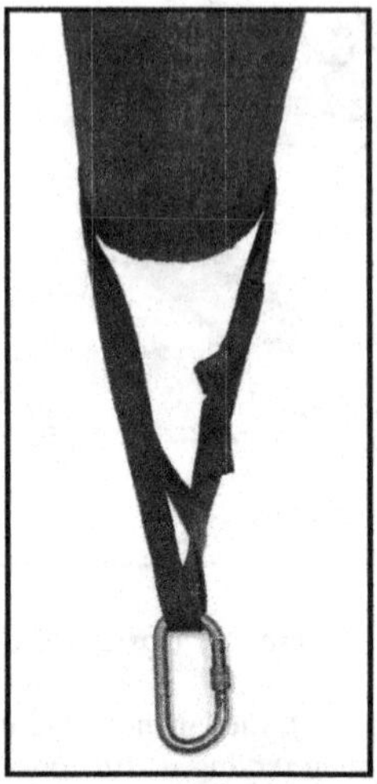

Figure 5-8. Wrap.

c. **Girth.** Tie the sling around the anchor with a girth hitch (Figure 5-9). Although a girth hitch reduces the strength of the sling, it allows the sling to remain in position and not slide on the anchor.

Figure 5-9. Girth.

Section II. ANCHORING WITH THE ROPE

The climbing or installation rope can be tied directly to the anchor using several different techniques. This requires less equipment, but also sacrifices some rope length to tie the anchor. The rope can be tied to the anchor using an appropriate anchor knot such as a bowline or a rerouted figure eight. Round turns can be used to help keep the rope in position on the anchor. A tensionless anchor can be used in high-load installations where tension on the attachment point and knot is undesirable.

5-8. ROPE ANCHOR

When tying the climbing or installation rope around an anchor, the knot should be placed approximately the same distance away from the anchor as the diameter of the anchor (Figure 5-10). The knot shouldn't be placed up against the anchor because this can stress and distort the knot under tension.

Figure 5-10. Rope tied to anchor with anchor knot.

5-9. TENSIONLESS ANCHOR

The tensionless anchor is used to anchor the rope on high-load installations such as bridging and traversing (Figure 5-11, page 5-8). The wraps of the rope around the anchor absorb the tension of the installation and keep the tension off the knot and carabiner. The anchor is usually tied with a minimum of four wraps, more if necessary, to absorb the tension. A smooth anchor may require several wraps, whereas a rough barked tree might only require a few. The rope is wrapped from top to bottom. A fixed loop is placed into the end of the rope and attached loosely back onto the rope with a carabiner.

Figure 5-11. Tensionless anchor.

Section III. ARTIFICIAL ANCHORS

Using artificial anchors becomes necessary when natural anchors are unavailable. The art of choosing and placing good anchors requires a great deal of practice and experience. Artificial anchors are available in many different types such as pitons, chocks, hexcentrics, and SLCDs. Anchor strength varies greatly; the type used depends on the terrain, equipment, and the load to be placed on it.

5-10. DEADMAN

A "deadman" anchor is any solid object buried in the ground and used as an anchor.

a. An object that has a large surface area and some length to it works best. (A hefty timber, such as a railroad tie, would be ideal.) Large boulders can be used, as well as a bundle of smaller tree limbs or poles. As with natural anchors, ensure timbers and tree limbs are not dead or rotting and that boulders are solid. Equipment, such as skis, ice axes, snowshoes, and ruck sacks, can also be used if necessary.

b. In extremely hard, rocky terrain (where digging a trench would be impractical, if not impossible) a variation of the deadman anchor can be constructed by building above the ground. The sling is attached to the anchor, which is set into the ground as deeply as possible. Boulders are then stacked on top of it until the anchor is strong enough for the load. Though normally not as strong as when buried, this method can work well for light-load installations as in anchoring a hand line for a stream crossing.

Note: Artificial anchors, such as pitons and bolts, are not widely accepted for use in all areas because of the scars they leave on the rock and the environment. Often they are left in place and become unnatural, unsightly fixtures in the natural environment. For training planning, local laws and courtesies should be taken into consideration for each area of operation.

5-11. PITONS

Pitons have been in use for over 100 years. Although still available, pitons are not used as often as other types of artificial anchors due primarily to their impact on the environment. Most climbers prefer to use chocks, SLCDs and other artificial anchors rather than pitons because they do not scar the rock and are easier to remove. Eye protection should always be worn when driving a piton into rock.

Note: The proper use and placement of pitons, as with any artificial anchor, should be studied, practiced, and tested while both feet are firmly on the ground and there is no danger of a fall.

a. **Advantages.** Some advantages in using pitons are:

- Depending on type and placement, pitons can support multiple directions of pull.
- Pitons are less complex than other types of artificial anchors.
- Pitons work well in thin cracks where other types of artificial anchors do not.

b. **Disadvantages.** Some disadvantages in using pitons are:

- During military operations, the distinct sound created when hammering pitons is a tactical disadvantage.
- Due to the expansion force of emplacing a piton, the rock could spread apart or break causing an unsafe condition.
- Pitons are more difficult to remove than other types of artificial anchors.
- Pitons leave noticeable scars on the rock.
- Pitons are easily dropped if not tied off when being used.

c. **Piton Placement.** The proper positioning or placement of pitons is critical. (Figure 5-12, page 5-10, shows examples of piton placement.) Usually a properly sized piton for a rock crack will fit one half to two thirds into the crack before being driven with the piton hammer. This helps ensure the depth of the crack is adequate for the size piton selected. As pitons are driven into the rock the pitch or sound that is made will change with each hammer blow, becoming higher pitched as the piton is driven in.

(1) Test the rock for soundness by tapping with the hammer. Driving pitons in soft or rotten rock is not recommended. When this type of rock must be used, clear the loose rock, dirt, and debris from the crack before driving the piton completely in.

(2) While it is being driven, attach the piton to a sling with a carabiner (an old carabiner should be used, if available) so that if the piton is knocked out of the crack, it will not be lost. The greater the resistance overcome while driving the piton, the firmer the anchor will be. The holding power depends on the climber placing the piton in a sound crack, and on the type of rock. The piton should not spread the rock, thereby loosening the emplacement.

Note: Pitons that have rings as attachment points might not display much change in sound as they are driven in as long as the ring moves freely.

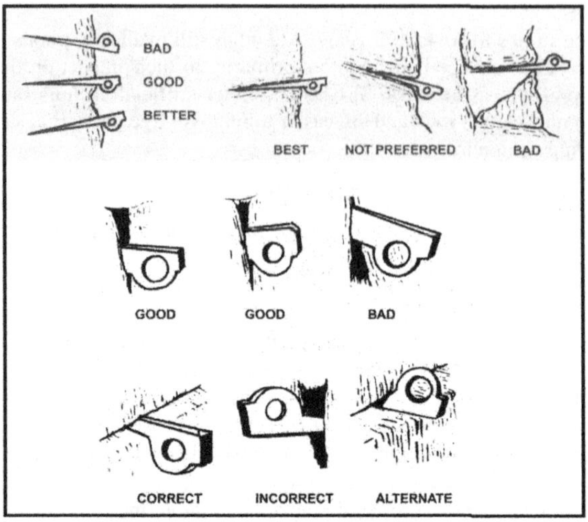

Figure 5-12. Examples of piton placements.

(3) Military mountaineers should practice emplacing pitons using either hand. Sometimes a piton cannot be driven completely into a crack, because the piton is too long. Therefore, it should be tied off using a hero-loop (an endless piece of webbing) (Figure 5-13). Attach this loop to the piton using a girth hitch at the point where the piton enters the rock so that the girth hitch is snug against the rock. Clip a carabiner into the loop.

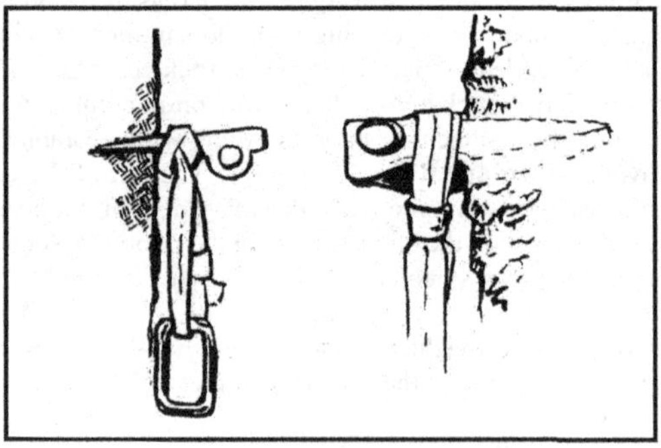

Figure 5-13. Hero-loop.

d. **Testing.** To test pitons pull up about 1 meter of slack in the climbing rope or use a sling. Insert this rope into a carabiner attached to the piton, then grasp the rope at least 1/2 meter from the carabiner. Jerk vigorously upward, downward, to each side, and then outward while observing the piton for movement. Repeat these actions as many times as necessary. Tap the piton to determine if the pitch has changed. If the pitch has changed greatly, drive the piton in as far as possible. If the sound regains its original pitch, the emplacement is probably safe. If the piton shows any sign of moving or if, upon driving it, there is any question of its soundness, drive it into another place. Try to be in a secure position before testing. This procedure is intended for use in testing an omni-directional anchor (one that withstands a pull in any direction). When a directional anchor (pull in one direction) is used, as in most free and direct-aid climbing situations, and when using chocks, concentrate the test in the direction that force will be applied to the anchor.

e. **Removing Pitons.** Attach a carabiner and sling to the piton before removal to eliminate the chance of dropping and losing it. Tap the piton firmly along the axis of the crack in which it is located. Alternate tapping from both sides while applying steady pressure. Pulling out on the attached carabiner eventually removes the piton (Figure 5-14).

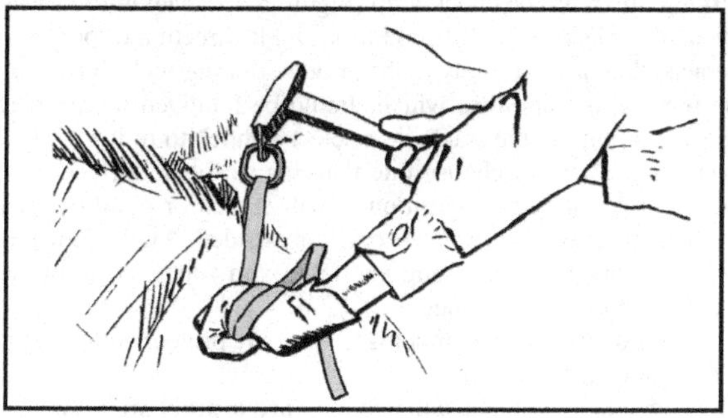

Figure 5-14. Piton removal.

f. **Reusing Pitons.** Soft iron pitons that have been used, removed, and straightened may be reused, but they must be checked for strength. In training areas, pitons already in place should not be trusted since weather loosens them in time. Also, they may have been driven poorly the first time. Before use, test them as described above and drive them again until certain of their soundness.

5-12. CHOCKS

Chock craft has been in use for many decades. A natural chockstone, having fallen and wedged in a crack, provides an excellent anchor point. Sometimes these chockstones are in unstable positions, but can be made into excellent anchors with little adjustment. Chock craft is an art that requires time and technique to master—simple in theory, but complex in practice. Imagination and resourcefulness are key principles to chock craft. The skilled climber must understand the application of mechanical advantage, vectors, and other forces that affect the belay chain in a fall.

a. **Advantages.** The advantages of using chocks are:

- Tactically quiet installation and recovery.
- Usually easy to retrieve and, unless severely damaged, are reusable.
- Light to carry.
- Easy to insert and remove.
- Minimal rock scarring as opposed to pitons.
- Sometimes can be placed where pitons cannot (expanding rock flakes where pitons would further weaken the rock).

b. **Disadvantages.** The disadvantages of using chocks are:

- May not fit in thin cracks, which may accept pitons.
- Often provide only one direction of pull.
- Practice and experience necessary to become proficient in proper placement.

c. **Placement.** The principles of placing chocks are to find a crack with a constriction at some point, place a chock of appropriate size above and behind the constriction, and set the chock by jerking down on the chock loop (Figure 5-15). Maximum surface contact with a tight fit is critical. Chocks are usually good for a single direction of pull.

(1) Avoid cracks that have crumbly (soft) or deteriorating rock, if possible. Some cracks may have loose rock, grass, and dirt, which should be removed before placing the chock. Look for a constriction point in the crack, then select a chock to fit it.

(2) When selecting a chock, choose one that has as much surface area as possible in contact with the rock. A chock resting on one small crystal or point of rock is likely to be unsafe. A chock that sticks partly out of the crack is avoided. Avoid poor protection. Ensure that the chock has a wire or runner long enough; extra ropes, cord, or webbing may be needed to extend the length of the runner.

(3) End weighting of the placement helps to keep the protection in position. A carabiner often provides enough weight

(4) Parallel-sided cracks without constrictions are a problem. Chocks designed to be used in this situation rely on camming principles to remain emplaced. Weighting the emplacement with extra hardware is often necessary to keep the chocks from dropping out.

(a) Emplace the wedge-shaped chock above and behind the constriction; seat it with a sharp downward tug.

(b) Place a camming chock with its narrow side into the crack, then rotate it to the attitude it will assume under load; seat it with a sharp downward tug.

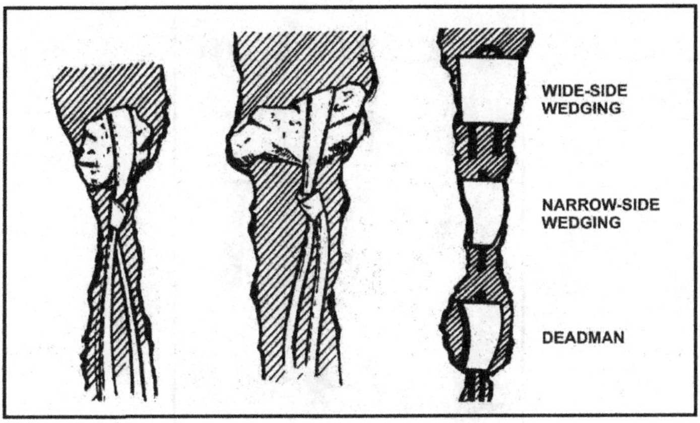

Figure 5-15. Chock placements.

d. **Testing.** After seating a chock, test it to ensure it remains in place. A chock that falls out when the climber moves past it is unsafe and offers no protection. To test it, firmly pull the chock in every anticipated direction of pull. Some chock placements fail in one or more directions; therefore, use pairs of chocks in opposition.

5-13. SPRING-LOADED CAMMING DEVICE

The SLCD offers quick and easy placement of artificial protection. It is well suited in awkward positions and difficult placements, since it can be emplaced with one hand. It can usually be placed quickly and retrieved easily (Figure 5-16, page 5-14).

a. To emplace an SLCD hold the device in either hand like a syringe, pull the retractor bar back, place the device into a crack, and release the retractor bar. The SLCD holds well in parallel-sided hand- and fist-sized cracks. Smaller variations are available for finger-sized cracks.

b. Careful study of the crack should be made before selecting the device for emplacement. It should be placed so that it is aligned in the direction of force applied to it. It should not be placed any deeper than is needed for secure placement, since it may be impossible to reach the extractor bar for removal. An SLCD should be extended with a runner and placed so that the direction of pull is parallel to the shaft; otherwise, it may rotate and pull out. The versions that have a semi-rigid wire cable shaft allow for greater flexibility and usage, without the danger of the shaft snapping off in a fall.

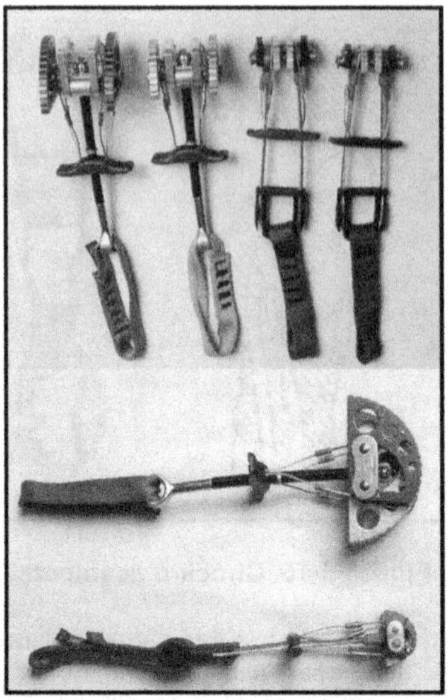

Figure 5-16. SLCD placements.

5-14. BOLTS

Bolts are often used in fixed-rope installations and in aid climbing where cracks are not available.

a. Bolts provide one of the most secure means of establishing protection. The rock should be inspected for evidence of crumbling, flaking, or cracking, and should be tested with a hammer. Emplacing a bolt with a hammer and a hand drill is a time-consuming and difficult process that requires drilling a hole in the rock deeper than the length of the bolt. This normally takes more than 20 minutes for one hole. Electric or even gas-powered drills can be used to greatly shorten drilling time. However, their size and weight can make them difficult to carry on the climbing route.

b. A hanger (carrier) and nut are placed on the bolt, and the bolt is inserted and then driven into the hole. A climber should never hammer on a bolt to test or "improve" it, since this permanently weakens it. Bolts should be used with carriers, carabiners, and runners.

c. When using bolts, the climber uses a piton hammer and hand drill with a masonry bit for drilling holes. Some versions are available in which the sleeve is hammered and turned into the rock (self-drilling), which bores the hole. Split bolts and expanding sleeves are common bolts used to secure hangers and carriers (Figure 5-17). Surgical tubing is useful in blowing dust out of the holes. Nail type bolts are emplaced by driving the nail with a hammer to expand the sleeve against the wall of the drilled hole. Safety glasses should always be worn when emplacing bolts.

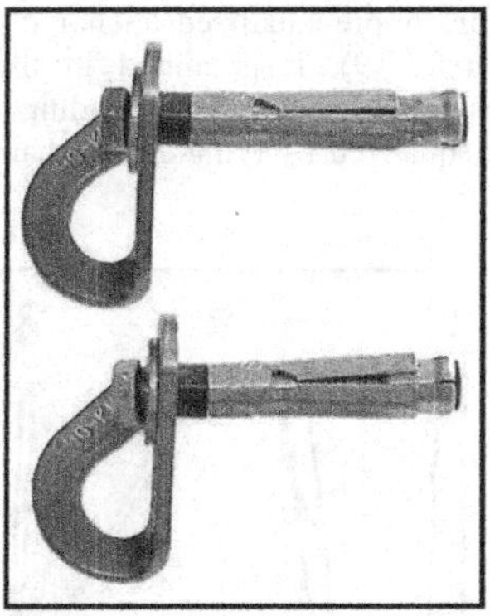

Figure 5-17. Bolt with expanding sleeve.

5-15. EQUALIZING ANCHORS

Equalizing anchors are made up of more than one anchor point joined together so that the intended load is shared equally. This not only provides greater anchor strength, but also adds redundancy or backup because of the multiple points.

a. **Self-equalizing Anchor.** A self-equalizing anchor will maintain an equal load on each individual point as the direction of pull changes (Figure 5-18). This is sometimes used in rappelling when the route must change left or right in the middle of the rappel. A self-equalizing anchor should only be used when necessary because if any one of the individual points fail, the anchor will extend and shock-load the remaining points or even cause complete anchor failure.

Figure 5-18. Self-equalizing anchors.

b. **Pre-equalized Anchor.** A pre-equalized anchor distributes the load equally to each individual point (Figure 5-19). It is aimed in the direction of the load. A pre-equalized anchor prevents extension and shock-loading of the anchor if an individual point fails. An anchor is pre-equalized by tying an overhand or figure-eight knot in the webbing or sling.

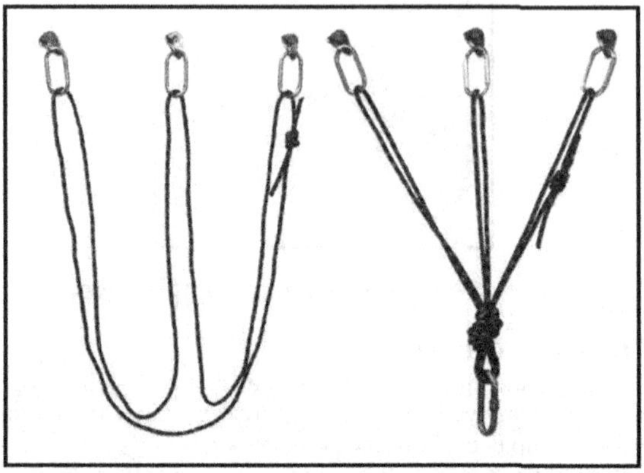

Figure 5-19. Pre-equalized anchor.

Note: When using webbing or slings, the angles of the webbing or slings directly affect the load placed on an anchor. An angle greater than 90 degrees can result in anchor failure (Figure 5-20).

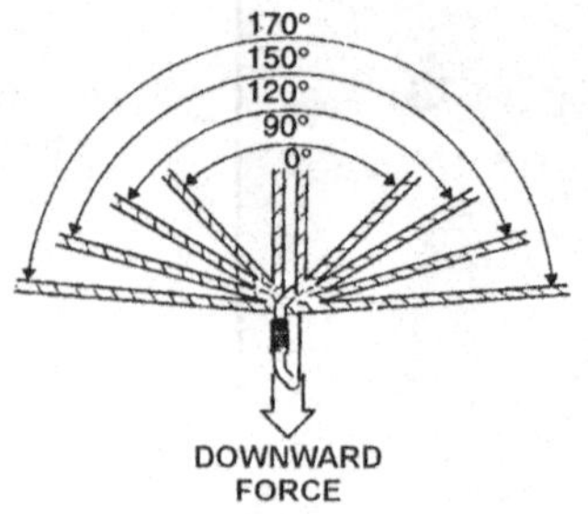

TENSION IN MULTIPLE ANCHOR RIGGING	
ANGLE	RESULTING LEG TENSION*
170°	1150%
150°	200%
120°	100%
90°	70%
0°	50%

On each leg relative to downward force.

Figure 5-20. Effects of angles on an anchor.

CHAPTER 6
CLIMBING

A steep rock face is a terrain feature that can be avoided most of the time through prior planning and good route selection. Rock climbing can be time consuming, especially for a larger unit with a heavy combat load. It can leave the climbing party totally exposed to weather, terrain hazards, and the enemy for the length of the climb.

Sometimes steep rock cannot be avoided. Climbing relatively short sections of steep rock (one or two pitches) may prove quicker and safer than using alternate routes. A steep rock route would normally be considered an unlikely avenue of approach and, therefore, might be weakly defended or not defended at all.

All personnel in a unit preparing for deployment to mountainous terrain should be trained in the basics of climbing. Forward observers, reconnaissance personnel, and security teams are a few examples of small units who may require rock climbing skills to gain their vantage points in mountainous terrain. Select personnel demonstrating the highest degree of skill and experience should be trained in roped climbing techniques. These personnel will have the job of picking and "fixing" the route for the rest of the unit.

Rock climbing has evolved into a specialized "sport" with a wide range of varying techniques and styles. This chapter focuses on the basics most applicable to military operations.

Section I. CLIMBING FUNDAMENTALS

A variety of refined techniques are used to climb different types of rock formations. The foundation for all of these styles is the art of climbing. Climbing technique stresses climbing with the weight centered over the feet, using the hands primarily for balance. It can be thought of as a combination of the balanced movement required to walk a tightrope and the technique used to ascend a ladder. No mountaineering equipment is required; however, the climbing technique is also used in roped climbing.

6-1. ROUTE SELECTION

The experienced climber has learned to climb with the "eyes." Even before getting on the rock, the climber studies all possible routes, or "lines," to the top looking for cracks, ledges, nubbins, and other irregularities in the rock that will be used for footholds and handholds, taking note of any larger ledges or benches for resting places. When picking the line, he mentally climbs the route, rehearsing the step-by-step sequence of movements that will be required to do the climb, ensuring himself that the route has an adequate number of holds and the difficulty of the climb will be well within the limit of his ability.

6-2. TERRAIN SELECTION FOR TRAINING

Route selection for military climbing involves picking the easiest and quickest possible line for all personnel to follow. However, climbing skill and experience can only be developed by increasing the length and difficulty of routes as training progresses. In the training environment, beginning lessons in climbing should be performed CLOSE to the ground on lower-angled rock with plenty of holds for the hands and feet. Personnel not climbing can act as "spotters" for those climbing. In later lessons, a "top-rope" belay can be used for safety, allowing the individual to increase the length and difficulty of the climb under the protection of the climbing rope.

6-3. PREPARATION

In preparation for climbing, the boot soles should be dry and clean. A small stick can be used to clean out dirt and small rocks that might be caught between the lugs of the boot sole. If the soles are wet or damp, dry them off by stomping and rubbing the soles on clean, dry rock. All jewelry should be removed from the fingers. Watches and bracelets can interfere with hand placements and may become damaged if worn while climbing. Helmets should be worn to protect the head from injury if an object, such as a rock or climbing gear, falls from climbers above. Most climbing helmets are not designed to provide protection from impact to the head if the wearer falls, but will provide a minimal amount of protection if a climber comes in contact with the rock during climbing.

CAUTION

Rings can become caught on rock facial features and or lodged into cracks, which could cause injuries during a slip or fall.

6-4. SPOTTING

Spotting is a technique used to add a level of safety to climbing without a rope. A second man stands below and just outside of the climbers fall path and helps (spots) the climber to land safely if he should fall. Spotting is only applicable if the climber is not going above the spotters head on the rock. Beyond that height a roped climbing should be conducted. If an individual climbs beyond the effective range of the spotter(s), he has climbed TOO HIGH for his own safety. The duties of the spotter are to help prevent the falling climber from impacting the head and or spine, help the climber land feet first, and reduce the impact of a fall.

CAUTION

The spotter should not catch the climber against the rock because additional injuries could result. If the spotter pushes the falling climber into the rock, deep abrasions of the skin or knee may occur. Ankle joints could be twisted by the fall if the climber's foot remained high on the rock. The spotter might be required to fully support the weight of the climber causing injury to the spotter.

6-5. CLIMBING TECHNIQUE

Climbing involves linking together a series of movements based on foot and hand placement, weight shift, and movement. When this series of movements is combined correctly, a smooth climbing technique results. This technique reduces excess force on the limbs, helping to minimize fatigue. The basic principle is based on the five body parts described here.

a. **Five Body Parts.** The five body parts used for climbing are the right hand, left hand, right foot, left foot, and body (trunk). The basic principle to achieve smooth climbing is to move only one of the five body parts at a time. The trunk is not moved in conjunction with a foot or in conjunction with a hand, a hand is not moved in conjunction with a foot, and so on. Following this simple technique forces both legs to do all the lifting simultaneously.

b. **Stance or Body Position.** Body position is probably the single most important element to good technique. A relaxed, comfortable stance is essential. (Figure 6-1 shows a correct climbing stance, and Figure 6-2, page 6-4, shows an incorrect stance.) The body should be in a near vertical or erect stance with the weight centered over the feet. Leaning in towards the rock will cause the feet to push outward, away from the rock, resulting in a loss of friction between the boot sole and rock surface. The legs are straight and the heels are kept low to reduce fatigue. Bent legs and tense muscles tire quickly. If strained for too long, tense muscles may vibrate uncontrollably. This vibration, known as "Elvis-ing" or "sewing-machine leg" can be cured by straightening the leg, lowering the heel, or moving on to a more restful position. The hands are used to maintain balance. Keeping the hands between waist and shoulder level will reduce arm fatigue.

Figure 6-1. Correct climbing stance—balanced over both feet.

Figure 6-2. Incorrect stance—stretched out.

(1) Whenever possible, three points of contact are maintained with the rock. Proper positioning of the hips and shoulders is critical. When using two footholds and one handhold, the hips and shoulders should be centered over both feet. In most cases, as the climbing progresses, the body is resting on one foot with two handholds for balance. The hips and shoulders must be centered over the support foot to maintain balance, allowing the "free" foot to maneuver.

(2) The angle or steepness of the rock also determines how far away from the rock the hips and shoulders should be. On low-angle slopes, the hips are moved out away from the rock to keep the body in balance with the weight over the feet. The shoulders can be moved closer to the rock to reach handholds. On steep rock, the hips are pushed closer to the rock. The shoulders are moved away from the rock by arching the back. The body is still in balance over the feet and the eyes can see where the hands need to go. Sometimes, when footholds are small, the hips are moved back to increase friction between the foot and the rock. This is normally done on quick, intermediate holds. It should be avoided in the rest position as it places more weight on the arms and hands. When weight must be placed on handholds, the arms should be kept straight to reduce fatigue. Again, flexed muscles tire quickly.

c. **Climbing Sequence.** The steps defined below provide a complete sequence of events to move the entire body on the rock. These are the basic steps to follow for a smooth climbing technique. Performing these steps in this exact order will not always be necessary because the nature of the route will dictate the availability of hand and foot placements. The basic steps are weight, shift, and movement (movement being either the foot, hand, or body). (A typical climbing sequence is shown in Figure 6-3, pages 6-6 through 6-8.)

STEP ONE: Shift the weight from both feet to one foot. This will allow lifting of one foot with no effect on the stance.

STEP TWO:	Lift the unweighted foot and place it in a new location, within one to two feet of the starting position, with no effect on body position or balance (higher placement will result in a potentially higher lift for the legs to make, creating more stress, and is called a high step) The trunk does not move during foot movement.
STEP THREE:	Shift the weight onto both feet. (Repeat steps 1 through 3 for remaining foot.)
STEP FOUR:	Lift the body into a new stance with both legs.
STEP FIVE:	Move one hand to a new position between waist and head height. During this movement, the trunk should be completely balanced in position and the removed hand should have no effect on stability.
STEP SIX:	Move the remaining hand as in Step 5.

Now the entire body is in a new position and ready to start the process again. Following these steps will prevent lifting with the hands and arms, which are used to maintain stance and balance. If both legs are bent, leg extension can be performed as soon as one foot has been moved. Hand movements can be delayed until numerous foot movements have been made, which not only creates shorter lifts with the legs, but may allow a better choice for the next hand movements because the reach will have increased.

Figure 6-3. Typical climbing sequence.

Figure 6-3. Typical climbing sequence (continued).

Figure 6-3. Typical climbing sequence (continued).

(1) Many climbers will move more than one body part at a time, usually resulting in lifting the body with one leg or one leg and both arms. This type of lifting is inefficient, requiring one leg to perform the work of two or using the arms to lift the body. Proper climbing technique is lifting the body with the legs, not the arms, because the legs are much stronger.

(2) When the angle of the rock increases, these movements become more critical. Holding or pulling the body into the rock with the arms and hands may be necessary as the angle increases (this is still not lifting with the arms). Many climbing routes have angles greater than ninety degrees (overhanging) and the arms are used to support partial body weight. The same technique applies even at those angles.

(3) The climber should avoid moving on the knees and elbows. Other than being uncomfortable, even painful, to rest on, these bony portions of the limbs offer little friction and "feel" on the rock.

6-6. SAFETY PRECAUTIONS

The following safety precautions should be observed when rock climbing.

a. While ascending a seldom or never traveled route, you may encounter precariously perched rocks. If the rock will endanger your second, it may be possible to remove it from the route and trundle it, tossing it down. This is extremely dangerous to climbers below and should not be attempted unless you are absolutely sure no men are below. If you are not sure that the flight path is clear, do not do it. Never dislodge loose rocks carelessly. Should a rock become loose accidentally, immediately shout the warning "ROCK" to alert climbers below. Upon hearing the warning, personnel should seek immediate cover behind any rock bulges or overhangs available, or flatten themselves against the rock to minimize exposure.

b. Should a climber fall, he should do his utmost to maintain control and not panic. If on a low-angle climb, he may be able to arrest his own fall by staying in contact with the rock, grasping for any possible hold available. He should shout the warning "FALLING" to alert personnel below.

CAUTION
Grasping at the rock in a fall can result in serious injuries to the upper body. If conducting a roped climb, let the rope provide protection.

c. When climbing close to the ground and without a rope, a spotter can be used for safety. The duties of the spotter are to ensure the falling climber does not impact the head or spine, and to reduce the impact of a fall.

d. Avoid climbing directly above or below other climbers (with the exception of spotters). When personnel must climb at the same time, following the same line, a fixed rope should be installed.

e. Avoid climbing with gloves on because of the decreased "feel" for the rock. The use of gloves in the training environment is especially discouraged, while their use in the mountains is often mandatory when it is cold. A thin polypropylene or wool glove is best for rock climbing, although heavier cotton or leather work gloves are often used for belaying.

f. Be extremely careful when climbing on wet or moss-covered rock; friction on holds is greatly reduced.

g. Avoid grasping small vegetation for handholds; the root systems can be shallow and will usually not support much weight.

6-7. MARGIN OF SAFETY

Besides observing the standard safety precautions, the climber can avoid catastrophe by climbing with a wide margin of safety. The margin of safety is a protective buffer the climber places between himself and potential climbing hazards. Both subjective (personnel-related) and objective (environmental) hazards must be considered when applying the margin of safety. The leader must apply the margin of safety taking into account the strengths and weaknesses of the entire team or unit.

a. When climbing, the climber increases his margin of safety by selecting routes that are well within the limit of his ability. When leading a group of climbers, he selects a route well within the ability of the weakest member.

b. When the rock is wet, or when climbing in other adverse weather conditions, the climber's ability is reduced and routes are selected accordingly. When the climbing becomes difficult or exposed, the climber knows to use the protection of the climbing rope and belays. A lead climber increases his margin of safety by placing protection along the route to limit the length of a potential fall.

Section II. USE OF HOLDS

The climber should check each hold before use. This may simply be a quick, visual inspection if he knows the rock to be solid. When in doubt, he should grab and tug on the hold to test it for soundness BEFORE depending on it. Sometimes, a hold that appears weak can actually be solid as long as minimal force is applied to it, or the force is applied in a direction that strengthens it. A loose nubbin might not be strong enough to support the climber's weight, but it may serve as an adequate handhold. Be especially careful when climbing on weathered, sedimentary-type rock.

6-8. CLIMBING WITH THE FEET

"Climb with the feet and use the hands for balance" is extremely important to remember. In the early learning stages of climbing, most individuals will rely heavily on the arms, forgetting to use the feet properly. It is true that solid handholds and a firm grip are needed in some combination techniques; however, even the most strenuous techniques require good footwork and a quick return to a balanced position over one or both feet. Failure to climb any route, easy or difficult, is usually the result of poor footwork.

a. The beginning climber will have a natural tendency to look up for handholds. Try to keep the hands low and train your eyes to look down for footholds. Even the smallest irregularity in the rock can support the climber once the foot is positioned properly and weight is committed to it.

b. The foot remains on the rock as a result of friction. Maximum friction is obtained from a correct stance over a properly positioned foot. The following describes a few ways the foot can be positioned on the rock to maximize friction.

(1) ***Maximum Sole Contact.*** The principle of using full sole contact, as in mountain walking, also applies in climbing. Maximum friction is obtained by placing as much of the boot sole on the rock as possible. Also, the leg muscles can relax the most when the entire foot is placed on the rock. (Figure 6-4 shows examples of maximum and minimum sole contact.)

(a) Smooth, low-angled rock (slab) and rock containing large "bucket" holds and ledges are typical formations where the entire boot sole should be used.

(b) On some large holds, like bucket holds that extend deep into the rock, the entire foot cannot be used. The climber may not be able to achieve a balanced position if the foot is stuck too far underneath a bulge in the rock. In this case, placing only part of the foot on the hold may allow the climber to achieve a balanced stance. The key is to use as much of the boot sole as possible. Remember to keep the heels low to reduce strain on the lower leg muscles.

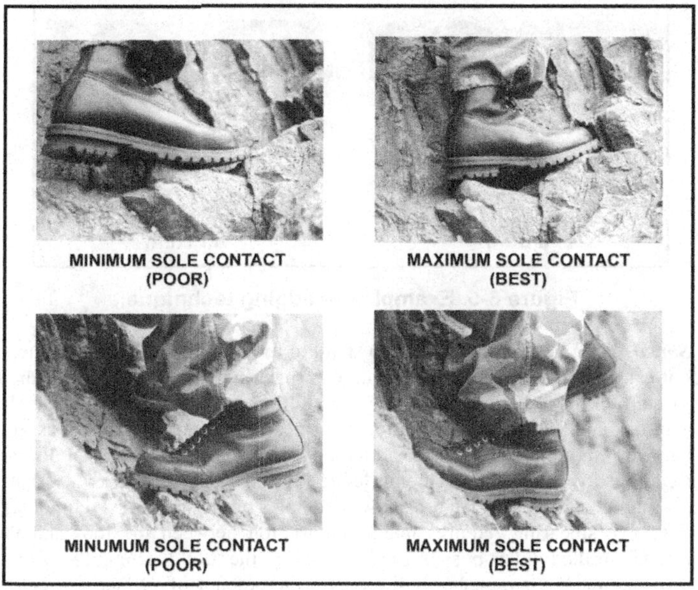

Figure 6-4. Examples of maximum and minimum sole contact.

(2) ***Edging.*** The edging technique is used where horizontal crack systems and other irregularities in the rock form small, well-defined ledges. The edge of the boot sole is placed on the ledge for the foothold. Usually, the inside edge of the boot or the edge area around the toes is used. Whenever possible, turn the foot sideways and use the entire inside edge of the boot. Again, more sole contact equals more friction and the legs can rest more when the heel is on the rock. (Figure 6-5, page 6-12, shows examples of the edging technique.)

(a) On smaller holds, edging with the front of the boot, or toe, may be used. Use of the toe is most tiring because the heel is off the rock and the toes support the climber's weight. Remember to keep the heel low to reduce fatigue. Curling and stiffening the toes in the boot increases support on the hold. A stronger position is usually obtained on small ledges by turning the foot at about a 45-degree angle, using the strength of the big toe and the ball of the foot.

(b) Effective edging on small ledges requires stiff-soled footwear. The stiffer the sole, the better the edging capability. Typical mountain boots worn by the US military have a relatively flexible lugged sole and, therefore, edging ability on smaller holds will be somewhat limited.

Figure 6-5. Examples of edging technique.

(3) ***Smearing.*** When footholds are too small to use a good edging technique, the ball of the foot can be "smeared" over the hold. The smearing technique requires the boot to adhere to the rock by deformation of the sole and by friction. Rock climbing shoes are specifically designed to maximize friction for smearing; some athletic shoes also work well. The Army mountain boot, with its softer sole, usually works better for smearing than for edging. Rounded, down-sloping ledges and low-angled slab rock often require good smearing technique. (Figure 6-6 shows examples of the smearing technique.)

(a) Effective smearing requires maximum friction between the foot and the rock. Cover as much of the hold as possible with the ball of the foot. Keeping the heel low will not only reduce muscle strain, but will increase the amount of surface contact between the foot and the rock.

(b) Sometimes flexing the ankles and knees slightly will place the climber's weight more directly over the ball of the foot and increase friction; however, this is more tiring and should only be used for quick, intermediate holds. The leg should be kept straight whenever possible.

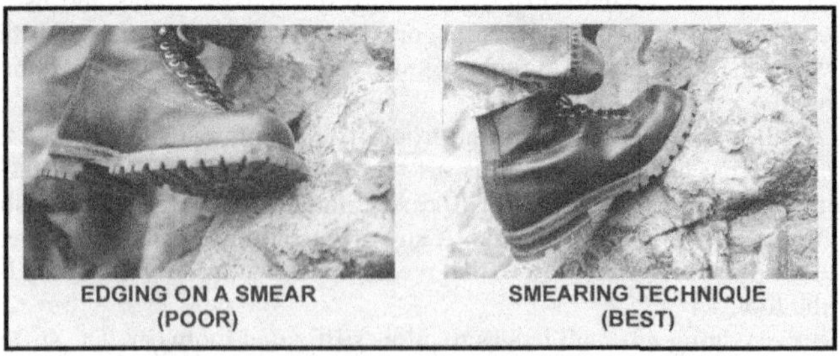

Figure 6-6. Examples of the smearing technique.

(4) ***Jamming.*** The jamming technique works on the same principal as chock placement. The foot is set into a crack in such a way that it "jams" into place, resisting a downward pull. The jamming technique is a specialized skill used to climb vertical or

near vertical cracks when no other holds are available on the rock face. The technique is not limited to just wedging the feet; fingers, hands, arms, even the entire leg or body are all used in the jamming technique, depending on the size of the crack. Jam holds are described in this text to broaden the range of climbing skills. Jamming holds can be used in a crack while other hand/foot holds are used on the face of the rock. Many cracks will have facial features, such as edges, pockets, and so on, inside and within reach. Always look or feel for easier to use features. (Figure 6-7 shows examples of jamming.)

(a) The foot can be jammed in a crack in different ways. It can be inserted above a constriction and set into the narrow portion, or it can be placed in the crack and turned, like a camming device, until it locks in place tight enough to support the climber's weight. Aside from these two basic ideas, the possibilities are endless. The toes, ball of the foot, or the entire foot can be used. Try to use as much of the foot as possible for maximum surface contact. Some positions are more tiring, and even more painful on the foot, than others. Practice jamming the foot in various ways to see what offers the most secure, restful position.

(b) Some foot jams may be difficult to remove once weight has been committed to them, especially if a stiffer sole boot is used. The foot is less likely to get stuck when it is twisted or "cammed" into position. When removing the boot from a crack, reverse the way it was placed to prevent further constriction.

FOOT JAM

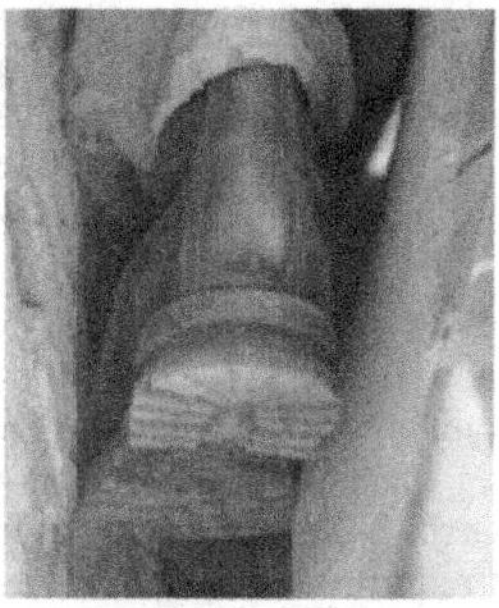

FOOT JAMMED INTO POSITION, HEEL IS PUSHED TO THE RIGHT

BOTH FEET JAMMED

Figure 6-7. Examples of jamming.

6-9. USING THE HANDS

The hands can be placed on the rock in many ways. Exactly how and where to position the hands and arms depends on what holds are available, and what configuration will best support the current stance as well as the movement to the next stance. Selecting handholds between waist and shoulder level helps in different ways. Circulation in the arms and hands is best when the arms are kept low. Secondly, the climber has less tendency to "hang" on his arms when the handholds are at shoulder level and below. Both of these contribute to a relaxed stance and reduce fatigue in the hands and arms.

a. As the individual climbs, he continually repositions his hands and arms to keep the body in balance, with the weight centered over the feet. On lower-angled rock, he may simply need to place the hands up against the rock and extend the arm to maintain balance; just like using an ice ax as a third point of contact in mountain walking. Sometimes, he will be able to push directly down on a large hold with the palm of the hand. More often though, he will need to "grip" the rock in some fashion and then push or pull against the hold to maintain balance.

b. As stated earlier, the beginner will undoubtedly place too much weight on the hands and arms. If we think of ourselves climbing a ladder, our body weight is on our legs. Our hands grip, and our arms pull on each rung only enough to maintain our balance and footing on the ladder. Ideally, this is the amount of grip and pull that should be used in climbing. Of course, as the size and availability of holds decreases, and the steepness of the rock approaches the vertical, the grip must be stronger and more weight might be placed on the arms and handholds for brief moments. The key is to move quickly from the smaller, intermediate holds to the larger holds where the weight can be placed back on the feet allowing the hands and arms to relax. The following describes some of the basic handholds and how the hand can be positioned to maximize grip on smaller holds.

(1) ***Push Holds.*** Push holds rely on the friction created when the hand is pushed against the rock. Most often a climber will use a push hold by applying "downward pressure" on a ledge or nubbin. This is fine, and works well; however, the climber should not limit his use of push holds to the application of down pressure. Pushing sideways, and on occasion, even upward on less obvious holds can prove quite secure. Push holds often work best when used in combination with other holds. Pushing in opposite directions and "push-pull" combinations are excellent techniques. (Figure 6-8 shows examples of push holds.)

(a) An effective push hold does not necessarily require the use of the entire hand. On smaller holds, the side of the palm, the fingers, or the thumb may be all that is needed to support the stance. Some holds may not feel secure when the hand is initially placed on them. The hold may improve or weaken during the movement. The key is to try and select a hold that will improve as the climber moves past it.

(b) Most push holds do not require much grip; however, friction might be increased by taking advantage of any rough surfaces or irregularities in the rock. Sometimes the strength of the hold can be increased by squeezing, or "pinching," the rock between the thumb and fingers (see paragraph on pinch holds).

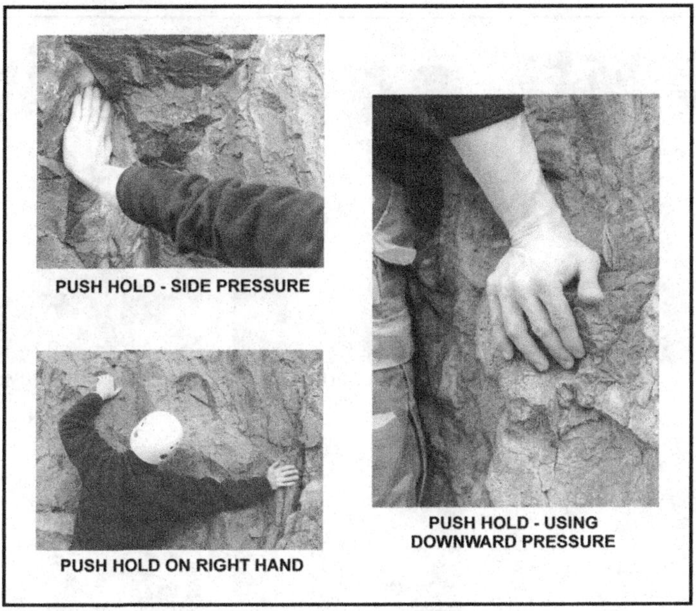

Figure 6-8. Examples of push holds.

(2) ***Pull Holds.*** Pull holds, also called "cling holds," which are grasped and pulled upon, are probably the most widely used holds in climbing. Grip plays more of a role in a pull hold, and, therefore, it normally feels more secure to the climber than a push hold. Because of this increased feeling of security, pull holds are often overworked. These are the holds the climber has a tendency to hang from. Most pull holds do not require great strength, just good technique. Avoid the "death grip" syndrome by climbing with the feet. (Figure 6-9, page 6-16, shows examples of pull holds.)

(a) Like push holds, pressure on a pull hold can be applied straight down, sideways, or upward. Again, these are the holds the climber tends to stretch and reach for, creating an unbalanced stance. Remember to try and keep the hands between waist and shoulder level, making use of intermediate holds instead of reaching for those above the head.

(b) Pulling sideways on vertical cracks can be very secure. There is less tendency to hang from "side-clings" and the hands naturally remain lower. The thumb can often push against one side of the crack, in opposition to the pull by the fingers, creating a stronger hold. Both hands can also be placed in the same crack, with the hands pulling in opposite directions. The number of possible combinations is limited only by the imagination and experience of the climber.

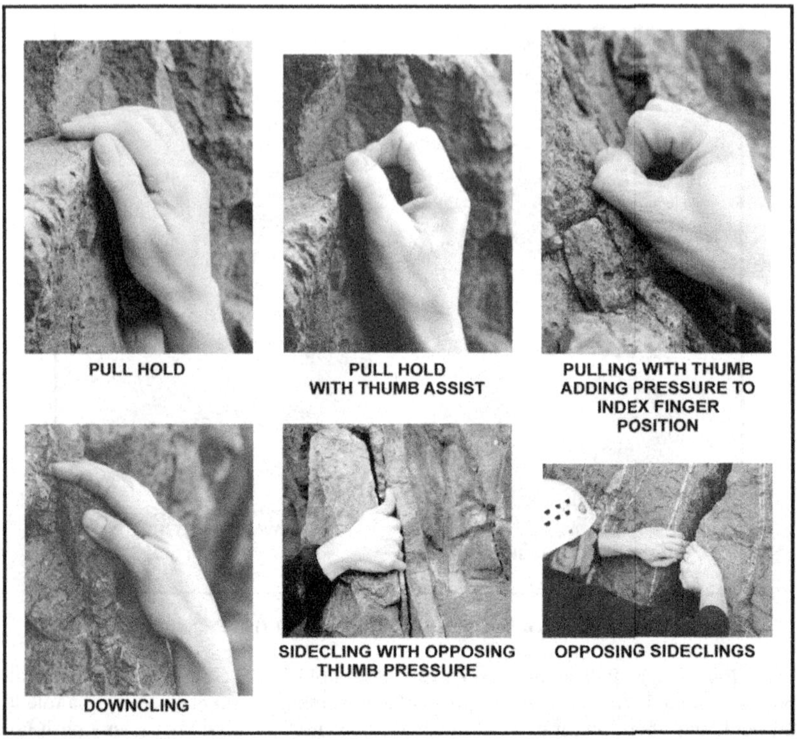

Figure 6-9. Examples of pull holds.

(c) Friction and strength of a pull hold can be increased by the way the hand grips the rock. Normally, the grip is stronger when the fingers are closed together; however, sometimes more friction is obtained by spreading the fingers apart and placing them between irregularities on the rock surface. On small holds, grip can often be improved by bending the fingers upward, forcing the palm of the hand to push against the rock. This helps to hold the finger tips in place and reduces muscle strain in the hand. Keeping the forearm up against the rock also allows the arm and hand muscles to relax more.

(d) Another technique that helps to strengthen a cling hold for a downward pull is to press the thumb against the side of the index finger, or place it on top of the index finger and press down. This hand configuration, known as a "ring grip," works well on smaller holds.

(3) ***Pinch Holds.*** Sometimes a small nubbin or protrusion in the rock can be "squeezed" between the thumb and fingers. This technique is called a pinch hold. Friction is applied by increasing the grip on the rock. Pinch holds are often overlooked by the

novice climber because they feel insecure at first and cannot be relied upon to support much body weight. If the climber has his weight over his feet properly, the pinch hold will work well in providing balance. The pinch hold can also be used as a gripping technique for push holds and pull holds. (Figure 6-10 shows examples of pinch holds.)

Figure 6-10. Examples of pinch holds.

(4) ***Jam Holds.*** Like foot jams, the fingers and hands can be wedged or cammed into a crack so they resist a downward or outward pull. Jamming with the fingers and hands can be painful and may cause minor cuts and abrasions to tender skin. Cotton tape can be used to protect the fingertips, knuckles, and the back of the hand; however, prolonged jamming technique requiring hand taping should be avoided. Tape also adds friction to the hand in jammed position. (Figure 6-11, page 6-18, shows examples of jam holds.)

(a) The hand can be placed in a crack a number of ways. Sometimes an open hand can be inserted and wedged into a narrower portion of the crack. Other times a clenched fist will provide the necessary grip. Friction can be created by applying cross pressure between the fingers and the back of the hand. Another technique for vertical cracks is to place the hand in the crack with the thumb pointed either up or down. The hand is then clenched as much as possible. When the arm is straightened, it will twist the hand and tend to cam it into place. This combination of clenching and camming usually produces the most friction, and the most secure hand jam in vertical cracks.

(b) In smaller cracks, only the fingers will fit. Use as many fingers as the crack will allow. The fingers can sometimes be stacked in some configuration to increase friction. The thumb is usually kept outside the crack in finger jams and pressed against the rock to increase friction or create cross pressure. In vertical cracks it is best to insert the fingers with the thumb pointing down to make use of the natural camming action of the fingers that occurs when the arm is twisted towards a normal position.

(c) Jamming technique for large cracks, or "off widths," requiring the use of arm, leg, and body jams, is another technique. To jam or cam an arm, leg, or body into an off width, the principle is the same as for fingers, hands, or feet—you are making the jammed appendage "fatter" by folding or twisting it inside the crack. For off widths, you may place your entire arm inside the crack with the arm folded and the palm pointing outward. The leg can be used, from the calf to the thigh, and flexed to fit the crack. Routes requiring this type of climbing should be avoided as the equipment normally used for protection might not be large enough to protect larger cracks and openings. However, sometimes a narrower section may be deeper in the crack allowing the use of "normal" size protection.

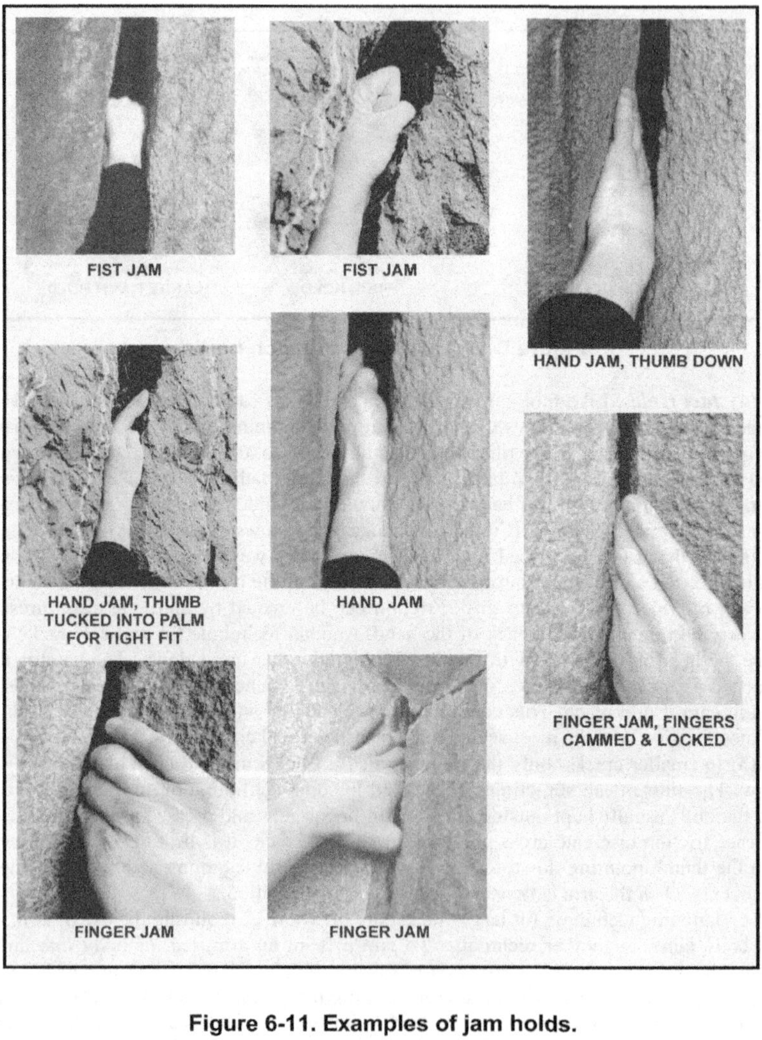

Figure 6-11. Examples of jam holds.

6-10. COMBINATION TECHNIQUES

The positions and holds previously discussed are the basics and the ones most common to climbing. From these fundamentals, numerous combination techniques are possible. As the climber gains experience, he will learn more ways to position the hands, feet, and body in relation to the holds available; however, he should always strive to climb with his weight on his feet from a balanced stance.

a. Sometimes, even on an easy route, the climber may come upon a section of the rock that defies the basic principles of climbing. Short of turning back, the only alternative is to figure out some combination technique that will work. Many of these type problems require the hands and feet to work in opposition to one another. Most will place more weight on the hands and arms than is desirable, and some will put the climber in an "out of balance" position. To make the move, the climber may have to "break the rules" momentarily. This is not a problem and is done quite frequently by experienced climbers. The key to using these type of combination techniques is to plan and execute them deliberately, without lunging or groping for holds, yet quickly, before the hands, arms, or other body parts tire. Still, most of these maneuvers require good technique more than great strength, though a certain degree of hand and arm strength certainly helps.

b. Combination possibilities are endless. The following is a brief description of some of the more common techniques.

(1) ***Change Step.*** The change step, or hop step, can be used when the climber needs to change position of the feet. It is commonly used when traversing to avoid crossing the feet, which might put the climber in an awkward position. To prevent an off balance situation, two solid handholds should be used. The climber simply places his weight on his handholds while he repositions the feet. He often does this with a quick "hop," replacing the lead foot with the trail foot on the same hold. Keeping the forearms against the rock during the maneuver takes some of the strain off the hands, while at the same time strengthening the grip on the holds.

(2) ***Mantling.*** Mantling is a technique that can be used when the distance between the holds increases and there are no immediate places to move the hands or feet. It does require a ledge (mantle) or projection in the rock that the climber can press straight down upon. (Figure 6-12, page 6-20, shows the mantling sequence.)

(a) When the ledge is above head height, mantling begins with pull holds, usually "hooking" both hands over the ledge. The climber pulls himself up until his head is above the hands, where the pull holds become push holds. He elevates himself until the arms are straight and he can lock the elbows to relax the muscles. Rotating the hands inward during the transition to push holds helps to place the palms more securely on the ledge. Once the arms are locked, a foot can be raised and placed on the ledge. The climber may have to remove one hand to make room for the foot. Mantling can be fairly strenuous; however, most individuals should be able to support their weight, momentarily, on one arm if they keep it straight and locked. With the foot on the ledge, weight can be taken off the arms and the climber can grasp the holds that were previously out of reach. Once balanced over the foot, he can stand up on the ledge and plan his next move.

(b) Pure mantling uses arm strength to raise the body; however, the climber can often smear the balls of the feet against the rock and "walk" the feet up during the maneuver to take some of the weight off the arms. Sometimes edges will be available for short steps in the process.

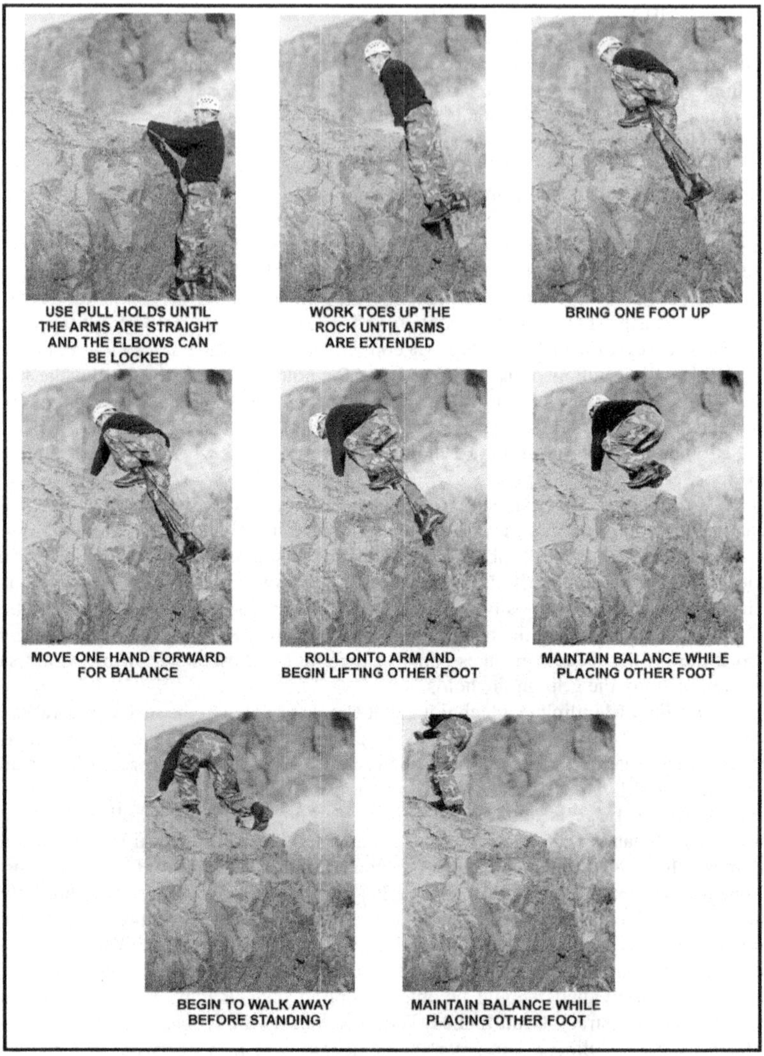

Figure 6-12. Mantling sequence.

(3) ***Undercling.*** An "undercling" is a classic example of handholds and footholds working in opposition (Figure 6-13). It is commonly used in places where the rock

projects outward, forming a bulge or small overhang. Underclings can be used in the tops of buckets, also. The hands are placed "palms-up" underneath the bulge, applying an upward pull. Increasing this upward pull creates a counterforce, or body tension, which applies more weight and friction to the footholds. The arms and legs should be kept as straight as possible to reduce fatigue. The climber can often lean back slightly in the undercling position, enabling him to see above the overhang better and search for the next hold.

Figure 6-13. Undercling.

(4) ***Lieback.*** The "lieback" is another good example of the hands working in opposition to the feet. The technique is often used in a vertical or diagonal crack separating two rock faces that come together at, more or less, a right angle (commonly referred to as a dihedral). The crack edge closest to the body is used for handholds while the feet are pressed against the other edge. The climber bends at the waist, putting the body into an L-shaped position. Leaning away from the crack on two pull holds, body tension creates friction between the feet and the hands. The feet must be kept relatively high to maintain weight, creating maximum friction between the sole and the rock surface. Either full sole contact or the smearing technique can be used, whichever seems to produce the most friction.

(a) The climber ascends a dihedral by alternately shuffling the hands and feet upward. The lieback technique can be extremely tiring, especially when the dihedral is near vertical. If the hands and arms tire out before completing the sequence, the climber will likely fall. The arms should be kept straight throughout the entire maneuver so the climber's weight is pulling against bones and ligaments, rather than muscle. The legs should be straightened whenever possible.

(b) Placing protection in a lieback is especially tiring. Look for edges or pockets for the feet in the crack or on the face for a better position to place protection from, or for a rest position. Often, a lieback can be avoided with closer examination of the available face features. The lieback can be used alternately with the jamming technique, or vice versa, for variation or to get past a section of a crack with difficult or nonexistent jam possibilities. The lieback can sometimes be used as a face maneuver (Figure 6-14).

Figure 6-14. Lieback on a face.

(5) ***Stemming.*** When the feet work in opposition from a relatively wide stance, the maneuver is known as stemming. The stemming technique can sometimes be used on faces, as well as in a dihedral in the absence of solid handholds for the lieback (Figure 6-15).

Figure 6-15. Stemming on a face.

(a) The classic example of stemming is when used in combination with two opposing push holds in wide, parallel cracks, known as chimneys. Chimneys are cracks in which the walls are at least 1 foot apart and just big enough to squeeze the body into. Friction is created by pushing outward with the hands and feet on each side of the crack. The climber ascends the chimney by alternately moving the hands and feet up the crack (Figure 6-16). Applying pressure with the back and bottom is usually necessary in wider chimneys. Usually, full sole contact of the shoes will provide the most friction, although smearing may work best in some instances. Chimneys that do not allow a full stemming position can be negotiated using the arms, legs, or body as an integral contact point. This technique will often feel more secure since there is more body to rock contact.

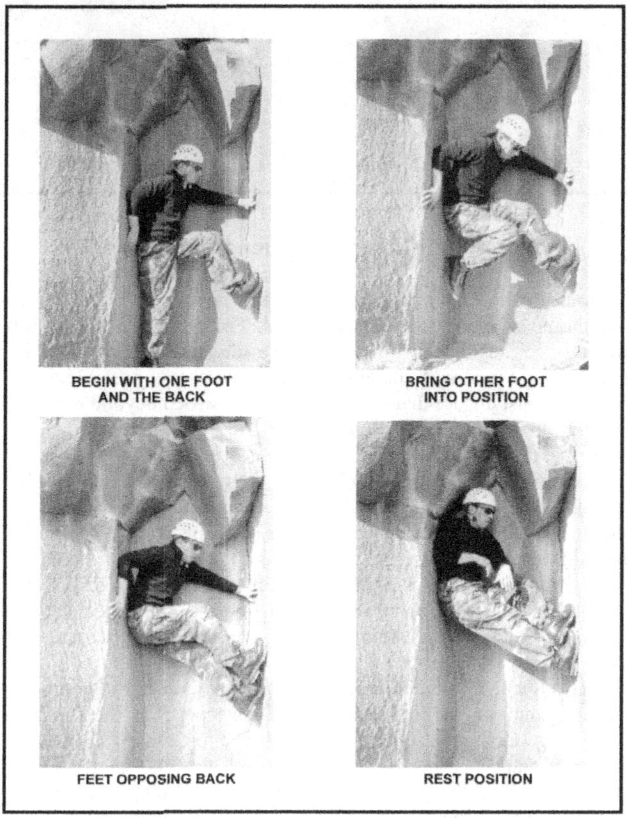

Figure 6-16. Chimney sequence.

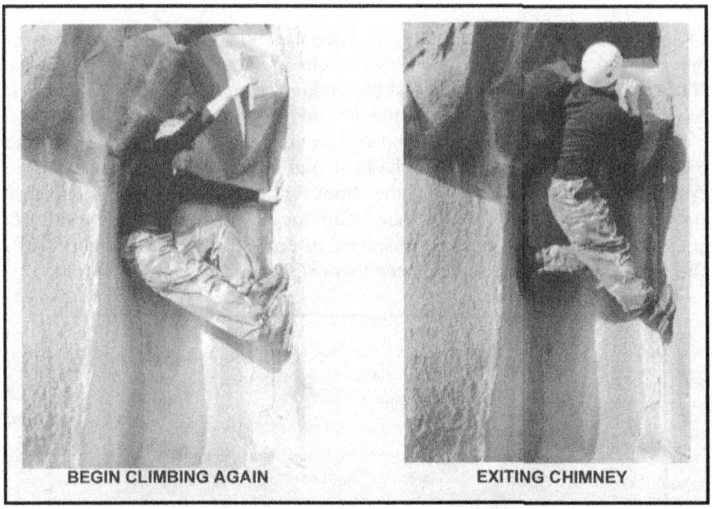

Figure 6-16. Chimney sequence (continued).

(b) The climber can sometimes rest by placing both feet on the same side of the crack, forcing the body against the opposing wall. The feet must be kept relatively high up under the body so the force is directed sideways against the walls of the crack. The arms should be straightened with the elbows locked whenever possible to reduce muscle strain. The climber must ensure that the crack does not widen beyond the climbable width before committing to the maneuver. Remember to look for face features inside chimneys for more security in the climb.

(c) Routes requiring this type of climbing should be avoided as the equipment normally used for protection might not be large enough to protect chimneys. However, face features, or a much narrower crack in one or both corners, may sometimes be found deeper in the chimney allowing the use of normal size protection.

(6) ***Slab Technique.*** A slab is a relatively smooth, low-angled rock formation that requires a slightly modified climbing technique (Figure 6-17). Since slab rock normally contains few, if any holds, the technique requires maximum friction and perfect balance over the feet.

(a) On lower-angled slab, the climber can often stand erect and climb using full sole contact and other mountain walking techniques. On steeper slab, the climber will need to apply good smearing technique. Often, maximum friction cannot be attained on steeper slab from an erect stance. The climber will have to flex the ankles and knees so his weight is placed more directly over the balls of the feet. He may then have to bend at the waist to place the hands on the rock, while keeping the hips over his feet.

(b) The climber must pay attention to any changes in slope angle and adjust his body accordingly. Even the slightest change in the position of the hips over the feet can mean the difference between a good grip or a quick slip. The climber should also take advantage of any rough surfaces, or other irregularities in the rock he can place his hands or feet on, to increase friction.

Figure 6-17. Slab technique.

(7) ***Down Climbing.*** Descending steep rock is normally performed using a roped method; however, the climber may at some point be required to down climb a route. Even if climbing ropes and related equipment are on hand, down climbing easier terrain is often quicker than taking the time to rig a rappel point. Also, a climber might find himself confronted with difficulties part way up a route that exceed his climbing ability, or the abilities of others to follow. Whatever the case may be, down climbing is a skill well worth practicing.

CAUTIONS

1. Down climbing can inadvertently lead into an unforeseen dangerous position on a descent. When in doubt, use a roped descent.
2. Down climbing is accomplished at a difficulty level well below the ability of the climber. When in doubt, use a roped descent.

(a) On easier terrain, the climber can face outward, away from the rock, enabling him to see the route better and descend quickly. As the steepness and difficulty increase, he can often turn sideways, still having a good view of the descent route, but being better able to use the hands and feet on the holds available. On the steepest terrain, the climber will have to face the rock and down climb using good climbing techniques.

(b) Down climbing is usually more difficult than ascending a given route. Some holds will be less visible when down climbing, and slips are more likely to occur. The climber must often lean well away from the rock to look for holds and plan his movements. More weight is placed on the arms and handholds at times to accomplish this, as well as to help lower the climber to the next foothold. Hands should be moved to holds as low as waist level to give the climber more range of movement with each step. If the handholds are too high, he may have trouble reaching the next foothold. The climber must be careful not to overextend himself, forcing a release of his handholds before reaching the next foothold.

CAUTION

Do not drop from good handholds to a standing position. A bad landing could lead to injured ankles or a fall beyond the planned landing area.

(c) Descending slab formations can be especially tricky. The generally lower angle of slab rock may give the climber a false sense of security, and a tendency to move too quickly. Down climbing must be slow and deliberate, as in ascending, to maintain perfect balance and weight distribution over the feet. On lower-angle slab the climber may be able to stand more or less erect, facing outward or sideways, and descend using good flat foot technique. The climber should avoid the tendency to move faster, which can lead to uncontrollable speed.

(d) On steeper slab, the climber will normally face the rock and down climb, using the same smearing technique as for ascending. An alternate method for descending slab is to face away from the rock in a "crab" position (Figure 6-18). Weight is still concentrated over the feet, but may be shifted partly onto the hands to increase overall friction. The climber is able to maintain full sole contact with the rock and see the entire descent route. Allowing the buttocks to "drag behind" on the rock will decrease the actual weight on the footholds, reducing friction, and leading to the likelihood of a slip. Facing the rock, and down-climbing with good smearing technique, is usually best on steeper slab.

Figure 6-18. Descending slab in the crab position.

Section III. ROPED CLIMBING

When the angle, length, and difficulty of the proposed climbing route surpasses the ability of the climbers' safety margin (possibly on class 4 and usually on class 5 terrain), ropes must be used to proceed. Roped climbing is only safe if accomplished correctly. Reading this manual does not constitute skill with ropes—much training and practice is necessary. Many aspects of roped climbing take time to understand and learn. Ropes are normally not used in training until the basic principles of climbing are covered.

Note: A rope is completely useless for climbing unless the climber knows how to use it safely.

6-11. TYING-IN TO THE CLIMBING ROPE

Over the years, climbers have developed many different knots and procedures for tying-in to the climbing rope. Some of the older methods of tying directly into the rope require minimal equipment and are relatively easy to inspect; however, they offer little support to the climber, may induce further injuries, and may even lead to strangulation in a severe fall. A severe fall, where the climber might fall 20 feet or more and be left dangling on the end of the rope, is highly unlikely in most instances, especially for most personnel involved in military climbing. Tying directly into the rope is perfectly safe for many roped party climbs used in training on lower-angled rock. All climbers should know how to properly tie the rope around the waist in case a climbing harness is unavailable.

6-12. PRESEWN HARNESSES

Although improvised harnesses are made from readily available materials and take little space in the pack or pocket, presewn harnesses provide other aspects that should be considered. No assembly is required, which reduces preparation time for roped movement. All presewn harnesses provide a range of adjustability. These harnesses have a fixed buckle that, when used correctly, will not fail before the nylon materials connected to it. However, specialized equipment, such as a presewn harness, reduce the flexibility of gear. Presewn harness are bulky, also.

a. **Seat Harness.** Many presewn seat harnesses are available with many different qualities separating them, including cost.

(1) The most notable difference will be the amount and placement of padding. The more padding the higher the price and the more comfort. Gear loops sewn into the waist belt on the sides and in the back are a common feature and are usually strong enough to hold quite a few carabiners and or protection. The gear loops will vary in number from one model/manufacturer to another.

(2) Although most presewn seat harnesses have a permanently attached belay loop connecting the waist belt and the leg loops, the climbing rope should be run around the waist belt and leg loop connector. The presewn belay loop adds another link to the chain of possible failure points and only gives one point of security whereas running the rope through the waist belt and leg loop connector provides two points of contact.

(3) If more than two men will be on the rope, connect the middle position(s) to the rope with a carabiner routed the same as stated in the previous paragraph.

(4) Many manufactured seat harnesses will have a presewn loop of webbing on the rear. Although this loop is much stronger than the gear loops, it is not for a belay anchor. It is a quick attachment point to haul an additional rope.

b. **Chest Harness.** The chest harness will provide an additional connecting point for the rope, usually in the form of a carabiner loop to attach a carabiner and rope to. This type of additional connection will provide a comfortable hanging position on the rope, but otherwise provides no additional protection from injury during a fall (if the seat harness is fitted correctly).

(1) A chest harness will help the climber remain upright on the rope during rappelling or ascending a fixed rope, especially while wearing a heavy pack. (If rappelling or ascending long or multiple pitches, let the pack hang on a drop cord below the feet and attached to the harness tie-in point.)

(2) The presewn chest harnesses available commercially will invariably offer more comfort or performance features, such as padding, gear loops, or ease of adjustment, than an improvised chest harness.

c. **Full-Body Harness.** Full-body harnesses incorporate a chest and seat harness into one assembly. This is the safest harness to use as it relocates the tie-in point higher, at the chest, reducing the chance of an inverted position when hanging on the rope. This is especially helpful when moving on ropes with heavy packs. A full-body harness only affects the body position when hanging on the rope and will not prevent head injury in a fall.

CAUTION

This type of harness does not prevent the climber from falling head first. Body position during a fall is affected only by the forces that generated the fall, and this type of harness promotes an upright position only when hanging on the rope from the attachment point.

6-13. IMPROVISED HARNESSES

Without the use of a manufactured harness, many methods are still available for attaching oneself to a rope. Harnesses can be improvised using rope or webbing and knots.

a. **Swami Belt.** The swami belt is a simple, belt-only harness created by wrapping rope or webbing around the waistline and securing the ends. One-inch webbing will provide more comfort. Although an effective swami belt can be assembled with a minimum of one wrap, at least two wraps are recommended for comfort, usually with approximately ten feet of material. The ends are secured with an appropriate knot.

b. **Bowline-on-a-Coil.** Traditionally, the standard method for attaching oneself to the climbing rope was with a bowline-on-a-coil around the waist. The extra wraps distribute the force of a fall over a larger area of the torso than a single bowline would, and help prevent the rope from riding up over the rib cage and under the armpits. The knot must be tied snugly around the narrow part of the waist, just above the bony portions of the hips (pelvis). Avoid crossing the wraps by keeping them spread over the waist area. "Sucking in the gut" a bit when making the wraps will ensure a snug fit.

(1) The bowline-on-a-coil can be used to tie-in to the end of the rope (Figure 6-19). The end man should have a minimum of four wraps around the waist before completing the knot.

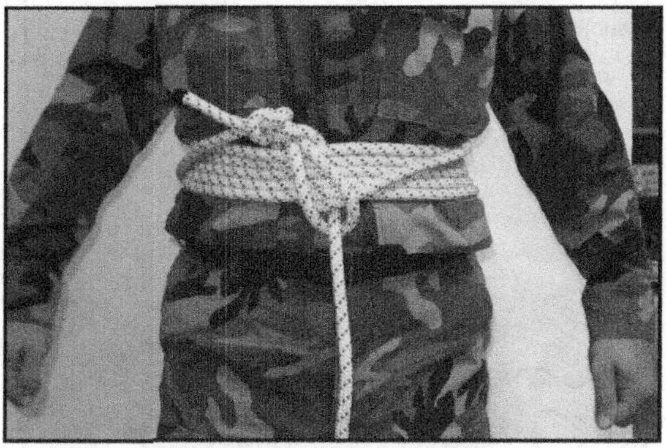

Figure 6-19. Tying-in with a bowline-on-a-coil.

(2) The bowline-on-a-coil is a safe and effective method for attaching to the rope when the terrain is low-angled, WITHOUT THE POSSIBILITY OF A SEVERE FALL. When the terrain becomes steeper, a fall will generate more force on the climber and this will be felt through the coils of this type of attachment. A hard fall will cause the coils to ride up against the ribs. In a severe fall, any tie-in around the waist only could place a "shock load" on the climber's lower back. Even in a relatively short fall, if the climber ends up suspended in mid-air and unable to regain footing on the rock, the rope around the waist can easily cut off circulation and breathing in a relatively short time.

(3) The climbing harness distributes the force of a fall over the entire pelvic region, like a parachute harness. Every climber should know how to tie some sort of improvised climbing harness from sling material. A safe, and comfortable, seat/chest combination harness can be tied from one-inch tubular nylon.

c. **Improvised Seat Harness.** A seat harness can be tied from a length of webbing approximately 25 feet long (Figure 6-20, page 6-30).

(1) Locate the center of the rope. Off to one side, tie two fixed loops approximately 6 inches apart (overhand loops). Adjust the size of the loops so they fit snugly around the thigh. The loops are tied into the sling "off center" so the remaining ends are different lengths. The short end should be approximately 4 feet long (4 to 5 feet for larger individuals).

(2) Slip the leg loops over the feet and up to the crotch, with the knots to the front. Make one complete wrap around the waist with the short end, wrapping to the outside, and hold it in place on the hip. Keep the webbing flat and free of twists when wrapping.

(3) Make two to three wraps around the waist with the long end in the opposite direction (wrapping to the outside), binding down on the short end to hold it in place. Grasping both ends, adjust the waist wraps to a snug fit. Connect the ends with the

appropriate knot between the front and one side so you will be able to see what you are doing.

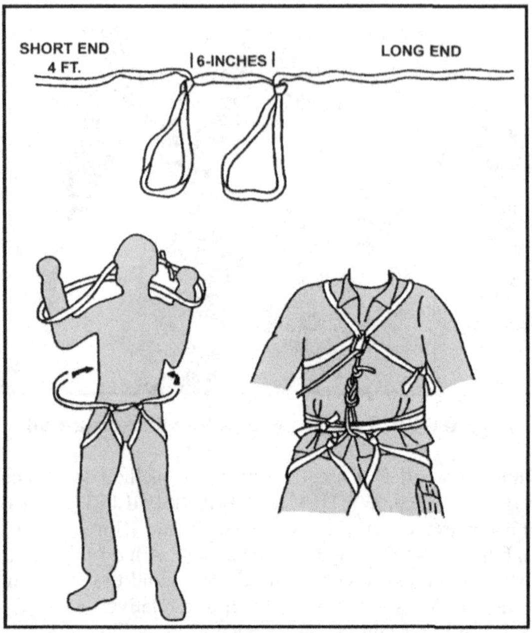

Figure 6-20. Improvised seat and chest harness.

d. **Improvised Chest Harness.** The chest harness can be tied from rope or webbing, but remember that with webbing, wider is better and will be more comfortable when you load this harness. Remember as you tie this harness that the remaining ends will need to be secured so choose the best length. Approximately 6 to 10 feet usually works.

(1) Tie the ends of the webbing together with the appropriate knot, making a sling 3 to 4 feet long.

(2) Put a single twist into the sling, forming two loops.

(3) Place an arm through each loop formed by the twist, just as you would put on a jacket, and drape the sling over the shoulders. The twist, or cross, in the sling should be in the middle of the back.

(4) Join the two loops at the chest with a carabiner. The water knot should be set off to either side for easy inspection (if a pack is to be worn, the knot will be uncomfortable if it gets between the body and the pack). The chest harness should fit just loose enough to allow necessary clothing and not to restrict breathing or circulation. Adjust the size of the sling if necessary.

e. **Improvised Full-Body Harness.** Full-body harnesses incorporate a chest and seat harness into one assembly.

(1) The full-body harness is the safest harness because it relocates the tie-in point higher, at the chest, reducing the chance of an inverted hanging position on the rope. This is especially helpful when moving on ropes with heavy packs. A full-body harness affects the body position only when hanging on the rope.

CAUTION

A full-body harness does not prevent falling head first; body position in a fall is caused by the forces that caused the fall.

(2) Although running the rope through the carabiner of the chest harness does, in effect, create a type of full-body harness, it is not a true full-body harness until the chest harness and the seat harness are connected as one piece. A true full-body harness can be improvised by connecting the chest harness to the seat harness, but not by just tying the rope into both—the two harnesses must be "fixed" as one harness. Fix them together with a short loop of webbing or rope so that the climbing rope can be connected directly to the chest harness and your weight is supported by the seat harness through the connecting material.

f. **Attaching the Rope to the Improvised Harness.** The attachment of the climbing rope to the harness is a CRITICAL LINK. The strength of the rope means nothing if it is attached poorly, or incorrectly, and comes off the harness in a fall. The climber ties the end of the climbing rope to the seat harness with an appropriate knot. If using a chest harness, the standing part of the rope is then clipped into the chest harness carabiner. The seat harness absorbs the main force of the fall, and the chest harness helps keep the body upright.

CAUTION

The knot must be tied around all the waist wraps and the 6-inch length of webbing between the leg loops.

(1) A middleman must create a fixed loop to tie in to. A rethreaded figure-eight loop tied on a doubled rope or the three loop bowline can be used. If using the three loop bowline, ensure the end, or third loop formed in the knot, is secured around the tie-in loops with an overhand knot. The standing part of the rope going to the lead climber is clipped into the chest harness carabiner.

Note: The climbing rope is not clipped into the chest harness when belaying.

(2) The choice of whether to tie-in with a bowline-on-a-coil or into a climbing harness depends entirely on the climber's judgment, and possibly the equipment

available. A good rule of thumb is: "Wear a climbing harness when the potential for severe falls exists and for all travel over snow-covered glaciers because of the crevasse fall hazard."

(3) Under certain conditions many climbers prefer to attach the rope to the seat harness with a locking carabiner, rather than tying the rope to it. This is a common practice for moderate snow and ice climbing, and especially for glacier travel where wet and frozen knots become difficult to untie.

CAUTION
Because the carabiner gate may be broken or opened by protruding rocks during a fall, tie the rope directly to the harness for maximum safety.

Section IV. BELAY TECHNIQUES

Tying-in to the climbing rope and moving as a member of a rope team increases the climber's margin of safety on difficult, exposed terrain. In some instances, such as when traveling over snow-covered glaciers, rope team members can often move at the same time, relying on the security of a tight rope and "team arrest" techniques to halt a fall by any one member. On steep terrain, however, simultaneous movement only helps to ensure that if one climber falls, he will jerk the other rope team members off the slope. For the climbing rope to be of any value on steep rock climbs, the rope team must incorporate "belays" into the movement.

Belaying is a method of managing the rope in such a way that, if one person falls, the fall can be halted or "arrested" by another rope team member (belayer). One person climbs at a time, while being belayed from above or below by another. The belayer manipulates the rope so that friction, or a "brake," can be applied to halt a fall. Belay techniques are also used to control the descent of personnel and equipment on fixed rope installations, and for additional safety on rappels and stream crossings.

Belaying is a skill that requires practice to develop proficiency. Setting up a belay may at first appear confusing to the beginner, but with practice, the procedure should become "second nature." If confronted with a peculiar problem during the setup of a belay, try to use common sense and apply the basic principles stressed throughout this text. Remember the following key points:

- Select the best possible terrain features for the position and use terrain to your advantage.
- Use a well braced, sitting position whenever possible.
- Aim and anchor the belay for all possible load directions.
- Follow the "minimum" rule for belay anchors—2 for a downward pull, 1 for an upward pull.
- Ensure anchor attachments are aligned, independent, and snug.
- Stack the rope properly.
- Choose a belay technique appropriate for the climbing.

- Use a guide carabiner for rope control in all body belays.
- Ensure anchor attachments, guide carabiner (if applicable), and rope running to the climber are all on the guidehand side.
- The brake hand remains on the rope when belaying.

CAUTION

Never remove the brake hand from the rope while belaying. If the brake hand is removed, there is no belay.

- Ensure you are satisfied with your position before giving the command "BELAY ON."
- The belay remains in place until the climber gives the command "OFF BELAY."

CAUTION

The belay remains in place the from the time the belayer commands *"BELAY ON"* until the climber commands *"OFF BELAY."*

6-14. PROCEDURE FOR MANAGING THE ROPE

A number of different belay techniques are used in modern climbing, ranging from the basic "body belays" to the various "mechanical belays," which incorporate some type of friction device.

a. Whether the rope is wrapped around the body, or run through a friction device, the rope management procedure is basically the same. The belayer must be able to perform three basic functions: manipulate the rope to give the climber slack during movement, take up rope to remove excess slack, and apply the brake to halt a fall.

b. The belayer must be able to perform all three functions while maintaining "total control" of the rope at all times. Total control means the brake hand is NEVER removed from the rope. When giving slack, the rope simply slides through the grasp of the brake hand, at times being fed to the climber with the other "feeling" or guide hand. Taking up rope, however, requires a certain technique to ensure the brake hand remains on the rope at all times. The following procedure describes how to take up excess rope and apply the brake in a basic body belay.

(1) Grasping the rope with both hands, place it behind the back and around the hips. The hand on the section of rope between the belayer and the climber would be the guide hand. The other hand is the brake hand.

(2) Take in rope with the brake hand until the arm is fully extended. The guide hand can also help to pull in the rope (Figure 6-21, step 1, page 6-34).

(3) Holding the rope in the brake hand, slide the guide hand out, extending the arm so the guide hand is father away from the body than the brake hand (Figure 6-21, step 2).

(4) Grasp both parts of the rope, to the front of the brake hand, with the guide hand (Figure 6-21, step 3).

(5) Slide the brake hand back towards the body (Figure 6-21, step 4).

(6) Repeat step 5 of Figure 6-21. The brake can be applied at any moment during the procedure. It is applied by wrapping the rope around the front of the hips while increasing grip with the brake hand (Figure 6-21, step 6).

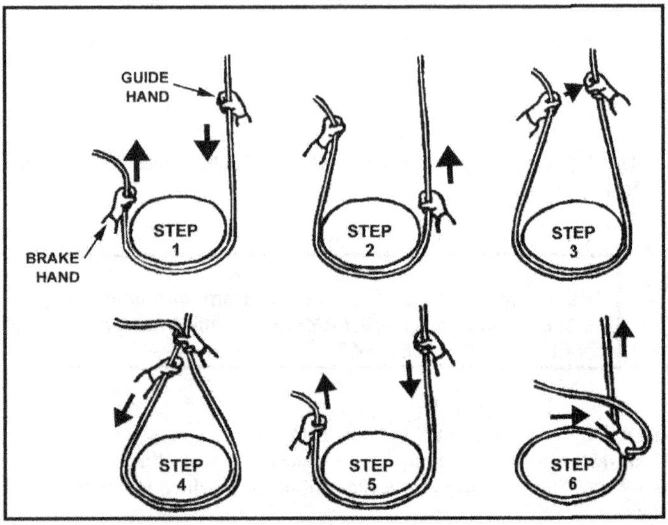

Figure 6-21. Managing the rope.

6-15. CHOOSING A BELAY TECHNIQUE

The climber may choose from a variety of belay techniques. A method that works well in one situation may not be the best choice in another. The choice between body belays and mechanical belays depends largely on equipment available, what the climber feels most comfortable with, and the amount of load, or fall force, the belay may have to absorb. The following describes a few of the more widely used techniques, and the ones most applicable to military mountaineering.

a. **Body Belay.** The basic body belay is the most widely used technique on moderate terrain. It uses friction between the rope and the clothed body as the rope is pressured across the clothing. It is the simplest belay, requiring no special equipment, and should be the first technique learned by all climbers. A body belay gives the belayer the greatest "feel" for the climber, letting him know when to give slack or take up rope. Rope management in a body belay is quick and easy, especially for beginners, and is effective in snow and ice climbing when ropes often become wet, stiff, and frozen. The body belay, in its various forms, will hold low to moderate impact falls well. It has been known to arrest some severe falls, although probably not without inflicting great pain on the belayer.

CAUTION
The belayer must ensure he is wearing adequate clothing to protect his body from rope burns when using a body belay. Heavy duty cotton or leather work gloves can also be worn to protect the hands.

(1) ***Sitting Body Belay.*** The sitting body belay is the preferred position and is usually the most secure (Figure 6-22). The belayer sits facing the direction where the force of a fall will likely come from, using terrain to his advantage, and attempts to brace both feet against the rock to support his position. It is best to sit in a slight depression, placing the buttocks lower than the feet, and straightening the legs for maximum support. When perfectly aligned, the rope running to the climber will pass between the belayer's feet, and both legs will equally absorb the force of a fall. Sometimes, the belayer may not be able to sit facing the direction he would like, or both feet cannot be braced well. The leg on the "guide hand" side should then point towards the load, bracing the foot on the rock when possible. The belayer can also "straddle" a large tree or rock nubbin for support, as long as the object is solid enough to sustain the possible load.

Figure 6-22. Sitting body belay.

(2) ***Standing Body Belay.*** The standing body belay is used on smaller ledges where there is no room for the belayer to sit (Figure 6-23). What appears at first to be a fairly unstable position can actually be quite secure when belay anchors are placed at or above shoulder height to support the stance when the force will be downward.

Figure 6-23. Standing body belay.

(a) For a body belay to work effectively, the belayer must ensure that the rope runs around the hips properly, and remains there under load when applying the brake. The rope should run around the narrow portion of the pelvic girdle, just below the bony high points of the hips. If the rope runs too high, the force of a fall could injure the belayer's midsection and lower rib cage. If the rope runs too low, the load may pull the rope below the buttocks, dumping the belayer out of position. It is also possible for a strong upward or downward pull to strip the rope away from the belayer, rendering the belay useless.

(b) To prevent any of these possibilities from happening, the belay rope is clipped into a carabiner attached to the guide hand side of the seat harness (or bowline-on-a-coil). This "guide carabiner" helps keep the rope in place around the hips and prevents loss of control in upward or downward loads (Figure 6-24).

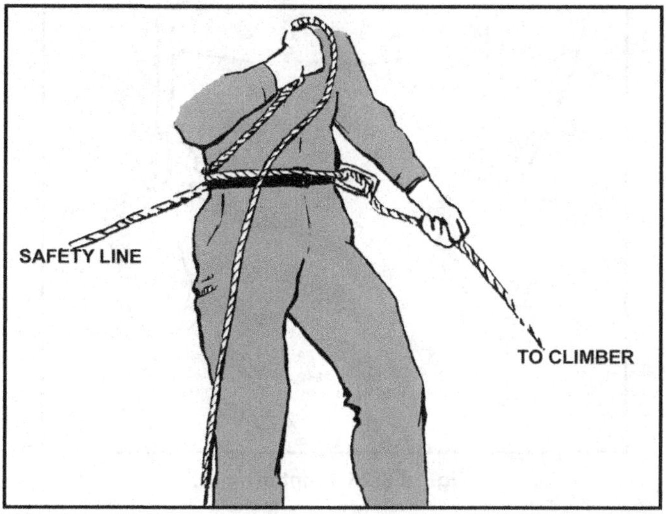

Figure 6-24. Guide carabiner for rope control in a body belay.

b. **Mechanical Belay.** A mechanical belay **must** be used whenever there is potential for the lead climber to take a severe fall. The holding power of a belay device is vastly superior to any body belay under high loads. However, rope management in a mechanical belay is more difficult to master and requires more practice. For the most part, the basic body belay should be totally adequate on a typical military route, as routes used during military operations should be the easiest to negotiate.

(1) ***Munter Hitch.*** The Munter hitch is an excellent mechanical belay technique and requires only a rope and a carabiner (Figure 6-25, page 6-38). The Munter is actually a two-way friction hitch. The Munter hitch will flip back and forth through the carabiner as the belayer switches from giving slack to taking up rope. The carabiner must be large enough, and of the proper design, to allow this function. The locking pear-shaped carabiner, or pearabiner, is designed for the Munter hitch.

(a) The Munter hitch works exceptionally well as a lowering belay off the anchor. As a climbing belay, the carabiner should be attached to the front of the belayer's seat harness. The hitch is tied by forming a loop and a bight in the rope, attaching both to the carabiner. It's fairly easy to place the bight on the carabiner backwards, which forms an obvious, useless hitch. Put some tension on the Munter to ensure it is formed correctly, as depicted in the following illustrations.

(b) The Munter hitch will automatically "lock-up" under load as the brake hand grips the rope. The brake is increased by pulling the slack rope away from the body, towards the load. The belayer must be aware that flipping the hitch DOES NOT change the function of the hands. The hand on the rope running to the climber, or load, is always the guide hand.

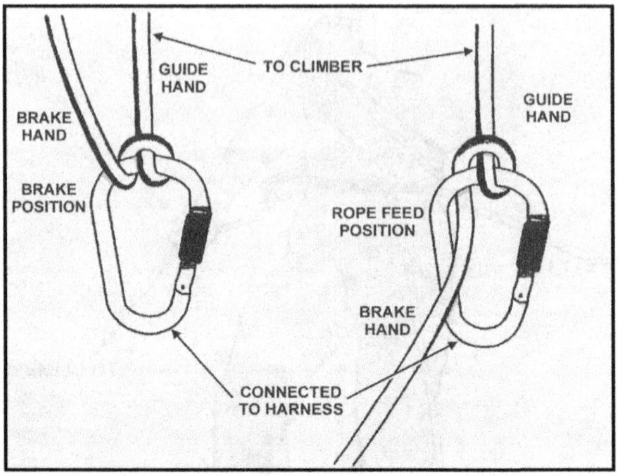

Figure 6-25. Munter hitch.

(2) ***Figure-Eight Device.*** The figure-eight device is a versatile piece of equipment and, though developed as a rappel device, has become widely accepted as an effective mechanical belay device (Figure 6-26). The advantage of any mechanical belay is friction required to halt a fall is applied on the rope through the device, rather than around the belayer's body. The device itself provides rope control for upward and downward pulls and excellent friction for halting severe falls. The main principle behind the figure-eight device in belay mode is the friction developing on the rope as it reaches and exceeds the 90-degree angle between the rope entering the device and leaving the device. As a belay device, the figure-eight works well for both belayed climbing and for lowering personnel and equipment on fixed-rope installations.

(a) As a climbing belay, a bight placed into the climbing rope is run through the "small eye" of the device and attached to a locking carabiner at the front of the belayer's seat harness. A short, small diameter safety rope is used to connect the "large eye" of the figure eight to the locking carabiner for control of the device. The guide hand is placed on the rope running to the climber. Rope management is performed as in a body belay. The brake is applied by pulling the slack rope in the brake hand towards the body, locking the rope between the device and the carabiner.

(b) As a lowering belay, the device is normally attached directly to the anchor with the rope routed as in rappelling.

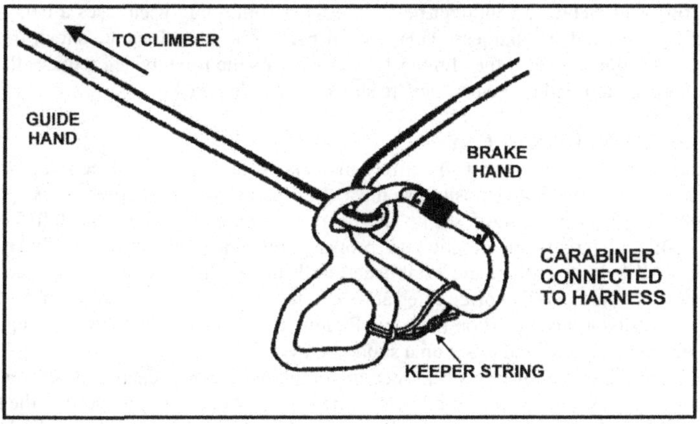

Figure 6-26. Figure-eight device.

Note: Some figure-eight descenders should not be used as belay devices due to their construction and design. Always refer to manufacturer's specifications and directions before use.

(3) ***Mechanical Camming Device.*** The mechanical camming device has an internal camming action that begins locking the rope in place as friction is increased. Unlike the other devices, the mechanical camming device can stop a falling climber without any input from the belayer. A few other devices perform similarly to this, but have no moving parts. Some limitations to these type devices are minimum and maximum rope diameters.

(4) ***Other Mechanical Belay Devices.*** There are many other commercially available mechanical belay devices. Most of these work with the same rope movement direction and the same braking principle. The air traffic controller (ATC), slotted plate, and other tube devices are made in many different shapes. These all work on the same principle as the figure-eight device—friction increases on the rope as it reaches and exceeds the 90-degree angle between the rope entering the device and leaving the device.

6-16. ESTABLISHING A BELAY

A belay can be established using either a direct or indirect connection. Each type has advantages and disadvantages. The choice will depend on the intended use of the belay.

a. **Direct Belay.** The direct belay removes any possible forces from the belayer and places this force completely on the anchor. Used often for rescue installations or to bring a second climber up to a new belay position in conjunction with the Munter hitch, the belay can be placed above the belayer's stance, creating a comfortable position and ease of applying the brake. Also, if the second falls or weights the rope, the belayer is not locked into a position. Direct belays provide no shock-absorbing properties from the belayer's attachment to the system as does the indirect belay; therefore, the belayer is apt to pay closer attention to the belaying process.

b. **Indirect Belay.** An indirect belay, the most commonly used, uses a belay device attached to the belayer's harness. This type of belay provides dynamic shock or weight absorption by the belayer if the climber falls or weights the rope, which reduces the direct force on the anchor and prevents a severe shock load to the anchor.

6-17. SETTING UP A BELAY

In rock climbing, climbers must sometimes make do with marginal protection placements along a route, but belay positions must be made as "bombproof" as possible. Additionally, the belayer must set up the belay in relation to where the fall force will come from and pay strict attention to proper rope management for the belay to be effective. All belay positions are established with the anchor connection to the front of the harness. If the belay is correctly established, the belayer will feel little or no force if the climber falls or has to rest on the rope. Regardless of the actual belay technique used, five basic steps are required to set up a sound belay.

a. **Select Position and Stance.** Once the climbing line is picked, the belayer selects his position. It's best if the position is off to the side of the actual line, putting the belayer out of the direct path of a potential fall or any rocks kicked loose by the climber. The position should allow the belayer to maintain a comfortable, relaxed stance, as he could be in the position for a fairly long time. Large ledges that allow a well braced, sitting stance are preferred. Look for belay positions close to bombproof natural anchors. The position must at least allow for solid artificial placements.

b. **Aim the Belay.** With the belay position selected, the belay must now be "aimed." The belayer determines where the rope leading to the climber will run and the direction the force of a fall will likely come from. When a lead climber begins placing protection, the fall force on the belayer will be in some upward direction, and in line with the first protection placement. If this placement fails under load, the force on the belay could be straight down again. The belayer must aim his belay for all possible load directions, adjusting his position or stance when necessary. The belay can be aimed through an anchor placement to immediately establish an upward pull; however, the belayer must always be prepared for the more severe downward fall force in the event intermediate protection placements fail.

c. **Anchor the Belay.** For a climbing belay to be considered bombproof, the belayer must be attached to a solid anchor capable of withstanding the highest possible fall force. A solid natural anchor would be ideal, but more often the belayer will have to place pitons or chocks. A single artificial placement should never be considered adequate for anchoring a belay (except at ground level). Multiple anchor points capable of supporting both upward and downward pulls should be placed. The rule of thumb is to place two anchors for a downward pull and one anchor for an upward pull as a MINIMUM. The following key points also apply to anchoring belays.

(1) Each anchor must be placed in line with the direction of pull it is intended to support.

(2) Each anchor attachment must be rigged "independently" so a failure of one will not shock load remaining placements or cause the belayer to be pulled out of position.

(3) The attachment between the anchor and the belayer must be snug to support the stance. Both belayer's stance and belay anchors should absorb the force of a fall.

(4) It is best for the anchors to be placed relatively close to the belayer with short attachments. If the climber has to be tied-off in an emergency, say after a severe fall, the belayer can attach a Prusik sling to the climbing rope, reach back, and connect the sling to one of the anchors. The load can be placed on the Prusik and the belayer can come out of the system to render help.

(5) The belayer can use either a portion of the climbing rope or slings of the appropriate length to connect himself to the anchors. It's best to use the climbing rope whenever possible, saving the slings for the climb. The rope is attached using either figure eight loops or clove hitches. Clove hitches have the advantage of being easily adjusted. If the belayer has to change his stance at some point, he can reach back with the guide hand and adjust the length of the attachment through the clove hitch as needed.

(6) The anchor attachments should also help prevent the force of a fall from "rotating" the belayer out of position. To accomplish this, the climbing rope must pass around the "guide-hand side" of the body to the anchors. Sling attachments are connected to the belayer's seat harness (or bowline-on-a-coil) on the guide-hand side.

(7) Arrangement of rope and sling attachments may vary according to the number and location of placements. Follow the guidelines set forth and remember the key points for belay anchors; "in line", "independent", and "snug". Figure 6-27 shows an example of a common arrangement, attaching the rope to the two "downward" anchors and a sling to the "upward" anchor. Note how the rope is connected from one of the anchors back to the belayer. This is not mandatory, but often helps "line-up" the second attachment.

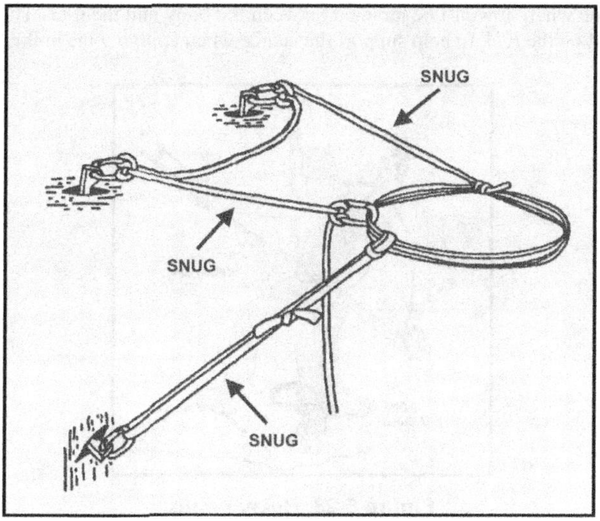

Figure 6-27. Anchoring a belay.

d. **Stack the Rope.** Once the belayer is anchored into position, he must stack the rope to ensure it is free of twists and tangles that might hinder rope management in the belay. The rope should be stacked on the ground, or on the ledge, where it will not get caught in cracks or nubbins as it is fed out to the climber.

(1) On small ledges, the rope can be stacked on top of the anchor attachments if there is no other place to lay it, but make sure to stack it carefully so it won't tangle with the anchored portion of the rope or other slings. The belayer must also ensure that the rope will not get tangled around his legs or other body parts as it "feeds" out.

(2) The rope should never be allowed to hang down over the ledge. If it gets caught in the rock below the position, the belayer may have to tie-off the climber and come out of the belay to free the rope; a time-consuming and unnecessary task. The final point to remember is the rope must be stacked "from the belayer's end" so the rope running to the climber comes off the "top" of the stacked pile.

e. **Attach the Belay.** The final step of the procedure is to attach the belay. With the rope properly stacked, the belayer takes the rope coming off the top of the pile, removes any slack between himself and the climber, and applies the actual belay technique. If using a body belay, ensure the rope is clipped into the guide carabiner.

(1) The belayer should make one quick, final inspection of his belay. If the belay is set up correctly, the anchor attachments, guide carabiner if applicable, and the rope running to the climber will all be on the "guide hand" side, which is normally closest to the rock (Figure 6-28). If the climber takes a fall, the force, if any, should not have any negative effect on the belayer's involvement in the system. The brake hand is out away from the slope where it won't be jammed between the body and the rock. The guide hand can be placed on the rock to help support the stance when applying the brake.

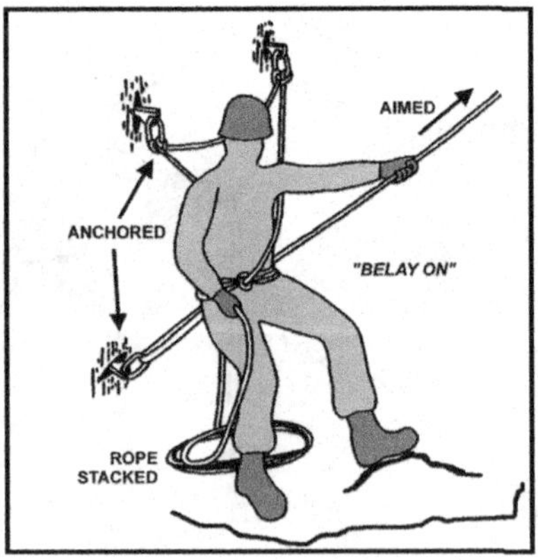

Figure 6-28. Belay setup.

(2) When the belayer is satisfied with his position, he gives the signal, "BELAY ON!". When belaying the "second", the same procedure is used to set up the belay. Unless the belay is aimed for an upward pull, the fall force is of course downward and the

belayer is usually facing away from the rock, the exception being a hanging belay on a vertical face. If the rope runs straight down to the climber and the anchors are directly behind the position, the belayer may choose to brake with the hand he feels most comfortable with. Anchor attachments, guide carabiner, and rope running to the climber through the guide hand must still be aligned on the same side to prevent the belayer from being rotated out of position, unless the belayer is using an improvised harness and the anchor attachment is at the rear.

6-18. TOP-ROPE BELAY

A "top-rope" is a belay setup used in training to protect a climber while climbing on longer, exposed routes. A solid, bombproof anchor is required at the top of the pitch. The belayer is positioned either on the ground with the rope running through the top anchor and back to the climber, or at the top at the anchor. The belayer takes in rope as the climber proceeds up the rock. If this is accomplished with the belayer at the bottom, the instructor can watch the belayer while he coaches the climber through the movements.

CAUTION

Do not use a body belay for top-rope climbing. The rope will burn the belayer if the climber has to be lowered.

Section V. CLIMBING COMMANDS

Communication is often difficult during a climb. As the distance between climber and belayer increases, it becomes harder to distinguish one word from another and the shortest sentence may be heard as nothing more than jumbled syllables. A series of standard voice commands were developed over the years to signal the essential rope management functions in a belayed climb. Each command is concise and sounds a bit different from another to reduce the risk of a misunderstanding between climber and belayer. They must be pronounced clearly and loudly so they can be heard and understood in the worst conditions.

6-19. VERBAL COMMANDS

Table 6-1, page 6-44, lists standard rope commands and their meanings in sequence as they would normally be used on a typical climb. (Note how the critical "BELAY" commands are reversed so they sound different and will not be confused.)

BELAYER	CLIMBER	MEANING/ACTION TAKEN
"BELAY ON"		The belay is on; you may climb when ready; the rope will be managed as needed.
	"CLIMBING" (as a courtesy)	I am ready to climb.
"CLIMB" (as a courtesy)		Proceed, and again, the rope will be managed as necessary.
"ROCK"	"ROCK"	**PROTECT YOURSELF FROM FALLING OBJECTS.** Signal will be echoed by all climbers in the area. If multipitch climbing, ensure climbers below hear.
	"TAKE ROPE"	Take in excess rope between us without pulling me off the route. Belayer takes in rope.
	"SLACK"	Release all braking/tension on the rope so I can have slack without pulling the rope. Belayer removes brake/tension.
	"TENSION"	Take all the slack, apply brake, and hold me. My weight will be on the rope. Belayer removes slack and applies brake.
	"FALLING"	I am falling. Belayer applies brake to arrest the fall.
"TWEN-TY-FIVE"		You have approximately 25 feet of rope left. Start looking for the next belay position. Climber selects a belay position.
"FIF-TEEN"		You have approximately 15 feet of rope left. Start looking for the next belay position. Climber selects a belay position within the next few feet.
"FIVE"	Set up the belay.	You have 5 feet of rope left. Set up the belay position. You have no more rope. Climber sets up the belay.
Removes the belay, remains anchored. Prepares to climb.	"OFF BELAY"	I have finished climbing and I am anchored. You may remove the belay. Belayer removes the belay and, remaining anchored, prepares to climb.

Table 6-1. Rope commands.

6-20. ROPE TUG COMMANDS

Sometimes the loudest scream cannot be heard when the climber and belayer are far apart. This is especially true in windy conditions, or when the climber is around a corner, above an overhang, or at the back of a ledge. It may be necessary to use a series of "tugs" on the rope in place of the standard voice commands. To avoid any possible confusion with interpretation of multiple rope tug commands, use only one.

a. While a lead climb is in progress, the most important command is "BELAY ON." This command is given only by the climber when the climber is anchored and is prepared for the second to begin climbing. With the issue of this command, the second knows the climber is anchored and the second prepares to climb.

b. For a rope tug command, the leader issues three distinct tugs on the rope AFTER anchoring and putting the second on belay. This is the signal for "BELAY ON" and signals the second to climb when ready. The new belayer keeps slack out of the rope.

Section VI. ROPED CLIMBING METHODS

In military mountaineering, the primary mission of a roped climbing team is to "fix" a route with some type of rope installation to assist movement of less trained personnel in the unit. This duty falls upon the most experienced climbers in the unit, usually working in two- or three-man groups or teams called assault climbing teams. Even if the climbing is for another purpose, roped climbing should be performed whenever the terrain becomes difficult and exposed.

6-21. TOP-ROPED CLIMBING

Top-roped climbing is used for training purposes only. This method of climbing is not used for movement due to the necessity of pre-placing anchors at the top of a climb. If you can easily access the top of a climb, you can easily avoid the climb itself.

a. For training, top-roped climbing is valuable because it allows climbers to attempt climbs above their skill level and or to hone present skills without the risk of a fall. Top-roped climbing may be used to increase the stamina of a climber training to climb longer routes as well as for a climber practicing protection placements.

b. The belayer is positioned either at the base of a climb with the rope running through the top anchor and back to the climber or at the top at the anchor. The belayer takes in rope as the climber moves up the rock, giving the climber the same protection as a belay from above. If this is accomplished with the belayer at the bottom, the instructor is able to keep an eye on the belayer while he coaches the climber through the movements.

6-22. LEAD CLIMBING

A lead climb consists of a belayer, a leader or climber, rope(s), and webbing or hardware used to establish anchors or protect the climb. As he climbs the route, the leader emplaces "intermediate" anchors, and the climbing rope is connected to these anchors with a carabiner. These "intermediate" anchors protect the climber against a fall-thus the term "protecting the climb."

Note: Intermediate anchors are commonly referred to as "protection," "pro," "pieces," "pieces of pro," "pro placements," and so on. For standardization within this publication, these specific anchors will be referred to as "protection;" anchors established for other purposes, such as rappel points, belays, or other rope installations, will be referred to as "anchors."

CAUTION

During all lead climbing, each climber in the team is either anchored or being belayed.

a. Lead climbing with two climbers is the preferred combination for movement on technically difficult terrain. Two climbers are at least twice as fast as three climbers, and are efficient for installing a "fixed rope," probably the most widely used rope installation in the mountains. A group of three climbers are typically used on moderate snow, ice, and snow-covered glaciers where the rope team can often move at the same time, stopping occasionally to set up belays on particularly difficult sections. A group or team of three climbers is sometimes used in rock climbing because of an odd number of personnel, a shortage of ropes (such as six climbers and only two ropes), or to protect and assist an individual who has little or no experience in climbing and belaying. Whichever technique is chosen, a standard roped climbing procedure is used for maximum speed and safety.

b. When the difficulty of the climbing is within the "leading ability" of both climbers, valuable time can be saved by "swinging leads." This is normally the most efficient method for climbing multipitch routes. The second finishes cleaning the first pitch and continues climbing, taking on the role of lead climber. Unless he requires equipment from the other rack or desires a break, he can climb past the belay and immediately begin leading. The belayer simply adjusts his position, re-aiming the belay once the new leader begins placing protection. Swinging leads, or "leap frogging," should be planned before starting the climb so the leader knows to anchor the upper belay for both upward and downward pulls during the setup.

c. The procedures for conducting a lead climb with a group of two are relatively simple. The most experienced individual is the "lead" climber or leader, and is responsible for selecting the route. The leader must ensure the route is well within his ability and the ability of the second. The lead climber carries most of the climbing equipment in order to place protection along the route and set up the next belay. The leader must also ensure that the second has the necessary equipment, such as a piton hammer, nut tool, etc., to remove any protection that the leader may place.

(1) The leader is responsible for emplacing protection frequently enough and in such a manner that, in the event that either the leader or the second should fall, the fall will be neither long enough nor hard enough to result in injury. The leader must also ensure that the rope is routed in a way that will allow it to run freely through the protection placements, thus minimizing friction, or "rope drag".

(2) The other member of the climbing team, the belayer (sometimes referred to as the "second"), is responsible for belaying the leader, removing the belay anchor, and retrieving the protection placed by the leader between belay positions (also called "cleaning the pitch").

(3) Before the climb starts, the second will normally set up the first belay while the leader is arranging his rack. When the belay is ready, the belayer signals, "BELAY ON", affirming that the belay is "on" and the rope will be managed as necessary. When the leader is ready, he double checks the belay. The leader can then signal, "CLIMBING", only as a courtesy, to let the belayer know he is ready to move. The belayer can reply with "CLIMB", again, only as a courtesy, reaffirming that the belay is "on" and the rope will be managed as necessary. The leader then begins climbing.

(4) While belaying, the second must pay close attention to the climber's every move, ensuring that the rope runs free and does not inhibit the climber's movements. If he cannot see the climber, he must "feel" the climber through the rope. Unless told otherwise by the climber, the belayer can slowly give slack on the rope as the climber proceeds on the route. The belayer should keep just enough slack in the rope so the climber does not have to pull it through the belay. If the climber wants a tighter rope, it can be called for. If the belayer notices too much slack developing in the rope, the excess

rope should be taken in quickly. It is the belayer's responsibility to manage the rope, whether by sight or feel, until the climber tells him otherwise.

(5) As the leader protects the climb, slack will sometimes be needed to place the rope through the carabiner (clipping), in a piece of protection above the tie-in point on the leaders harness. In this situation, the leader gives the command "SLACK" and the belayer gives slack, (if more slack is needed the command will be repeated). The leader is able to pull a bight of rope above the tie-in point and clip it into the carabiner in the protection above. When the leader has completed the connection, or the clip, the command "TAKE ROPE" is given by the leader and the belayer takes in the remaining slack.

(6) The leader continues on the route until either a designated belay location is reached or he is at the end of or near the end of the rope. At this position, the leader sets an anchor, connects to the anchor and signals "OFF BELAY". The belayer prepares to climb by removing all but at least one of his anchors and secures the remaining equipment. The belayer remains attached to at least one anchor until the command "BELAY ON" is given.

d. When the leader selects a particular route, he must also determine how much, and what types, of equipment might be required to safely negotiate the route. The selected equipment must be carried by the leader. The leader must carry enough equipment to safely protect the route, additional anchors for the next belay, and any other items to be carried individually such as rucksacks or individual weapons.

(1) The leader will assemble, or "rack," the necessary equipment onto his harness or onto slings around the head and shoulder. A typical leader "rack" consists of:

- Six to eight small wired stoppers on a carabiner.
- Four to six medium to large wired stoppers on a carabiner.
- Assorted hexentrics, each on a separate carabiner.
- SLCDs of required size, each on a separate carabiner.
- Five to ten standard length runners, with two carabiners on each.
- Two to three double length runners, with two carabiners on each.
- Extra carabiners.
- Nut tool.

Note: The route chosen will dictate, to some degree, the necessary equipment. Members of a climbing team may need to consolidate gear to climb a particular route.

(2) The belayer and the leader both should carry many duplicate items while climbing.

- Short Prusik sling.
- Long Prusik sling.
- Cordellette.
- 10 feet of 1-inch webbing.
- 20 feet of 1-inch webbing.
- Belay device (a combination belay/rappel device is multifunctional).
- Rappel device (a combination belay/rappel device is multifunctional).
- Large locking carabiner (pear shape carabiners are multifunctional).

- Extra carabiners.
- Nut tool (if stoppers are carried).

Note: If using an over the shoulder gear sling, place the items in order from smallest to the front and largest to the rear.

e. Leading a difficult pitch is the most hazardous task in roped climbing. The lead climber may be exposed to potentially long, hard falls and must exercise keen judgment in route selection, placement of protection, and routing of the climbing rope through the protection. The leader should try to keep the climbing line as direct as possible to the next belay to allow the rope to run smoothly through the protection with minimal friction. Protection should be placed whenever the leader feels he needs it, and BEFORE moving past a difficult section.

> **CAUTION**
> The climber must remember he will fall twice the distance from his last piece of protection before the rope can even begin to stop him.

(1) ***Placing Protection.*** Generally, protection is placed from one stable position to the next. The anchor should be placed as high as possible to reduce the potential fall distance between placements. If the climbing is difficult, protection should be placed more frequently. If the climbing becomes easier, protection can be placed farther apart, saving hardware for difficult sections. On some routes an extended diagonal or horizontal movement, known as a traverse, is required. As the leader begins this type of move, he must consider the second's safety as well as his own. The potential fall of the second will result in a pendulum swing if protection is not adequate to prevent this. The danger comes from any objects in the swinging path of the second.

> **CAUTION**
> Leader should place protection prior to, during, and upon completion of any traverse. This will minimize the potential swing, or pendulum, for both the leader and second if either should fall.

(2) ***Correct Clipping Technique.*** Once an anchor is placed, the climber "clips" the rope into the carabiner (Figure 6-29, page 6-50). As a carabiner hangs from the protection, the rope can be routed through the carabiner in two possible ways. One way will allow the rope to run smoothly as the climber moves past the placement; the other way will often create a dangerous situation in which the rope could become "unclipped" from the carabiner if the leader were to fall on this piece of protection. In addition, a

series of incorrectly clipped carabiners may contribute to rope drag. When placing protection, the leader must ensure the carabiner on the protection does not hang with the carabiner gate facing the rock; when placing protection in a crack ensure the carabiner gate is not facing into the crack.

- Grasp the rope with either hand with the thumb pointing down the rope towards the belayer
- Pull enough rope to reach the carabiner with a bight
- Note the direction the carabiner is hanging from the protection
- Place the bight into the carabiner so that, when released, the rope does not cause the carabiner to twist.

(a) If the route changes direction, clipping the carabiner will require a little more thought. Once leaving that piece of protection, the rope may force the carabiner to twist if not correctly clipped. If the clip is made correctly, a rotation of the clipped carabiner to ensure that the gate is not resting against the rock may be all that is necessary.

CAUTION
Ensure the carabiner gate is not resting against a protrusion or crack edge in the rock surface; the rock may cause the gate to open.

(b) Once the rope is clipped into the carabiner, the climber should check to see that it is routed correctly by pulling on the rope in the direction it will travel when the climber moves past that position.

(c) Another potential hazard peculiar to leading should be eliminated before the climber continues. The carabiner is attached to the anchor or runner with the gate facing away from the rock and opening down for easy insertion of the rope. However, in a leader fall, it is possible for the rope to run back over the carabiner as the climber falls below the placement. If the carabiner is left with the gate facing the direction of the route there is a chance that the rope will open the gate and unclip itself entirely from the placement. To prevent this possibility, the climber should ensure that after the clip has been made, the gate is facing away from the direction of the route. There are two ways to accomplish this: determine which direction the gate will face before the protection or runner is placed or once clipped, rotate the carabiner upwards 180 degrees. This problem is more apt to occur if bent gate carabiners are used. Straight gate ovals or "Ds" are less likely to have this problem and are stronger and are highly recommended. Bent gate carabiners are easier to clip the rope into and are used mostly on routes with bolts preplaced for protection. Bent gate carabiners are not recommended for many climbing situations.

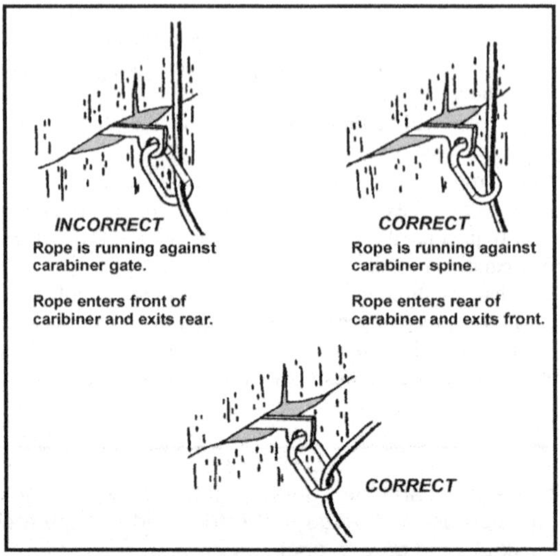

Figure 6-29. Clipping on to protection.

(3) ***Reducing Rope Drag; Using Runners***. No matter how direct the route, the climber will often encounter problems with "rope drag" through the protection positions. The friction created by rope drag will increase to some degree every time the rope passes through a carabiner, or anchor. It will increase dramatically if the rope begins to "zigzag" as it travels through the carabiners. To prevent this, the placements should be positioned so the rope creates a smooth, almost straight line as it passes through the carabiners (Figure 6-30). Minimal rope drag is an inconvenience; severe rope drag may actually pull the climber off balance, inducing a fall.

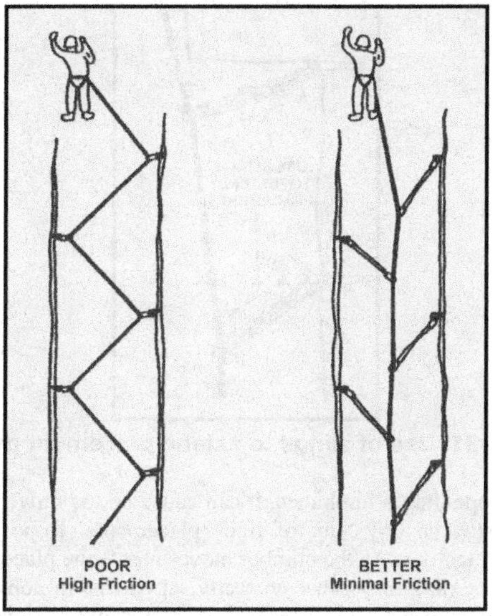

Figure 6-30. Use of slings on protection.

CAUTION

Rope drag can cause confusion when belaying the second or follower up to a new belay position. Rope drag can be mistaken for the climber, causing the belayer to not take in the necessary slack in the rope and possibly resulting in a serious fall.

(a) If it is not possible to place all the protection so the carabiners form a straight line as the rope moves through, you should "extend" the protection (Figure 6-31, page 6-52). Do this by attaching an appropriate length sling, or runner, to the protection to extend the rope connection in the necessary direction. The runner is attached to the protection's carabiner while the rope is clipped into a carabiner at the other end of the runner. Extending placements with runners will allow the climber to vary the route slightly while the rope continues to run in a relatively straight line.

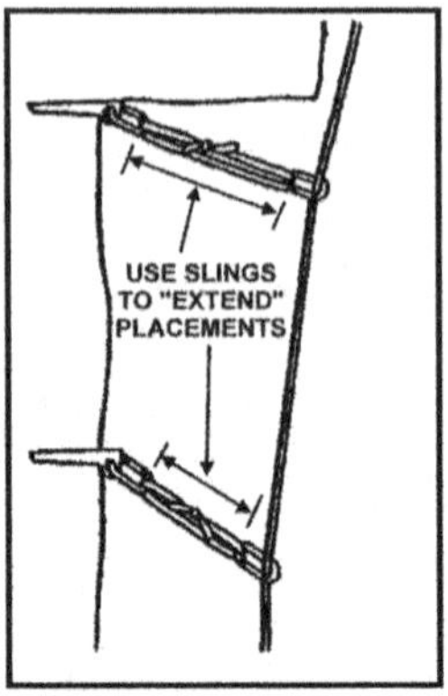

Figure 6-31. Use of slings to extend placement positions.

(b) Not only is rope drag a hindrance, it can cause undue movement of protection as the rope tightens between any "out of line" placements. Rope drag through chock placements can be dangerous. As the climber moves above the placements, an outward or upward pull from rope drag may cause correctly set chocks to pop out, even when used "actively". Most all chocks placed for leader protection should be extended with a runner, even if the line is direct to eliminate the possibility of movement.

(c) Wired chocks are especially prone to wiggling loose as the rope pulls on the stiff cable attachment. All wired chocks used for leader protection should be extended to reduce the chance of the rope pulling them out (Figure 6-32). Some of the larger chocks, such as roped Hexentrics and Tri-Cams, have longer slings pre-attached that will normally serve as an adequate runner for the placement. Chocks with smaller sling attachments must often be extended with a runner. Many of today's chocks are manufactured with pre-sewn webbing installed instead of cable.

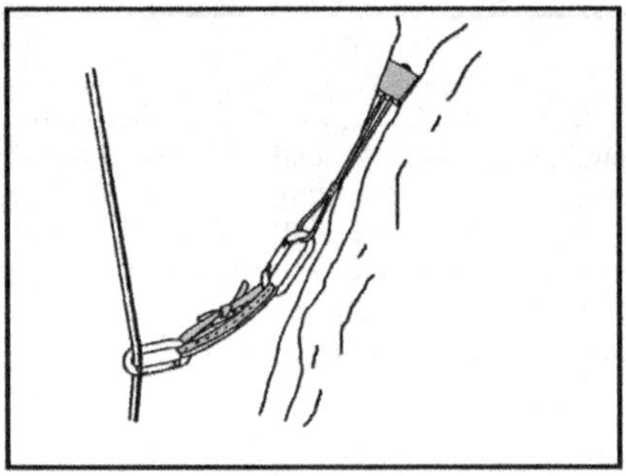

Figure 6-32. Use of sling on a wired stopper.

(d) When a correctly placed piton is used for protection, it will normally not be affected by rope drag. A correctly placed piton is generally a multi-directional anchor, therefore, rope drag through pitons will usually only affect the leader's movements but will continue to protect as expected.

(e) Rope drag will quite often move SLCDs out of position, or "walk" them deeper into the crack than initially placed, resulting in difficult removal or inability to remove them at all. Furthermore, most cases of SLCD movement result in the SLCD moving to a position that does not provide protection in the correct direction or no protection at all due to the lobes being at different angles from those at the original position.

Note: Any placement extended with a runner will increase the distance of a potential fall by the actual length of the sling. Try to use the shortest runners possible, ensuring they are long enough to function properly.

f. Belaying the follower is similar to belaying a top-roped climb in that the follower is not able to fall any farther than rope stretch will allow. This does not imply there is no danger in following. Sharp rocks, rock fall, and inadequately protected traverses can result in damage to equipment or injury to the second.

g. Following, or seconding, a leader has a variety of responsibilities. The second has to issue commands to the leader, as well as follow the leader's commands. Once the lead climber reaches a good belay position, he immediately establishes an anchor and connects to it. When this is completed he can signal "OFF BELAY" to the belayer. The second can now remove the leader's belay and prepare to climb. The second must remain attached to at least one of the original anchors while the leader is preparing the next belay position. The removed materials and hardware can be organized and secured on the second's rack in preparation to climb.

(1) When the leader has established the new belay position and is ready to belay the follower, the "new" belayer signals "BELAY ON." The second, now the climber, removes any remaining anchor hardware/materials and completes any final preparations. The belayer maintains tension on the rope, unless otherwise directed, while the final preparations are taking place, since removal of these remaining anchors can introduce slack into the rope. When the second is ready, he can, as a courtesy, signal "CLIMBING," and the leader can, again as a courtesy, reply with "CLIMB."

(2) Upon signaling "BELAY ON," the belayer must remove and keep all slack from the rope. (This is especially important as in many situations the belayer cannot see the follower. A long pitch induces weight and sometimes "drag" on the rope and the belayer above will have difficulty distinguishing these from a rope with no slack.)

h. When removing protection, the man cleaning the pitch should rack it properly to facilitate the exchange and or arrangement of equipment at the end of the pitch. When removing the protection, or "cleaning the pitch", SLCDs or chocks may be left attached to the rope to prevent loss if they are accidentally dropped during removal. If necessary, the hardware can remain on the rope until the second reaches a more secure stance. If removing a piton, the rope should be unclipped from the piton to avoid the possibility of damaging the rope with a hammer strike.

(1) The second may need to place full body weight on the rope to facilitate use of both hands for protection removal by giving the command "TENSION." The second

must also ensure that he does not climb faster than the rope is being taken in by the belayer. If too much slack develops, he should signal "TAKE ROPE" and wait until the excess is removed before continuing the climb. Once the second completes the pitch, he should immediately connect to the anchor. Once secured, he can signal "OFF BELAY." The leader removes the belay, while remaining attached to an anchor. The equipment is exchanged or organized in preparation for the next pitch or climb.

(2) When the difficulty of the climbing is within the "leading ability" of both climbers, valuable time can be saved by "swinging leads." This is normally the most efficient method for climbing multi-pitch routes. The second finishes cleaning the first pitch and continues climbing, taking on the role of lead climber. Unless he requires equipment from the belayer or desires a break, he can climb past the belay and immediately begin leading. The belayer simply adjusts his position, re-aiming the belay once the new leader begins placing protection. Swinging leads, or "leap frogging," should be planned before starting the climb so the leader knows to anchor the upper belay for both upward and downward pulls during the setup.

6-23. AID CLIMBING

When a route is too difficult to free climb and is unavoidable, if the correct equipment is available you might aid climb the route. Aid climbing consists of placing protection and putting full body weight on the piece. This allows you to hang solely on the protection you place, giving you the ability to ascend more difficult routes than you can free climb. Clean aid consists of using SLCDs and chocks, and is the simplest form of aid climbing.

a. **Equipment**. Aid climbing can be accomplished with various types of protection. Regardless of the type of protection used, the method of aid climbing is the same. In addition to the equipment for free climbing, other specialized equipment will be needed.

(1) ***Pitons***. Pitons are used the same as for free climbing. Most piton placements will require the use of both hands. Piton usage will usually leave a scar in the rock just by virtue of the hardness of the piton and the force required to set it with a hammer. Swinging a hammer to place pitons will lead to climber fatigue sooner than clean aid. Since pitons are multidirectional, the strength of a well-placed piton is more secure than most clean aid protection. Consider other forms of protection when noise could be hazardous to tactics.

(2) ***Bolts***. Bolts are used when no other protection will work. They are a more permanent form of protection and more time is needed to place them. Placing bolts creates more noise whether drilled by hand or by motorized drill. Bolts used in climbing are a multi-part expanding system pounded into predrilled holes and then tightened to the desired torque with a wrench or other tool. Bolts are used in many ways in climbing today. The most common use is with a hanger attached and placed for anchors in face climbing. However, bolts can be used for aid climbing, with or without the hanger.

(a) Placing bolts for aid climbing takes much more time than using pitons or clean aid. Bolting for aid climbing consists of consecutive bolts about 2 feet apart. Drilling a deep enough hole takes approximately thirty minutes with a hand drill and up to two minutes with a powered hammer drill. A lot of time and work is expended in a short distance no matter how the hole is drilled. (The weight of a powered hammer drill becomes an issue in itself.) Noise will also be a factor in both applications. A constant pounding with a hammer on the hand drill or the motorized pounding of the powered drill may alert the enemy to the position. The typical climbing bolt/hanger combination normally is left in the hole where it was placed.

(b) Other items that can be used instead of the bolt/hanger combination are the removable and reusable "spring-loaded removable bolts" such as rivets (hex head threaded bolts sized to fit tightly into the hole and pounded in with a hammer), split-shaft rivets, and some piton sizes that can be pounded into the holes. When using rivets or bolts without a hanger, place a loop of cable over the head and onto the shaft of the rivet or bolt and attach a carabiner to the other end of the loop (a stopper with the chock slid back will suffice). Rivet hangers are available that slide onto the rivet or bolt after it is placed and are easily removed for reuse. Easy removal means a slight loss of security while in use.

(3) ***SLCDs.*** SLCDs are used the same as for free climbing, although in aid climbing, full body weight is applied to the SLCD as soon as it is placed.

(4) ***Chocks.*** Chocks are used the same as for free climbing, although in aid climbing, weight is applied to the chock as soon as it is placed.

(5) ***Daisy Chains.*** Daisy chains are tied or presewn loops of webbing with small tied or presewn loops approximately every two inches. The small loops are just large enough for two or three carabiners. Two daisy chains should be girth-hitched to the tie-in point in the harness.

(6) ***Etriers (or Aiders).*** Etriers (aiders) are tied or presewn webbing loops with four to six tied or presewn internal loops, or steps, approximately every 12 inches. The internal loops are large enough to easily place one booted foot into. At least two etriers (aiders) should be connected by carabiner to the free ends of the daisy chains.

(7) ***Fifi Hook.*** A fifi hook is a small, smooth-surfaced hook strong enough for body weight. The fifi hook should be girth-hitched to the tie-in point in the harness and is used in the small loops of the daisy chain. A carabiner can be used in place of the fifi hook, although the fifi hook is simpler and adequate.

(8) ***Ascenders.*** Ascenders are mechanical devices that will move easily in one direction on the rope, but will lock in place if pushed or pulled the other direction. (Prusiks can be used but are more difficult than ascenders.)

b. **Technique.** The belay will be the same as in normal lead climbing and the rope will be routed through the protection the same way also. The big difference is the movement up the rock. With the daisy chains, aiders, and fifi hook attached to the rope tie-in point of the harness as stated above, and secured temporarily to a gear loop or gear sling, the climb continues as follows:

(1) The leader places the first piece of protection as high as can safely be reached and attaches the appropriate sling/carabiner

(2) Attach one daisy chain/aider group to the newly placed protection

(3) Clip the rope into the protection, (the same as for normal lead climbing)

(4) Insure the protection is sound by weighting it gradually; place both feet, one at a time, into the steps in the aider, secure your balance by grasping the top of the aider with your hands.

(5) When both feet are in the aider, move up the steps until your waist is no higher than the top of the aider.

(6) Place the fifi hook (or substituted carabiner) into the loop of the daisy chain closest to the daisy chain/aider carabiner, this effectively shortens the daisy chain;

maintain tension on the daisy chain as the hook can fall out of the daisy chain loop if it is unweighted.

Note: Moving the waist higher than the top of the aider is possible, but this creates a potential for a fall to occur even though you are on the aider and "hooked" close to the protection with the daisy chain. As the daisy chain tie-in point on the harness moves above the top of the aider, you are no longer supported from above by the daisy chain, you are now standing above your support. From this height, the fifi hook can easily fall out of the daisy chain loop if it is unweighted. If this happens, you could fall the full length of the daisy chain resulting in a static fall on the last piece of protection placed.

(7) Release one hand from the aider and place the next piece of protection, again, as high as you can comfortably reach; if using pitons or bolts you may need both hands free- "lean" backwards slowly, and rest your upper body on the daisy chain that you have "shortened" with the fifi hook

(8) Clip the rope into the protection

(9) Attach the other daisy chain/aider group to the next piece of protection

(10) Repeat entire process until climb is finished

c. **Seconding.** When the pitch is completed, the belayer will need to ascend the route. To ascend the route, use ascenders instead of Prusiks, ascenders are much faster and safer than Prusiks. Attach each ascender to a daisy chain/aider group with carabiners. To adjust the maximum reach/height of the ascenders on the rope, adjust the effective length of the daisy chains with a carabiner the same as with the fifi hook; the typical height will be enough to hold the attached ascender in the hand at nose level. When adjusted to the correct height, the arms need not support much body weight. If the ascender is too high, you will have difficulty reaching and maintaining a grip on the handle.

(1) Unlike lead climbing, there will be a continuous load on the rope during the cleaning of the route, this would normally increase the difficulty of removing protection. To make this easier, as you approach the protection on the ascenders, move the ascenders, one at a time, above the piece. When your weight is on the rope above the piece, you can easily unclip and remove the protection.

CAUTION

If both ascenders should fail while ascending the pitch, a serious fall could result. To prevent this possibility, *tie-in short* on the rope every 10-20 feet by tying a figure eight loop and clipping it into the harness with a separate locking carabiner as soon as the ascent is started. After ascending another 20 feet, repeat this procedure. Do not unclip the previous figure eight until the new knot is attached to another locking carabiner. Clear each knot as you unclip it.

Notes: 1. Ensure the loops formed by the short tie-ins do not catch on anything below as you ascend.

2. If the nature of the rock will cause the "hanging loop" of rope, formed by tying in at the end of the rope, to get caught as you move upward, do not tie into the end of the rope.

(2) Seconding an aid pitch can be done in a similar fashion as seconding free-climbed pitches. The second can be belayed from above as the second "climbs" the protection. However, the rope is unclipped from the protection before the aider/daisy chain is attached.

d. **Seconding Through a Traverse.** While leading an aid traverse, the climber is hanging on the protection placed in front of the current position. If the second were to clean the section by hanging on the rope while cleaning, the protection will be pulled in more than one direction, possibly resulting in the protection failing. To make this safer and easier, the second should hang on the protection just as the leader did. As the second moves to the beginning of the traverse, one ascender/daisy chain/aider group is removed from the rope and clipped to the protection with a carabiner, (keep the ascenders attached to the daisy chain/aider group for convenience when the traverse ends). The second will negotiate the traverse by leapfrogging the daisy chain/aider groups on the next protection just as the leader did. Cleaning is accomplished by removing the protection as it is passed when all weight is removed from it. This is in effect a self-belay. The second maintains a shorter safety tie-in on the rope than for vertical movement to reduce the possibility of a lengthy pendulum if the protection should pull before intended.

e. **Clean Aid Climbing**. Clean aid climbing consists of using protection placed without a hammer or drill involvement: chocks, SLCDs, hooks, and other protection placed easily by hand. This type of aid climbing will normally leave no trace of the climb when completed. When climbing the aiders on clean aid protection, ensure the protection does not "move" from it's original position.

(1) Hooks are any device that rests on the rock surface without a camming or gripping action. Hooks are just what the name implies, a curved piece of hard steel with a hole in one end for webbing attachment. The hook blade shape will vary from one model to another, some have curved or notched "blades" to better fit a certain crystal shape on a face placement. These types of devices due to their passive application, are only secure while weighted by the climber.

(2) Some featureless sections of rock can be negotiated with hook use, although bolts can be used. Hook usage is faster and quieter but the margin of safety is not there unless hooks are alternated with more active forms of protection. If the last twenty foot section of a route is negotiated with hooks, a forty foot fall could result.

6-24. THREE-MAN CLIMBING TEAM

Often times a movement on steep terrain will require a team of more than two climbers, which involves more difficulties. A four-man team (or more) more than doubles the difficulty found in three men climbing together. A four-man team should be broken down into two groups of two unless prevented by a severe lack of gear.

a. Given one rope, a three-man team is at a disadvantage on a steep, belayed climb. It takes at least twice as long to climb an average length pitch because of the third climber and the extra belaying required. The distance between belay positions will be halved if only one rope is used because one climber must tie in at the middle of the rope. Two ropes are recommended for a team of three climbers.

Note: Time and complications will increase when a three-man team uses only one rope. For example: a 100-foot climb with a 150-foot rope would normally require two belays for two climbers; a 100-foot climb with a 150-foot rope would require six belays for three climbers.

b. At times a three-man climb may be unavoidable and personnel should be familiar with the procedure. Although a team of three may choose from many different methods, only two are described below. If the climb is only one pitch, the methods will vary.

CAUTION

When climbing with a team of three, protected traverses will require additional time. The equipment used to protect the traverse must be left in place to protect both the second and third climbers.

(1) The first method can be used when the belay positions are not large enough for three men. If using one rope, two climbers tie in at each end and the other at the mid point. When using two ropes, the second will tie in at one end of both ropes, and the other two climbers will each tie in to the other ends. The most experienced individual is the leader, or number 1 climber. The second, or number 2 climber, is the stronger of the remaining two and will be the belayer for both number 1 and number 3. Number 3 will be the last to climb. Although the number 3 climber does no belaying in this method, each climber should be skilled in the belay techniques required. The sequence for this method (in one pitch increments) is as follows (repeated until the climb is complete):

(a) Number 1 ascends belayed by number 2. Number 2 belays the leader up the first pitch while number 3 is simply anchored to the rock for security (unless starting off at ground level) and manages the rope between himself and number 2. When the leader completes the pitch, he sets up the next belay and belays number 2 up.

(b) Number 2 ascends belayed by number 1, and cleans the route (except for traverses). Number 2 returns the hardware to the leader and belays him up the next pitch. When the leader completes this pitch, he again sets up a new belay. When number 2 receives "OFF BELAY" from the leader, he changes ropes and puts number 3 on belay. He should not have to change anchor attachments because the position was already aimed for a downward as well as an upward pull when he belayed the leader.

(c) Number 3 ascends belayed by number 2. When number 3 receives "BELAY ON," he removes his anchor and climbs to number 2's position. When the pitch is completed he secures himself to one of number 2's belay anchors. When number 1's belay is ready, he brings up number 2 while number 3 remains anchored for security. Number 2 again cleans the pitch and the procedure is continued until the climb is completed.

(d) In this method, number 3 performs no belay function. He climbs when told to do so by number 2. When number 3 is not climbing, he remains anchored to the rock for security. The standard rope commands are used; however, the number 2 climber may include the trailing climber's name or number in the commands to avoid confusion as to who should be climbing.

(d) Normally, only one climber would be climbing at a time; however, the number 3 climber could ascend a fixed rope to number 2's belay position using proper ascending technique, with no effect on the other two members of the team. This would save time for a team of three, since number 2 would not have to belay number 3 and could be either belaying number 1 to the next belay or climbing to number 1. If number 3 is to ascend a fixed rope to the next belay position, the rope will be loaded with number 3's weight, and positioned directly off the anchors established for the belay. The rope should be located so it does not contact any sharp edges. The rope to the ascending number 3 could be secured to a separate anchor, but this would require additional time and gear.

(2) The second method uses either two ropes or a doubled rope, and number 2 and number 3 climb simultaneously. This requires either a special belay device that accepts two ropes, such as the tuber type, or with two Munter hitches. The ropes must travel through the belay device(s) without affecting each other.

(a) As the leader climbs the pitch, he will trail a second rope or will be tied in with a figure eight in the middle of a doubled rope. The leader reaches the next belay position and establishes the anchor and then places both remaining climbers on belay. One remaining climber will start the ascent toward the leader and the other will start when a gap of at least 10 feet is created between the two climbers. The belayer will have to remain alert for differences in rope movement and the climbers will have to climb at the same speed. One of the "second" climbers also cleans the pitch.

(b) Having at least two experienced climbers in this team will also save time. The belayer will have additional requirements to meet as opposed to having just one second. The possible force on the anchor will be twice that of one second. The second that is not cleaning the pitch can climb off route, but staying on route will usually prevent a possible swing if stance is not maintained.

CHAPTER 7

ROPE INSTALLATIONS

Obstacles on the battlefield today are inevitable. They can limit the battlefield and, even worse, prevent a unit from accomplishing its mission. However, with highly skilled personnel trained on rope installations, leaders can be assured that even a unit with limited mountain skills and experience will be able to successfully move and operate in terrain that would otherwise have been impassable.

Section I. FIXED ROPE

A fixed rope is a rope anchored in place to assist soldiers in movement over difficult terrain. Its simplest form is a rope tied off on the top of steep terrain. As terrain becomes steeper or more difficult, fixed rope systems may require intermediate anchors along the route. Moving on a fixed rope requires minimal equipment. The use of harnesses, ascenders, and other technical gear makes fixed rope movement easier, faster, and safer, but adds to total mission weight.

7-1. INSTALLATION

To install a fixed rope, two experienced climbers rope up for a roped climb. The leader must have the necessary equipment to rig the anchor at the top of the pitch. Although leader protection is usually not needed on a typical slope, additional hardware can be brought along and placed at the leader's discretion. The second will establish a belay if protection is being placed. Otherwise, he will stack and manage the rope. He ensures the rope runs smoothly up the slope and does not get tangled as the climber ascends. Upon reaching the end of the pitch, the leader will establish the top anchor. Once the anchor is rigged, the leader will take up any remaining slack between himself and the second. He will anchor the installation rope and remain tied into the rope. The second unties from his end of the rope and begins to climb. If the leader placed protection, the second will clean the pitch on his way up.

7-2. UTILIZATION

All personnel using the fixed rope grasp the rope with the palm downward and use it for assistance as they ascend the slope (Figure 7-1, page 7-2). An individual can easily prevent a long fall by attaching himself to the rope with a sling using a friction knot (for example, Prusik, autoblock). The knot is slid along the rope as the individual ascends. If the climber slips and loses control of the rope, the friction knot will grab the rope and arrest the fall. The friction knot used in this manner is referred to as a self-belay (Figure 7-2, page 7-2).

Figure 7-1. Using a fixed rope.

Figure 7-2. Using a self-belay.

7-3. RETRIEVAL

If the fixed rope is to be used on the descent, it can be left in place and recovered after the last rappel. If not, the last climber will tie into the rope and be belayed from above. The climber now can easily free the rope if it gets caught on anything as it is taken up from the belayer.

7-4. FIXED ROPE WITH INTERMEDIATE ANCHORS

Whenever the route varies from the fall line of the slope, the fixed rope must be anchored at intermediate anchor points (Figure 7-3). Intermediate anchor points should also be used on any long routes that exceed the length of a single rope. The use of intermediate anchor points creates independent sections and allows for changes in direction from one section to the next. The independent sections allow for more personnel to move on the fixed rope. This type of fixed rope is commonly used along exposed ridges and narrow mountain passes.

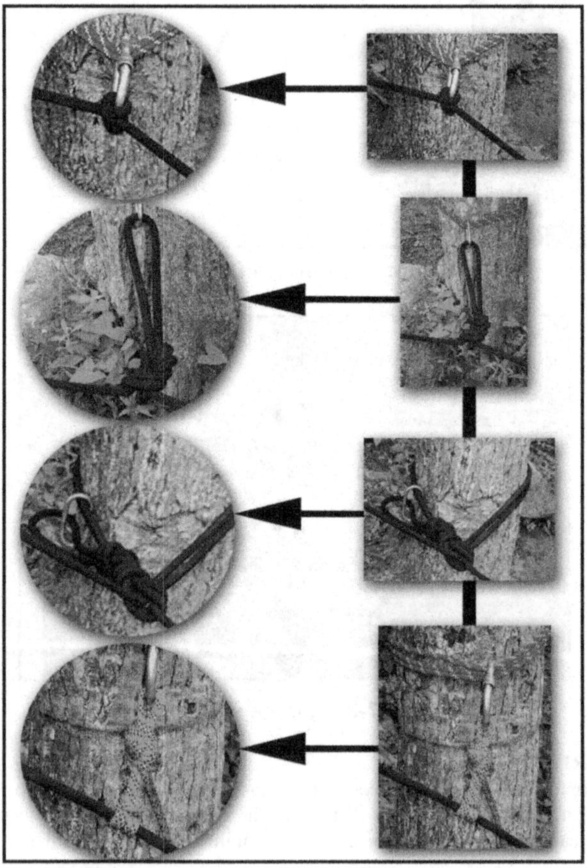

Figure 7-3. Fixed rope with intermediate anchors.

a. **Installation.** Two experienced climbers prepare for a roped climb. The leader will carry a typical rack with enough hardware to place an adequate number of intermediate anchor points. The second sets up a standard belay. The route they select must have the following characteristics:

- Most suitable location, ease of negotiation, avoids obstacles.
- Availability of anchors (natural and artificial).
- Area is safe from falling rock or ice.
- Tactical considerations are met.
- A rope routed between knee and chest height (waist high preferred).
- Rope crossovers should be avoided.

As the leader climbs the route, he will place the anchors and route the climbing rope as in a typical roped climb. The leader makes use of any available natural anchors.

Note: Sling attachments should be kept as short and snug as possible to ensure that a load on the fixed rope below the anchor is placed only on that anchor. This will prevent one section from affecting another section.

(1) The leader places an anchor at all points where a change of direction occurs. He also makes every attempt to route the rope so personnel will not have to cross back and forth over the rope between sections.

(2) When the leader reaches the end of the pitch, he temporarily anchors the rope. He should use a sling to anchor himself if there is any chance of slipping and falling. He then takes up any excess slack, and attaches the rope to the anchor.

Note: Enough slack must be left in the rope so the second can tie the knots necessary to fix the rope.

(3) The second unties from the rope and anchors it at the bottom. He attaches himself to the rope with a sling using a friction knot to create a self-belay. The self-belay will protect the second as he climbs and fixes the rope to the intermediate anchor points. When he reaches an anchor point, he unclips the climbing rope so he can advance the self-belay beyond the anchor point. He then takes the slack out of the section below the anchor point. He ensures that the fixed rope will be approximately knee to chest level as climbers negotiate the installation. He then attaches the rope using an anchor knot (for example, middle-of-the-rope clove hitch, double figure-eight). The second then moves to next anchor point and repeats the process.

(4) If a long runner is to be used at any anchor point, the second should adjust the section below it so the runner is oriented in the direction where the load or pull on the anchor will come from. This will help isolate the section.

(5) The sections are normally adjusted fairly snug between anchor points. A slack section may be necessary to move around obstacles in the route or large bulges in the terrain. If clove hitches are used, adjusting the clove hitches at each end of the section can leave any amount of slack.

(6) A middle-of-the-rope Prusik safetied with a figure eight may be used when utility ropes are available. These are used to adjust the rope height (either higher or lower).

(7) In addition to the fixed rope, the second could anchor etriers to be used as footholds.

(8) When the second reaches the end of the pitch, the rope is removed from the top anchor and the remaining slack is removed from the last section. The rope is reattached to the anchor. If additional fixed rope is required the procedure is repeated using another rope. The second will tie the ropes together before anchoring the next section, creating one continuous fixed rope.

b. **Utilization.** Personnel should be attached to the fixed rope during movement for safety reasons.

(1) If a self-belay is desired, a harness should be worn. A friction knot will be tied to the installation rope using a short sling. The sling will then be attached to the harness. Another short sling will be used as a safety line. One end of the sling will be attached to the harness and the other will have a carabiner inserted. This safety line is also attached to the fixed rope during movement. Once the climber reaches an anchor point, he removes his safety line and attaches it to the anchor or attaches it to the next section of rope. He will then untie the friction knot and tie another friction knot beyond the anchor point. The use of a mechanical ascender in the place of the friction knot could greatly speed up movement.

(2) There will be many situations where a self-belay may not be required. In these situations an individual may attach himself to the fixed rope using only a safety line. The individual will tie into the middle of a sling rope approximately 12 feet long. Fixed loops are tied into the running ends and a carabiner is attached into each of the fixed loops. The individual now has two points of attachments to the fixed rope. Upon reaching an anchor point, one safety line is removed and advanced beyond the anchor point onto the next section. Then the next safety is removed and placed on the next section. This way the individual is always secured to the fixed rope at all times.

(3) Personnel will move one at a time per section during the entire movement. Once an individual changes over to the next section he signals the next man to climb. When descending on the fixed rope, personnel can down climb using the installation for assistance. Another option would be to descend using a hasty rappel.

c. **Retrieval.** When the installation is retrieved, the next to last man on the system will untie the knots at the intermediate anchor points and reclips the rope as he ascends. He will be attached to rope using a self-belay. Once he reaches the top of the pitch, the rope should be running the same as when the leader initially placed it. The last man will untie the rope from the bottom anchor and tie into the rope. He will the clean the pitch as he climbs while being belayed from above.

Section II. RAPPELLING

When an individual or group must descend a vertical surface quickly, a rappel may be performed. Rappelling is a quick method of descent but it is extremely dangerous. These dangers include anchor failure, equipment failure, and individual error. Anchors in a mountainous environment should be selected carefully. Great care must be taken to load the anchor slowly and to ensure that no excessive stress is placed on the anchor. To ensure this, bounding rappels should be prohibited, and only walk down rappels used. Constant vigilance to every detail will guarantee a safe descent every time.

7-5. SELECTION OF A RAPPEL POINT

The selection of the rappel point depends on factors such as mission, cover, route, anchor points, and edge composition (loose or jagged rocks). There must be good anchors (primary and secondary). The anchor point should be above the rappeller's departure point. Suitable loading and off-loading platforms should be available.

7-6. INSTALLATION OF THE RAPPEL POINT

A rappel lane should have equal tension between all anchor points by establishing primary and secondary anchor points. The rappel rope should not extend if one anchor point fails. The following methods of establishing an anchor can be performed with a single or double rope. A double rope application should be used when possible for safety purposes.

a. If a rappel lane is less than half the rope length, the climber may apply one of the following techniques:

(1) Double the rope and tie a three-loop bowline around the primary anchor to include the primary anchor inside two loops and enough rope in the third loop to run to the secondary anchor (another three-loop bowline secured with an overhand knot).

(2) Bowline secured with an overhand knot (or any appropriate anchor knot).

(3) Double the rope and establish a self-equalizing anchor system with a three-loop bowline or any other appropriate anchor knot

b. If a rappel lane is greater than half the rope length, the climber may apply one of the following techniques:

(1) Use two ropes. With both ropes, tie a round turn anchor bowline around a primary anchor point. Take the remaining rope (the tail from the primary anchor bowline) and tie another round turn anchor bowline to a secondary anchor point. The secondary anchor point should be in a direct line behind the primary anchor point. The anchor can be either natural or artificial.

(2) Use two ropes. Establish a multi-point anchor system using a bowline on a bight or any other appropriate anchor knot.

c. Situations may arise where, due to the length of the rappel, the rappel rope cannot be tied to the anchor (if the rope is used to tie the knots, it will be too short to accomplish the rappel). The following techniques can be used:

(1) When using a natural anchor, tie a sling rope, piece of webbing, or another rope around the anchor using proper techniques for slinging natural anchors. The rappel rope will have a fixed loop tied in one end, which is attached to the anchor created.

(2) When using an artificial anchor, tie off a sling rope, piece of webbing, runner, or another rope to form a loop. Use this loop to create an equalizing or pre-equalized anchor, to which the rappel rope will be attached.

7-7. OPERATION OF THE RAPPEL POINT

Due to the inherent dangers of rappelling, special care must be taken to ensure a safe and successful descent.

a. **Communication.** Climbers at the top of a rappel point must be able to communicate with those at the bottom. During a tactical rappel, radios, hand signals, and rope signals are considered. For training situations use the commands shown in Table 7-1.

COMMAND	GIVEN BY	MEANING
LANE NUMBER ___, ON RAPPEL	Rappeller	I am ready to begin rappelling.
LANE NUMBER ___, ON BELAY	Belayer	I am on belay and you may begin your rappel.
LANE NUMBER ___, OFF RAPPEL	Rappeller	I have completed the rappel, cleared the rappel lane, and am off the rope.
LANE NUMBER ___, OFF BELAY	Belayer	I am off belay.

Table 7-1. Rappel commands.

Notes: 1. In a training environment, the lane number must be understood.
2. In a tactical situation, a series of tugs on the rope may be substituted for the oral commands to maintain noise discipline. The number of tugs used to indicate each command is IAW the unit SOP.

b. **Duties and Responsibilities.**

(1) Duties of the rappel point commander are as follows:

- Ensures that the anchors are sound and the knots are properly tied.
- Ensures that loose rock and debris are cleared from the loading platform.
- Allows only one man on the loading platform at a time and ensures that the rappel point is run orderly.
- Ensures that each man is properly prepared for the particular rappel: gloves on, sleeves down, helmet with chin strap fastened, gear prepared properly, and rappel seat and knots correct (if required). He also ensures that the rappeller is hooked up to the rope correctly and is aware of the proper braking position.
- Ensures that the proper signals or commands are used.
- Dispatches each man down the rope.
- Is the last man down the rope.

(2) Duties of the first rappeller down are as follows:

- Selects a smooth route, for the rope, that is clear of sharp rocks.
- Conducts a self-belay.
- Clears the route, placing loose rocks far enough back on ledges to be out of the way, which the rope may dislodge.
- Ensures the rope reaches the bottom or is at a place from which additional rappels can be made.
- Ensures that the rope will run freely around the rappel point when pulled from below.
- Clears the rappel lane by straightening all twists and tangles from the ropes.
- Belays subsequent rappellers down the rope or monitors subsequent belayers
- Takes charge of personnel as they arrive at the bottom (off-loading platform).

Note: A rappeller is always belayed from the bottom, except for the first man down. The first man belays himself down the rope using a self-belay attached to his rappel seat, which is hooked to the rappel rope with a friction knot. As the first man rappels down the rope, he "walks" the friction knot down with him.

(3) Each rappeller down clears the ropes, and shouts, "Off rappel," (if the tactical situation permits). After the rope is cleared and the rappeller is off rappel, he acts as the belayer for next rappeller.

(4) Soldiers wear gloves for all types of rappels to protect their hands from rope burns.

(5) Rappellers descend in a smooth, controlled manner.

(6) The body forms an L-shape with the feet shoulder-width apart, legs straight, and buttocks parallel to the ground. When carrying equipment or additional weight, a modified L-shape is used with the legs slightly lower than the buttocks to compensate for the additional weight. The rappeller's back is straight. He looks over the brake shoulder. The guide hand is extended on the rope with the elbow extended and locked. The rope slides freely through the guide hand. The guide hand is used to adjust equipment and assist balance during descent. The rappeller grasps the rope firmly with the brake hand and places it in the brake position. Releasing tension on the rope and moving the brake hand regulates the rate of descent. The rappeller never lets go of the ropes with his brake hand until the rappel is complete.

c. **Tying Off During the Rappel.** It may be necessary to stop during descent. This can be accomplished by passing the rope around the body and placing three or more wraps around the guide-hand-side leg, or by tying off using the appropriate knot for the rappel device.

7-8. RECOVERY OF THE RAPPEL POINT

After almost all personnel have descended, only two personnel will remain at the top of the rappel point. They will be responsible for establishing a retrievable rappel.

a. **Establishing the Retrievable Rappel.** To set up a retrievable rappel point, a climber must apply one of the following methods:

(1) Double the rope when the rappel is less than half the total length of the rope. Place the rope, with the bight formed by the midpoint, around the primary anchor. Join the tails of the rappel rope and throw the rope over the cliff. Tie a clove hitch around a carabiner, just below the anchor point, with the locking bar outside the carabiner away from the gate opening end and facing uphill. Snap the opposite standing portion into the carabiner. When the rappeller reaches the bottom, he pulls on that portion of the rope to which the carabiner is secured to allow the rope to slide around the anchor point.

(2) When the length of the rappel is greater than half the length of the rope used, join two ropes around the anchor point with an appropriate joining knot (except the square knot). Adjust the joining knot so that it is away from the anchor. Tie a clove hitch around a carabiner just below the anchor point with the locking bar outside the carabiner away from the gate opening end and facing uphill. Snap the opposite standing portion into the carabiner. Upon completion of the rappel, pull the rope to which the carabiner is secured to allow the rope to slide around the anchor point.

Notes: 1. When setting up a retrievable rappel, use only a primary point; care is taken in selecting the point.
2. Ensure the soldiers have a safety line when approaching the rappel point, with only the rappeller going near the edge.

b. **Retrieving the Rappel Rope.** The next to last rappeller will descend the lane, removing any twists, and routes the rope for easiest retrieval. Once he reaches the end of the rappel, he tests the rope for retrieval. If the rappel is retrievable, the last man will rappel down. Once he is off rappel, he pulls the lane down.

7-9. TYPES OF RAPPELS

During military mountaineering operations, many types of rappels may be used. The following paragraphs describe some these rappels.

a. **Hasty Rappel** (Figure 7-4). The hasty rappel is used only on moderate pitches. Its main advantage is that it is easier and faster than other methods. Gloves are worn to prevent rope burns.

(1) Facing slightly sideways to the anchor, the rappeller places the ropes horizontally across his back. The hand nearest to the anchor is his guide hand, and the other is the brake hand.

(2) To stop, the rappeller brings his brake hand across in front of his body locking the rope. At the same time, he turns to face up toward the anchor point.

Figure 7-4. Hasty rappel.

b. **Body Rappel** (Figure 7-5). The rappeller faces the anchor point and straddles the rope. He then pulls the rope from behind, and runs it around either hip, diagonally across the chest, and back over the opposite shoulder. From there, the rope runs to the brake hand, which is on the same side of the hip that the rope crosses (for example, the right hip to the left shoulder to the right hand). The rappeller leads with the brake hand down and faces slightly sideways. The foot corresponding to the brake hand precedes the guide hand at all times. The rappeller keeps the guide hand on the rope above him to guide himself--not to brake himself. He must lean out at a sharp angle to the rock. He keeps his legs spread well apart and relatively straight for lateral stability, and his back straight to reduce friction. The BDU collar is turned up to prevent rope burns on the neck. Gloves are worn, and other clothing may be used to pad the shoulders and buttocks. To brake, the rappeller leans back and faces directly toward the rock area so his feet are horizontal to the ground.

Figure 7-5. Body rappel.

Notes: 1. Hasty rappels and body rappels are not used on pitches that have overhangs; feet must maintain surface contact.

2. Hasty rappels and body rappels are not belayed from below.

c. **Seat-Hip Rappel** (Figure 7-6). The seat rappel differs from the body rappel in that the friction is absorbed by a carabiner that is inserted in a sling rope seat and fastened to the rappeller. This method provides a faster and more frictional descent than other methods. Gloves can be worn to prevent rope burns.

Figure 7-6. Seat-hip rappel.

(1) An alternate technique is to insert two carabiners opposite and opposed. Then insert a locking carabiner into the two carabiners with opening gate on brake hand side. Then run the rope through the single carabiner. This helps to keep the rappel rope away from the harness.

(2) To hook up for the seat-hip method, stand to one side of the rope. If using a right-hand brake, stand to the left of the rappel rope facing the anchor; if using a left-hand brake, stand to the right of the rappel rope. Place the rappel rope(s) into the locking carabiner; slack is taken between the locking carabiner and anchor point and wrapped around the shaft of the locking carabiner and placed into the gate so that a round turn is made around the shaft of the locking carabiner (Figure 7-7, page 7-12). Any remaining slack is pulled toward the uphill anchor point. If a single rope is used, repeat this process to place two round turns around the shaft of the locking carabiner. Face the anchor point and descend using the upper hand as the guide and the lower hand as the brake. This method has minimal friction, and is fast and safe. However, care is taken that the rope is hooked correctly into the carabiner to avoid the gate being opened by the rope. Loose clothing or equipment around the waist may be accidentally pulled into the locking carabiner and lock (stop) the rappel. For this reason, the rappeller must tuck in his shirt and keep his equipment out of the way during his descent.

Figure 7-7. Proper hookup using carabiner wrap.

d. **Figure-Eight Descender**. The figure-eight descender puts less kinks in the rope, and it can be used with one or two ropes (Figure 7-8).

(1) To use the figure-eight descender, pass a bight through the large eye and then over the small eye onto the neck. Place the small eye into a locking carabiner. To reduce the amount of friction on the figure-eight, place the original bight into the carabiner and not around the neck of the descender. (Less friction requires more braking force from the rappeller.)

(2) The guide hand goes on the rope that is running from the anchor. The brake hand goes on the slack rope. The brake is applied by moving the brake hand to the rear or downward.

Figure 7-8. Figure-eight descender.

d. **Other Devices**. Many different types of devices are similar in design and operation to the basic plate. These include slots or plates and tubers. Most of these devices can accommodate two ropes not greater than 7/16 of an inch in size. Follow manufacturer's directions for using these devices for rappelling.

e. **Extending the Rappel Device**. The rappel device can be extended using either a piece of webbing or cordage to move the device away from the body and the harness, preventing accidental damage (Figure 7-9, page 7-14). It also allows for easier self-belay.

f. **Self-Belay Techniques.** A friction knot can be used as a belay for a rappeller (Figure 7-9, page 7-14). The knot acts as the brake hand when the rappeller must work or negotiate an obstacle requiring the use of both hands. The knot acts as a belay if the rappeller loses control of the rope.

Figure 7-9. Extended hookup with self-belay.

Section III. ONE-ROPE BRIDGE

The one-rope bridge is constructed using a static rope. The rope is anchored with an anchor knot on the far side of the obstacle and is tied off at the near end with a tightening system. A one-rope bridge may be built many ways, depending upon the tactical situation and area to be crossed (crossing a gorge above the tree line may require constructing artificial anchors). However, they all share common elements to safely construct and use the bridge: two suitable anchors; good loading and unloading platforms; a rope about 1-meter (waist) high for loading and unloading; a tightening system; and a rope tight enough for ease of crossing. Which side the tightening system is utilized, or whether an anchor knot or retrievable bowline is used, depends on the technique.

7-10. SITE SELECTION

A suitable crossing site must have "bombproof" anchors on both the near side and far side. These anchors must be extremely strong due to the amount of tension that will be placed upon them. Natural anchors, such as large trees and solid rock formations, are always preferred. The site must also have suitable loading and off-loading platforms to facilitate safe personnel movement.

7-11. INSTALLATION USING TRANSPORT TIGHTENING SYSTEM

The transport tightening system provides a mechanical advantage without requiring additional equipment.

a. The rope must first be anchored on the far side of the obstacle. If crossing a stream, the swimmer must be belayed across. If crossing a ravine or gorge, crossing may involve rappelling and a roped climb. Once across, the swimmer/climber will temporarily anchor the installation rope.

b. One man on the near side ties a fixed-loop knot (for example, wireman's, figure-eight slip knot) approximately 3 feet from the near side anchor and places the carabiner into the loop of the knot. The opening gate must be up and away from the loop. If two carabiners are used, the gates will be opposing. At that time, soldiers route the remainder of the rope around the near side anchor point and hook the rope into the carabiner. This system is known as a transport-tightening system (Figure 7-10). The man on the far side pulls the knot out four to six feet from the near anchor.

c. Once the knot has been pulled out, the far side man anchors the rope using a tensionless anchor. The anchor should be waist high.

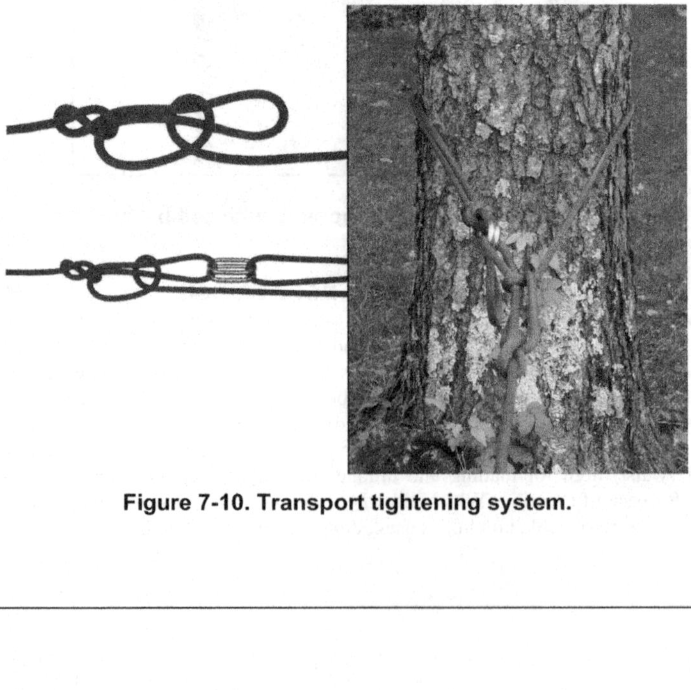

Figure 7-10. Transport tightening system.

d. A three-man pull team on the near side pulls the slack out of the installation rope. The knot should be close enough to the near side anchor to allow personnel to easily load the installation.

Note: No more than three personnel should be used to tighten the rope. Using more personnel can over-tighten the rope and bring the rope critically close to failure.

e. The rope the can be secured using one of three methods: transport knot (Figure 7-11), round turn around anchor and two half hitches on a bight (Figure 7-12), or a tensionless anchor knot (Figure 7-13).

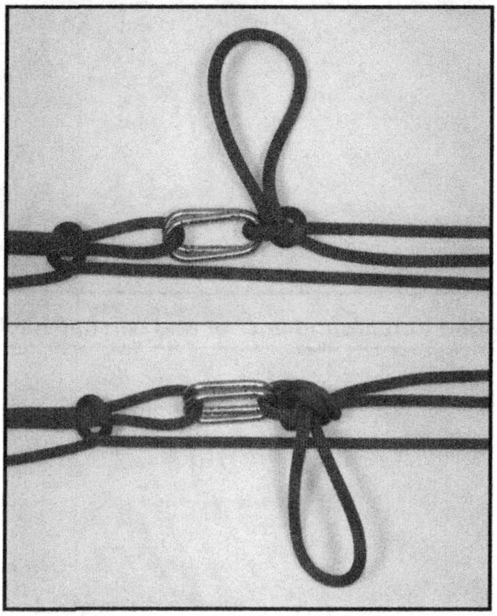

Figure 7-11. Transport knot.

Figure 7-12. Round turn around anchor and two half hitches on a bight.

Figure 7-13. Tensionless anchor knot.

Note: During training, a second static rope may be installed under less tension and alongside the tight rope to increase safety. An individual would clip into both ropes when crossing, thus having a backup in case of failure of the tighter rope.

7-12. INSTALLATION USING Z-PULLEY TIGHTENING SYSTEM

The Z-pulley tightening system (Figure 7-14) is another method for gaining a mechanical advantage.

a. The rope is brought across the obstacle the same way as discussed in paragraph 7-10.

b. Once across, the far side man anchors the rope.

c. One soldier ties a friction knot (autoblock, web wrap, Kleimheist) with a sling rope onto the bridging rope on the near side bank. Two steel carabiners are inserted with opposing gates into the friction knot.

d. The rope is routed around the near side anchor and through the carabiners, from inside to outside, and is run back to the near side anchor.

e. A second sling rope is tied to the bridge rope and then anchored to the near side anchor. This knot will be used as a progress capture device.

Figure 7-14. Z-pulley tightening system.

f. The three-man pull team on the near side then pulls on the rope, creating a pulley effect that tightens the system. As the rope is pulled tight, one man pushes the friction knot back toward the far side.

g. When the rope is tight, it is tied off with a tensionless anchor knot, transport knot, or round turn around anchor and two half hitches on a bight.

7-13. UTILIZATION

The rope bridge can be used to move personnel and equipment over obstacles. There are several methods of accomplishing this.

a. **Method of Crossing**. If dry crossing is impossible, soldiers will use the rope bridge as a hand line. Preferably, all soldiers will tie a safety line and attach it to the rope installation as they cross. If the soldier must cross with his rucksack, he may wear it over both shoulders, although the preferred method is to place another carabiner into the top of the rucksack frame, attach it to the bridge, and pull the rucksack across. Soldiers will always cross on the downstream side of the installation. If a dry crossing is possible soldiers will use one of three methods: commando crawl, monkey crawl, and Tyrolean traverse.

(1) ***Commando Crawl*** (Figure 7-15). The soldier lies on top of the rope with the upstream foot hooked on the rope and the knee bent close to the buttocks; the downstream leg hangs straight to maintain balance. He progresses by pulling with his hands and arms. To recover if he falls over, the soldier hooks one leg and the opposite arm over the rope, and then pushes down with the other hand to regain position.

Figure 7-15. Commando crawl.

Note: Only one man at a time is allowed on the bridge while conducting a commando crawl.

(2) ***Monkey Crawl*** (Figure 7-16, page 7-20). The soldier hangs below the rope suspended by his hands with both heels crossed over the rope. He pulls with his hands and arms, and pushes with his feet to make progress.

Figure 7-16. Monkey crawl.

(3) ***Rappel Seat Method*** (Figure 7-17). The soldier ties a rappel seat (or dons a seat harness) with the carabiner facing up and away from his body. He then faces the rope and clips into the rope bridge. He rotates under the rope and pulls with his hands and arms to make progress. The rappel seat method is the preferred method. If crossing with rucksacks, a carabiner is inserted into the frame and attached to the rope bridge. The soldier the places one or both legs through the shoulder carrying straps and pulls the rucksack across.

Figure 7-17. Rappel seat method.

b. **Rigging Special Equipment**. Any special equipment, such as crew-served weapons, ammunition, or supplies, must be rigged for movement across the rope bridge. A unit SOP may dictate the rigging of these items, but many expedient methods exist. The rigging should use various items that would be readily available to a deployed unit. Some of these items include tubular nylon webbing, cordage (various sizes), and carabiners.

(1) ***Machine Guns***. To rig machine guns, use a sling rope and tie a rerouted figure-eight around the spine of the front sight post. Then tie two evenly spaced fixed loops. Finally, anchor the sling rope to the buttstock of the machine gun. Additional tie downs may be necessary to prevent accidental disassembly of the weapon.

(2) ***ALICE Packs.*** ALICE packs can be joined together with a sling to facilitate moving more than one rucksack at one time.

7-14. HAULING LINE

A hauling line may be used to move rucksacks or casualties across the rope bridge (Figure 7-18).

a. **Construction**. An additional rope is brought across the rope bridge and anchored to the far side. The other end is anchored on the near side. All the slack is pulled to the near side, and a figure-eight slip knot is tied at the loading platform. A carabiner is inserted into the loop and clipped onto the rope bridge.

Figure 7-18. Hauling line.

b. **Moving Rucksacks**. Use carabiners to attach the rucksack frames to the rope bridge. Then clip the carabiner of the hauling line into the carabiner of the rucksack closest to the far side. Personnel on the far side pull the rucksacks across using the hauling line while personnel on the near side manages the slack at all times.

c. **Moving Litters**. The carabiner of the hauling line will remain on the rope bridge. On each side of this carabiner, using the hauling line tie a middle-of-the-rope clove hitch around both of the horizontal lift straps of the litter. Remove the slack between the carabiners. Then place the carabiners in each of the lift straps onto the rope bridge. The same technique used for the rucksacks is used to pull the litter across.

7-15. RETRIEVAL

Once all except two troops have crossed the rope bridge, the bridge team commander (BTC) chooses either the wet or dry method to dismantle the rope bridge.

a. If the BTC chooses the dry method, he should have anchored his tightening system with the transport knot.

(1) The BTC back-stacks all of the slack coming out of the transport knot, then ties a fixed loop and places a carabiner into the fixed loop.

(2) The next to last man to cross attaches the carabiner to his rappel seat or harness, and then moves across the bridge using the Tyrolean traverse method.

(3) The BTC then removes all knots from the system. The far side remains anchored. The rope should now only pass around the near side anchor.

(4) A three-man pull team, assembled on the far side, takes the end brought across by the next to last man and pulls the rope tight again and holds it.

(5) The BTC then attaches himself to the rope bridge and moves across.

(6) Once across, the BTC breaks down the far side anchor, removes the knots, and then pulls the rope across.

(b) If the BTC chooses a wet crossing, any method can be used to anchor the tightening system.

(1) All personnel cross except the BTC or the strongest swimmer.

(2) The BTC then removes all knots from the system.

(3) The BTC ties a fixed loop, inserts a carabiner, and attaches it to his rappel seat or harness. He then manages the rope as the slack is pulled to the far side.

(4) The BTC then moves across the obstacle while being belayed from the far side.

Section IV. SUSPENSION TRAVERSE

The suspension traverse is used to move personnel and equipment over rivers, ravines, chasms, and up or down a vertical rock face (Figure 7-19). The system may be established on a plane, varying from horizontal to near vertical.

Figure 7-19. Suspension traverse.

7-16. SITE SELECTION

The crossing site must have bombproof anchors at the near side and the far side, and suitable loading and off-loading platforms. If the anchors do not provide sufficient height to allow clearance, an A-frame must be used.

7-17. INSTALLATION

Installation of a suspension traverse can be time-consuming and equipment-intensive. All personnel must be well trained and well rehearsed in the procedures.

a. **A-frames**. Even in wooded mountainous terrain constructing an A-frame may be necessary due to the lack of height where the installation is needed. Site selection determines whether more height is needed; mission requirements determine site selection. The two main installations that use A-frames are the suspension traverse and vertical hauling line.

b. **Equipment**. Two sturdy poles are needed. The exact size of the poles depends on the type of load and location of the installation. The average size A-frame pole should be at least 3 inches in diameter and 9 to 12 feet long. Three to five 14-foot sling ropes are needed, depending on the size of the poles used for the A-frame.

c. **Construction**. Place two poles with the butt ends flush, and mark the apex on both poles.

(1) Ensure that proper height is attained and that the installation runs in a straight line between the two anchors. An A-frame placed out of proper alignment can cause the system to collapse. Try to find natural pockets in which to place the base of the A-frame poles.

(2) With a sling rope, tie a clove hitch around the left pole (standing at the base of the poles and facing the top) 3 inches above the apex marking, leaving about 18 inches of the sling rope free on top of the clove hitch. Place the locking bar on the outside edge of the pole. Make sure the rope end is pointing down as it is tied. (See Figure 7-20A, page 7-24.)

(3) Place the poles side by side and wrap the sling rope horizontally around both poles six to eight times, wrapping down from the clove hitch (Figure 7-20B, page 7-24). It may be necessary to join another sling rope to the first by using a square knot secured with overhand knots. Position this knot on the outside of one of the poles so as not to interfere with the vertical wraps. Make at least two additional wraps below the joining square knot. (See Figure 7-20C, page 7-24.)

(4) On the last horizontal wrap (ensure there are at least two wraps below the joining knot) to which the clove hitch is not tied, pass the rope between the poles below the wraps, and make four to six tight vertical wraps around the horizontal wraps (Figure 7-20D, page 7-24). Make the wraps as tight as possible. The vertical wraps must be as flat as possible next to each other. When starting the first vertical wrap, ensure it is in the same direction as the 18-inch tail on the top of the clove hitch. Insert a carabiner into the last two vertical wraps (Figure 7-20E, page 7-24).

(5) On the last vertical wrap, pass the rope between the poles above the horizontal wraps. Tie it off with a square knot in the section of rope coming from the clove hitch. Secure with overhand knots tied in the tails. (See Figure 7-20F, page 7-24.)

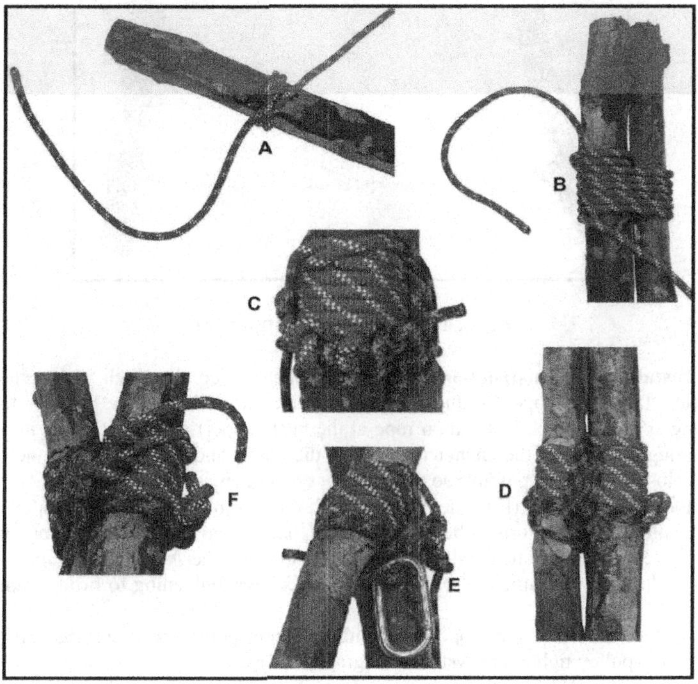

Figure 7-20. A-frame horizontal and vertical wraps.

(6) Use a spreader rope to prevent the A-frame from collapsing from pressure applied at the apex (Figure 7-21). If the ground is soft, dig the legs in about 6 inches. Tie a sling rope between the legs with a round turn with two half hitches around each leg. Remove all slack in the rope between the legs.

(7) If the ground is a hard surface, tie end-of-the-rope clove hitches with the locking portions facing to the rear, the direction of kick. Tie the tails off at a 45-degree angle with a round turn and two half hitches to a secondary anchor point. The spreader rope should be no more than 6 inches above ground level. The use of clove hitches and half hitches permits easy adjustment of the spreader rope. If more than one sling rope is needed, tie the two ropes together with a square knot and secure with half hitches or overhand knots.

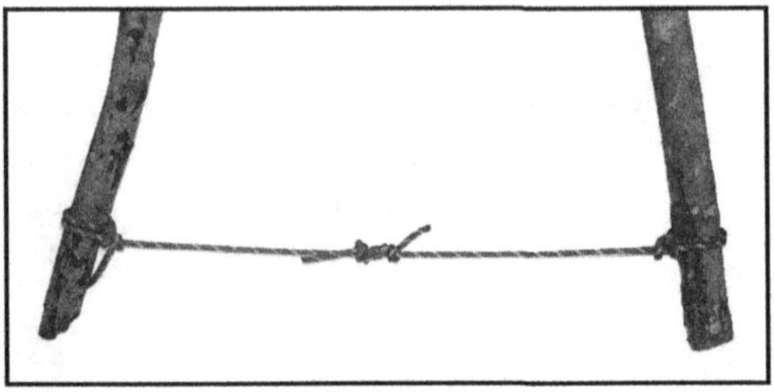

Figure 7-21. A-frame spreader.

d. **Installation Construction.** One man rappels down the pitch and secures two installation (traverse) ropes to the far anchor with an anchor knot. Place a transport tightening system in each installation rope at the near (upper) anchor. Run the installation ropes through or around the anchor in opposite directions and tie off. Anchor the traverse ropes as close together as possible so that the ropes do not cross.

(1) Place the A-frame (if needed) so that both traverse ropes run over the apex and the A-frame splits the angle formed between the near (upper) and far (lower) anchors, with the legs firmly emplaced or anchored with pitons. Ensure that the A-frame is in line with the anchors. Adjust the A-frame under the traverse ropes after tightening to firmly implant the A-frame.

(2) Tighten the installation ropes using either the transport tightening system (paragraph 7-11) or the z-pulley tightening system (paragraph 7-12).

(3) Anchor the A-frame to the traverse rope. Tie a clove hitch at the center of a sling rope. Place it over one of the poles above the apex and move down to the apex so that the locking bar of the clove hitch is to the inside of the A-frame. Secure each end of the sling rope to one of the tightened static lines with two Prusik knots-one forward and one to the rear of the A-frame on the same static line rope (Figure 7-22).

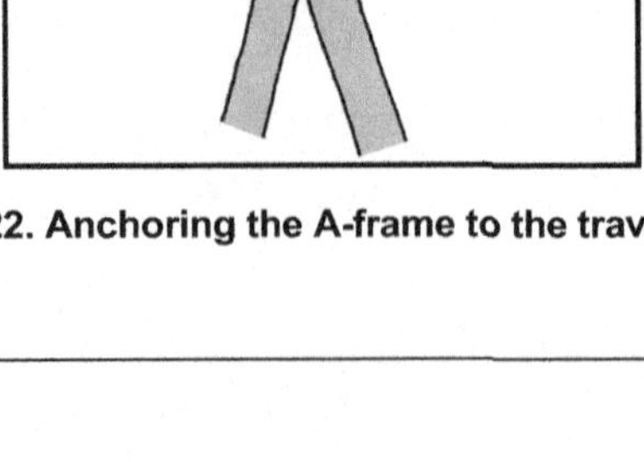

Figure 7-22. Anchoring the A-frame to the traverse rope.

Note: The A-frame should be positioned so that the angles created by the A-frame bisecting the installation rope are approximately equal on both sides. This creates downward pressure holding the A-frame in position, not forcing it in a lateral direction. It must also be placed in a straight line between the upper and lower anchor points.

(4) Use a carrying rope to attach loads to the traverse ropes (Figure 7-23). Join the ends of a 14-foot sling rope with a square knot and two overhand knots. Displace the knot one-third of the distance down the loop and tie an overhand knot both above and below the square knot. This forms two small loops and one large loop that is longer than the two small loops combined.

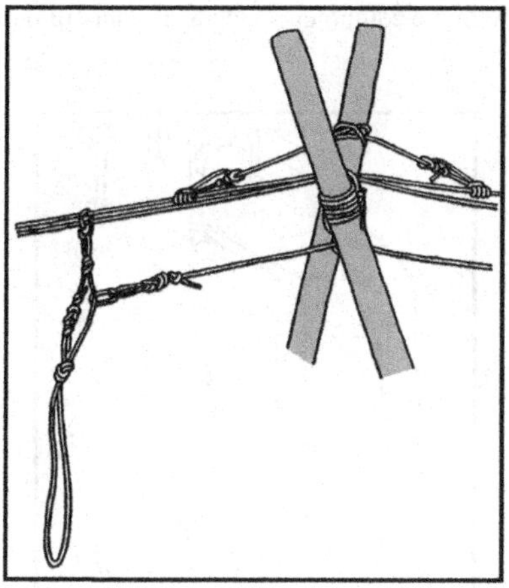

Figure 7-23. Carrying rope for use on a traverse.

(5) Attach the carrying rope to the traverse ropes with carabiners (or a pulley) that have the gates reversed and opening in opposite directions. Attach a belay rope to the center loop of the carrying rope using a fixed loop or locking carabiner on the side opposite the joining knot (Figure 7-23). When the suspension traverse is near horizontal, a second rope may be needed to pull the load across and should be attached to the carrying rope the same as the first.

(6) Insert second carabiner into the one placed into the wraps of the A-frame. This is where a belay rope will be attached

(7) With a sling rope, tie a six wrap middle-of-the-rope Prusik knot to both static ropes near the far side off-loading point. This acts as a stopper knot for the man descending, preventing him from hitting the lower anchor.

(8) Attach the load by running the long loop of the carrying rope through the load or through the soldier's harness and attaching the bottom loop to the traverse rope carabiner.

Descent must be belayed slowly and be controlled. Soldiers descending should hold onto the carrying rope and keep their feet high to avoid contact with the ground. Due to the constant tension maintained on the belayer, use a mechanical belay. If the belayer cannot view the entire descent route, use a relay man.

7-18. RETRIEVAL

The suspension traverse is not as readily retrievable as the one-rope bridge. Therefore, the installing unit should dismantle it after it is no longer needed.

Section V. VERTICAL HAULING LINE

The vertical hauling line is an installation used to move men and equipment up vertical or near-vertical slopes (Figure 7-24). It is often used with a fixed rope for personnel movement. The hauling line is used to move equipment, such as mortars or other crew-served weapons, rucksacks, or supplies

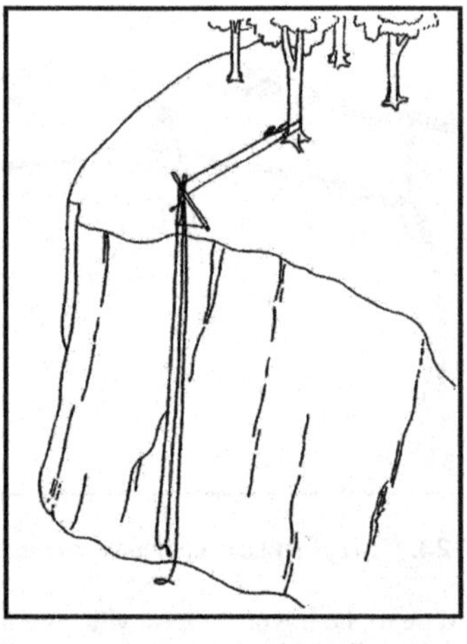

Figure 7-24. Vertical hauling line.

7-19. SITE SELECTION

The first and most important task is to determine where to construct the vertical hauling line. The site must have an appropriate top anchor that is secure enough to hold the system and load. Loading and unloading platforms should be easily accessible natural platforms that provide a safe working area. The ideal platform at the top allows construction of the vertical hauling line without the use of an A-frame. The site should also have sufficient clearance to allow for space between the slope and pulley rope for easy hauling of troops or equipment.

7-20. INSTALLATION

Construct an A-frame, if necessary, and anchor it. Double one installation rope, find the middle, and lay the middle of the installation rope over the apex of the A-frame; a 30-centimeter (12-inch) bight should hang below the apex.

a. To maintain the 12-inch bight, tie clove hitches above the A-frame lashing on each side of the apex with the installation rope, ensuring that the locking bars of the clove hitches are on the inside. Ensure that the portion of the rope that forms the bight comes out of the bottom of the clove hitch. (See Figure 7-25.)

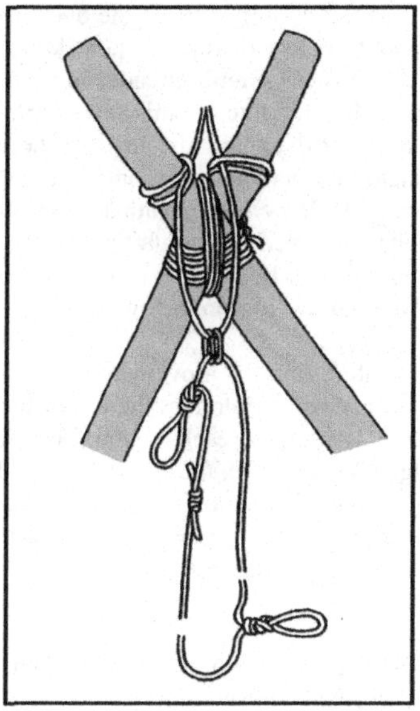

Figure 7-25. Attaching the anchor rope to the A-frame.

b. To anchor the A-frame, use a transport tightening system with the doubled rope, which is tied to the A-frame. Tie this off at an anchor point to the rear of the A-frame installation and adjust the angle of the A-frame so it leans out over the cliff edge. The angle should be 15 to 25 degrees unloaded. The A-frame should not lean outward more than 45 degrees once loaded since the legs can lose their position.

c. Tie the ends of another installation rope together with a joining knot to form the hauling line. Attach the rope to the system by two carabiners with gates up and opposed or one mountain rescue pulley with a locking steel carabiner in the 12-inch bight hanging from the apex of the A-frame. Tie fixed loops (wireman's, directional figure-eight, or single butterfly) on opposite sides of the endless rope at the loading and unloading platforms.

d. Attach equipment to the hauling line 12 inches above the joining knot by a carabiner in the fixed loop.

e. Additional fixed loops may be tied in the hauling line for more control over the object when moving large loads. Attach personnel to the hauling line by use of a rappel seat or seat harness.

Note: Mortar tubes and similar objects are attached to the line by two knots so that the tube stays parallel and as close to the hauling line as possible.

f. When personnel are moved using a vertical hauling line, make a knotted hand line; anchor it in line with, or to, the primary anchor (round turn with a bowline); and place it over the spreader on the legs of the A-frame. Space the overhand knots in the knotted hand line 12 inches apart, with about 20 feet of rope without knots at one end for the anchor. Throw the knotted hand line over the A-frame spreader rope and down the side of the cliff. Personnel ascending the vertical hauling line use this as a simple fixed rope.

g. Use as many men as needed to pull the load to the top by pulling on the rope opposite the load. If equipment and personnel are only being lowered, belay from the top using the hauling line. Station two climbers at the unloading platform to retrieve loads.

h. If only equipment is being hauled up, it is not necessary to use the knotted hand line rope, but it may be necessary to use a belay rope. To move materials or troops up on one side of the hauling line, pull the other side from below.

Note: Personnel using the hauling line for movement must apply all related principles of climbing. Always station two operators at the top of the vertical hauling line to aid men or to retrieve loads when they reach the top. They will always be safetied while working near the edge. When in use, the A-frame should lean slightly over the edge of the cliff to prevent excessive wear on the ropes that pass over sharp rocks. Reduce excessive friction on the system. Remove all obstacles and any loose objects that could be dislodged by personnel and equipment.

7-21. RETRIEVAL

The vertical hauling line is used along a main supply route. When it is no longer needed, the installing unit will return and dismantle the system.

Section VI. SIMPLE RAISING SYSTEMS

Moving heavy objects with limited manpower may be necessary in mountainous terrain. To reduce fatigue of those personnel moving the load, simple rigging techniques can be used to increase the mechanical advantage of the hauling system.

7-22. Z-PULLEY SYSTEM

The Z-pulley system is a simple, easily constructed hauling system (Figure 7-26, page 7-30).

a. **Considerations**. Anchors must be sturdy and able to support the weight of the load. Site selection is governed by different factors: tactical situation, weather, terrain, equipment, load weight, and availability of anchors.

b. **Theory.** Use carabiners as a substitute if pulleys are not available. The mechanical advantage obtained in theory is 3:1. The less friction involved the greater the mechanical advantage. Friction is caused by the rope running through carabiners, the load rubbing against the rock wall, and the rope condition.

c. **Construction.** Use the following procedures to construct a Z-pulley system.

(1) Establish an anchor (anchor pulley system [APS]). Place a carabiner on the runner at the anchor point, place a pulley into the carabiner, and run the hauling rope through the pulley.

(2) With a sling rope (preferably 7 millimeter), tie a middle-of-rope Prusik knot secured with a figure-eight knot on the load side of the pulley. This will be used as a progress capture device (PCD). A mechanical descender may be used in place of the Prusik knot. Take the tails exiting the figure-eight and tie a Munter hitch secured by a mule knot. Ensure the Munter hitch is loaded properly before tying the mule knot.

(3) At an angle away from the APS, establish a moveable pulley system (MPS) to create a "Z" in the hauling rope. Tie another Prusik knot on the load side of the hauling rope. Secure it with a figure-eight knot. Using the tails tie a double-double figure-eight knot. Insert a locking carabiner into the two loops formed, then place the working end into the carabiner. Mechanical ascenders should not be used as an MPS. Move the working end back on a parallel axis with the APS. Provide a pulling team on the working end with extra personnel to monitor the Prusik knots.

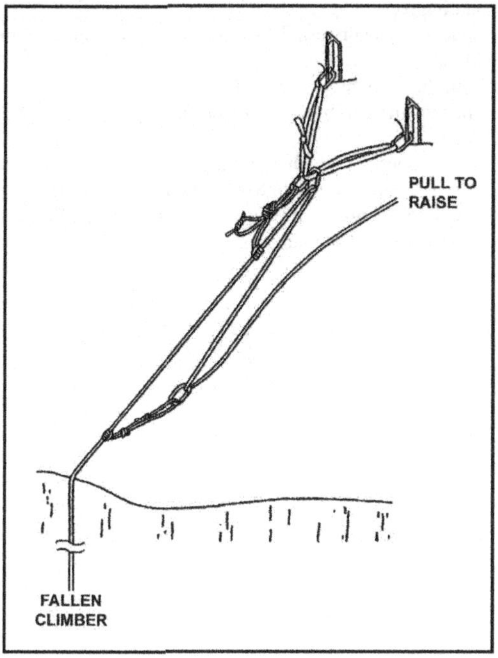

Figure 7-26. Z-pulley system.

d. **Other Factors**. If the two pulleys touch, the "Z" is lost along with the mechanical advantage. For greater efficiency, the main anchor should be well back from the edge and all ropes should pull parallel to the load.

Note: Avoid the possibility of overstressing the anchors. Be aware of reduced sensitivity to the load due to the mechanical advantage. Use belays and backup safeties. Protect the rope from edges and other abrasive parts of the rock.

7-23. U-PULLEY SYSTEM

The U-pulley system is another simple, easily-constructed hauling system (Figure 7-27, page 7-32).

a. **Considerations**. Anchors must be sturdy and able to support the weight of the load. Site selection is governed by different factors: tactical situation, weather, terrain, equipment, load weight, and availability of anchors.

b. **Theory.** Use carabiners as a substitute if pulleys are not available. The mechanical advantage obtained in theory is 2:1. The less friction involved the greater the mechanical advantage. Friction is caused by the rope running through carabiners, the load rubbing against the rock wall, and the rope condition.

c. **Construction.** Use the following procedures construct a U-pulley system.

(1) Anchor the hauling rope.

(2) Prepare the load or casualty for hauling. Place a locking carabiner the on to the harness or the rigged load.

(3) Lower a bight to the casualty or the load.

(4) Place the bight into the carabiner; or place the bight on to a pulley and then place pulley into the carabiner.

(5) Construct a second anchor. Attach a locking carabiner to the anchor.

(6) Tie a middle of the rope Prusik onto the haul rope exiting the pulley. Secure the Prusik with a double-double figure eight. This is the PCD. Place the fixed loops into the locking carabiner of the second anchor.

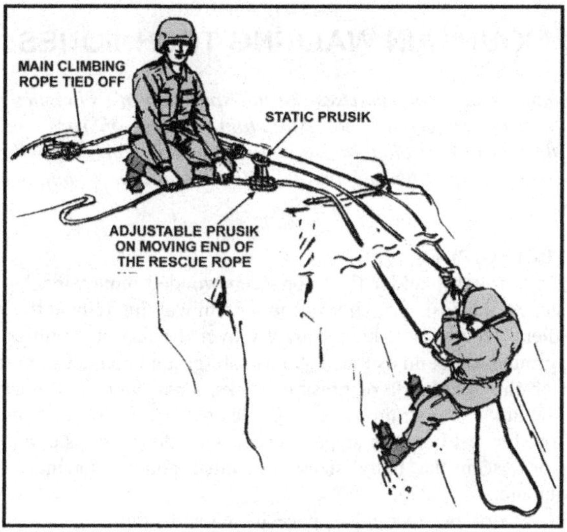

Figure 7-27. U-pulley system.

CHAPTER 8

MOUNTAIN WALKING TECHNIQUES

Mountain travel encompasses the full spectrum of techniques used to negotiate steep, rugged terrain. Mountain walking on rock and snow, technical rock and ice climbing, skiing or snow shoeing, rappelling, and stream crossing are the key travel skills a military mountaineer must possess.

8-1. BASIC PRINCIPLES

Up scree or talus, through boulder fields or steep wooded mountainsides, over snow or grass-covered slopes, the basic principles of mountain walking remain the same.

a. The soldier's weight is centered directly over the feet at all times. He places his foot flat on the ground to obtain as much (boot) sole-ground contact as possible. Then, he places his foot on the uphill side of grass tussocks, small talus and other level spots to avoid twisting the ankle and straining the Achilles tendon. He straightens the knee after each step to allow for rest between steps, and takes moderate steps at a steady pace. An angle of ascent or descent that is too steep is avoided, and any indentations in the slope are used to advantage.

b. In addition to proper technique, pace is adapted to conditions. The mountaineer sets a tempo, or number of steps per minute, according to the pace of the unit in which he is moving. (Physical differences mean that the tempos of two people moving at the same speed will not always be the same.) The soldier maintains tempo and compensates for changes of slope or terrain by adjusting the length of his stride. Tempo, pace, and rhythm are enhanced when an interval of three to five paces is kept between individuals. This interval helps lessen the "accordion" effect of people at the end of the file who must constantly stop and start.

c. The terrain, weather, and light conditions affect the rate of climb. The more adverse the conditions, the slower the pace. Moving too fast, even under ideal conditions, produces early fatigue, requires more rest halts, and results in loss of climbing time. A soldier can only move as fast as his lungs and legs will allow. The trained, conditioned and acclimatized soldier has greater endurance and moves more efficiently. Rest, good nutrition and hydration, conditioning, acclimatization, proper training, and the will to climb are key to successful mountain operations.

d. Breaks are kept to a minimum. When a moderate pace is set, the need for rest halts decreases, the chance of personnel overheating is lessened, and a unit can cover a given distance in a minimal time. If possible, rests should be taken on level ground avoiding steeper inclines.

(1) During the first half-hour of movement an adjustment halt should be taken. Soldiers will loosen or tighten bootlaces as needed, adjust packs and add or remove appropriate layers of clothing.

(2) Following the first halt, a well-conditioned party may take a short rest every 1 to 1.5 hours. If possible, soldiers lean against a tree, rock, or hillside to relieve the shoulders of pack weight, breathe deeply, hydrate, and snack on trail food. These halts are kept short enough to avoid muscles stiffening (one to two minutes).

(3) Later in the march longer halts may be necessary due to fatigue or mission requirements. At these halts soldiers should immediately put on additional clothing to avoid chilling—it is much easier to keep a warm body warm than to warm up a cold one.

(4) After a climb, a good rest is needed to revive tired muscles.

e. The rest step is used for steep slopes, snowfields, and higher elevations. It controls pace and limits fatigue by giving the lungs and legs a moment to recuperate between steps. Pace is kept slow and rhythmic.

(1) After each step forward, the soldier pauses briefly, relaxing the muscles of the forward leg while resting his entire bodyweight on the rear leg. The rear leg is kept straight with the knee locked so that bone, not muscle, supports the weight.

(2) Breathing is synchronized with the rest step. The number of breaths per step will change depending on the difficulty of the climb. Steeper slopes or higher elevations may require several breaths per step. When the air thins at altitude it is especially important to breathe deeply, using the "pressure breathing" technique. The soldier exhales strongly, enabling an easier, deeper inhale.

(3) This slow, steady, halting rest step is more efficient than spurts of speed, which are rapidly exhausting and require longer recovery.

f. Downhill walking uses less energy than uphill but is much harder on the body. Stepping down can hammer the full bodyweight onto the feet and legs. Blisters and blackened toenails, knee damage, and back pain may follow. To avoid these problems the soldier should start by tightening bootlaces to ensure a snug fit (also keep toenails trimmed). A ski pole, ice ax, or walking stick will help take some of the load and give additional stability. (Refer to Chapter 11 for techniques and use of the ice ax.) Keep a moderate pace and walk with knees flexed to absorb shock.

g. Side hill travel on any surface should be avoided whenever possible. Weighted down with a rucksack, the soldier is vulnerable to twisted ankles, back injury, and loss of balance. If side hill travel is necessary, try to switchback periodically, and use any lower angle flat areas such as rocks, animal trails, and the ground above grass or brush clumps to level off the route.

8-2. TECHNIQUES

Mountain walking techniques can be divided according to the general formation, surface, and ground cover such as walking on hard ground, on snow slopes and grassy slopes, through thick brush, and on scree and talus slopes.

a. **Hard Ground.** Hard ground is firmly compacted, rocky soil that does not give way under the weight of a soldier's step. It is most commonly found under mature forest canopy, in low brush or heather, and areas where animals have beaten out multiple trails.

(1) When ascending, employ the rest step to rest the leg muscles. Steep slopes can be traversed rather than climbed straight up. To turn at the end of each traverse, the soldier should step off in the new direction with the uphill foot. This prevents crossing the feet and possible loss of balance. While traversing, the full sole-to-ground principle is accomplished by rolling the ankle downhill on each step. For small stretches the herringbone step may be used—ascending straight up a slope with toes pointed out. A normal progression, as the slope steepens, would be from walking straight up, to a herringbone step, and then to a traverse on the steeper areas.

(2) Descending is best done by walking straight down the slope without traversing. The soldier keeps his back straight and bends at the knees to absorb the shock of each step. Body weight is kept directly over the feet and the full boot sole is placed on the ground with each step. Walking with a slight forward lean and with the feet in a normal position make the descent easier.

b. **Snow Slopes.** Snow-covered terrain can be encountered throughout the year above 1,500 meters in many mountainous areas. Talus and brush may be covered by hardened snowfields, streams made crossable with snowbridges. The techniques for ascending and descending moderate snow slopes are similar to walking on hard ground with some exceptions.

(1) ***Diagonal Traverse Technique.*** The diagonal traverse is the most efficient means to ascend snow. In conjunction with the ice ax it provides balance and safety for the soldier. This technique is a two-step sequence. The soldier performs a basic rest step, placing the leading (uphill) foot above and in front of the trailing (downhill) foot, and weighting the trail leg. This is the in-balance position. The ice ax, held in the uphill hand, is placed in the snow above and to the front. The soldier shifts his weight to the leading (uphill) leg and brings the unweighted trail (downhill) foot ahead of the uphill foot. He shifts weight to the forward (downhill) leg and then moves the uphill foot up and places it out ahead of the trail foot, returning to the in-balance position. At this point the ax is moved forward in preparation for the next step.

(2) ***Step Kicking.*** Step kicking is a basic technique used when crampons are not worn. It is best used on moderate slopes when the snow is soft enough to leave clear footprints. On softer snow the soldier swings his foot into the snow, allowing the leg's weight and momentum to carve the step. Fully laden soldiers will need to kick steps, which take half of the boot. The steps should be angled slightly into the slope for added security. Succeeding climbers will follow directly in the steps of the trailbreaker, each one improving the step as he ascends. Harder snow requires more effort to kick steps, and they will not be as secure. The soldier may need to slice the step with the side of his boot and use the diagonal technique to ascend.

(3) ***Descending Snow.*** If the snow is soft and the slope gentle, simply walk straight down. Harder snow or steeper slopes call for the plunge step, which must be done in a positive, aggressive manner. The soldier faces out, steps off, and plants his foot solidly, driving the heel into the snow while keeping his leg straight. He shifts his weight to the new foot plant and continues down with the other foot. On steeper terrain it may be necessary to squat on the weighted leg when setting the plunge step. The upper body should be kept erect or canted slightly forward.

(4) ***Tips on Snow Travel.*** The following are tips for travelling on snow.

(a) Often the best descent is on a different route than the ascent. When looking for a firmer travel surface, watch for dirty snow—this absorbs more heat and thus hardens faster than clean snow.

(b) In the Northern Hemisphere, slopes with southern and western exposures set up earlier in the season and quicker after storms, but are more prone to avalanches in the spring. These slopes generally provide firm surfaces while northern and eastern exposures remain unconsolidated.

(c) Travel late at night or early in the morning is best if daytime temperatures are above freezing and the sun heats the slopes. The night's cold hardens the snow surface.

(d) Avoid walking on snow next to logs, trees, and rocks as the subsurface snow has melted away creating hidden traps.

c. **Grassy Slopes.** Grassy slopes are usually composed of small tussocks of growth rather than one continuous field.

(1) When ascending, step on the upper side of each hummock or tussock, where the ground is more level.

(2) When descending a grassy slope, the traverse technique should be used because of the uneven nature of the ground. A climber can easily build up too much speed and fall if a direct descent is tried. The hop-skip step can be useful on this type of slope. In this technique, the lower leg takes all of the weight, and the upper leg is used only for balance. When traversing, the climber's uphill foot points in the direction of travel. The downhill foot points about 45 degrees off the direction of travel (downhill). This maintains maximum sole contact and prevents possible downhill ankle roll-out.

Note: Wet grass can be extremely slippery; the soldier must be aware of ground cover conditions.

d. **Thick Brush.** For the military mountaineer, brush is both a help and a hindrance. Brush-filled gullies can provide routes and rally points concealed from observation; on the other hand steep brushy terrain is hazardous to negotiate. Cliffs and steep ravines are hidden traps, and blow downs and thickets can obstruct travel as much as manmade obstacles. When brush must be negotiated take the most direct route across the obstacle; look for downed timber to use as raised paths through the obstacle; or create a tunnel through the obstacle by prying the brush apart, standing on lower branches and using upper limbs for support.

e. **Scree Slopes.** Slopes composed of the smallest rocks are called scree slopes. Scree varies in size from the smallest gravel to about the size of a man's fist.

(1) Ascending scree slopes is difficult and tiring and should be avoided, if possible. All principles of ascending hard ground and snow apply, but each step is carefully chosen so that the foot does not slide down when weighted. This is done by kicking in with the toe of the upper foot (similar to step-kicking in snow) so that a step is formed in the loose scree. After determining that the step is stable, weight is transferred to the upper leg, the soldier then steps up and repeats the process with the lower foot.

(2) The best method for descending scree slopes is to come straight down the slope using a short shuffling step with the knees bent, back straight, feet pointed downhill, and heels dug in. When several climbers descend a scree slope together, they should be as close together as possible (one behind the other at single arm interval) to prevent injury from dislodged rocks. Avoid running down scree as this can cause a loss of control. When the bottom of the slope (or run out zone) cannot be seen, use caution because drop-offs may be encountered.

(3) Scree slopes can be traversed using the ice ax as a third point of contact. Always keep the ax on the uphill side. When the herringbone or diagonal method is used to ascend scree, the ax can be used placing both hands on the top and driving the spike into the scree slope above the climber. The climber uses the ax for balance as he moves up to it, and then repeats the process.

f. **Talus Slopes.** Talus slopes are composed of rocks larger than a man's fist. When walking in talus, ascending or descending, climbers should always step on the uphill side of rocks and stay alert for movement underfoot. Disturbing unstable talus can cause rockslides. Climbers must stay in close columns while walking through talus so that dislodged rocks do not reach dangerous speeds before reaching lower soldiers. To prevent rock fall injuries, avoid traversing below other climbers. All other basics of mountain walking apply.

8-3. SAFETY CONSIDERATIONS

The mountain walking techniques presented here are designed to reduce the hazards of rock fall and loss of control leading to a fall. Carelessness can cause the failure of the best-planned missions.

a. Whenever a rock is kicked loose, the warning, "Rock!" is shouted immediately. Personnel near the bottom of the cliff immediately lean into the cliff to reduce their exposure, and do not look up. Personnel more than 3 meters away from the bottom of the cliff may look up to determine where the rock is heading and seek cover behind an obstacle. Lacking cover, personnel should anticipate which way the rock is falling and move out of its path to the left or right.

b. If a soldier slips or stumbles on sloping terrain (hard ground, grass, snow, or scree) he must immediately self-arrest, digging into the slope with hands, elbows, knees and toes. If he falls backwards and rolls over he must immediately try to turn over onto his stomach with his legs downhill and self-arrest with hands and toes.

c. When traveling through steep terrain, soldiers should be trained in the use of the ice ax for self-arrest. The ax can be used to arrest a fall on solid ground, grass and scree as well as snow. It may also be used as a third point of contact on difficult terrain. If not in use the ice ax is carried in or on the rucksack with its head down and secured.

8-4. NAVIGATION

Navigation is the process of determining one's present position, the location of a target objective, and selecting and following a route between these two points. Navigation consists of three distinct stages: orientation, navigation, and route finding.

- Orientation is simply figuring out exactly where one is. The use of the map, compass and identifiable terrain features, assisted by an altimeter and GPS, is the foundation of good navigation.
- Navigation includes the determination of the objective's location and the direction from the soldier's starting point to the objective. The same skills and equipment used in orientation are essential for good navigation.
- Route finding is picking the best line of travel that matches the equipment and capabilities of the team. Good route finding incorporates a comprehensive awareness of terrain, a solid base of mountaineering experience, good judgement and sound tactical instincts.

a. **Compasses.** The magnetic compass is the simplest and most widely used instrument for measuring directions and angles in the mountains. The lensatic compass is most commonly used in the military and can be employed in a variety of ways for either day or night navigation.

b. **Altimeters.** The altimeter is a vital piece of navigational equipment that can save valuable time in determining position through elevation.

(1) The standard altimeter is a modified barometer. A barometer is an instrument that measures the weight of a column of air above itself and displays the result on a scale marked in units of pressure, usually inches of mercury, millimeters of mercury, or millibars. Since air pressure drops uniformly as elevation is gained, it can be used to read altitude by means of the altimeter's scale, marked in feet or meters of elevation above sea level. By measuring air pressure, the altimeter/barometer gives the navigator new techniques for position finding, route planning, checking progress and terrain identification. It also gives the navigator valuable weather information specific to his immediate location.

(2) Changes in the weather are usually accompanied by air pressure changes, which are reflected in the altimeter. As the air pressure drops due to the approach of inclement weather for instance, the displayed elevation will rise by a corresponding amount. This means that a barometric pressure change of one inch of mercury equals roughly 1,000 feet of elevation. If the altimeter displays an elevation gain of 300 feet, a loss of barometric pressure of .3 inches has occurred, and bad weather should be expected.

(3) Altimeters come in two types: wrist-mounted digital altimeters and analog altimeters, usually attached to a cord.

(4) Because the altimeter is sensitive to changes in air pressure it must be recalibrated whenever a point of known elevation (summits, saddles, stream-trail intersections, survey monuments, and so forth) is reached. This is especially important when weather fronts are moving rapidly through the area.

(5) The altimeter may expand or contract because of changes in temperature. This can result in faulty elevation readings. Although some altimeters are temperature-compensated, rapid ascents or descents sometime overcome the adjustment, causing them to give poor readings.

(6) Keep the altimeter at a constant temperature. This is best accomplished by storing the altimeter (analog) in a pocket or on a cord around the neck, or on the wrist under the parka and hand gear (digital).

(7) Even though altimeters can be precise they are affected by both pressure and temperature changes and should be monitored carefully. The soldier should become familiar with the specific altimeter he employs and understand its capabilities and limitations.

c. **Global Positioning System.** The GPS is a space-based, global, all-weather, continuously available radio positioning navigation system. It is highly accurate in determining position location derived from a satellite constellation system. It can determine the latitude, longitude and elevation of the individual user. Location information is also displayed in military grid coordinates.

(1) The GPS provides precise steering information as well as position locations. The receiver can accept many checkpoints entered in any coordinate system by the user and convert them to the desired coordinate system. The user then calls up the desired checkpoint and the receiver will display direction and distance to the checkpoint. It can also compute travel time to the next checkpoint.

(2) Because the GPS does not need visible landmarks to operate, it can provide position (accurate up to 16 meters) in whiteouts or on featureless terrain. It also does not compound navigational errors as compass use can.

(3) During route planning, after choosing critical checkpoints, start point and objective, enter their coordinates as way points. The best use of the GPS is to verify these as they are reached, as a backup to terrain association and compass navigation.

(4) Since the 21-satellite constellation is not yet complete, coverage may be limited to specific hours of the day in certain areas of the world. The GPS navigational signals are similar to light rays, so anything that blocks light will reduce or block the effectiveness of the signals. The more unobstructed the view of the sky, the better the system performs. Although the GPS can be used in any terrain, it is performs best in more open areas such as the desert.

(5) Because the GPS requires horizon to horizon views for good satellite reception its use can be limited in the mountains. Canyons, deep valleys, saddles, and steep mountainsides are all problematic spots to use for shots. Ridgelines, spurs, summits, open valleys, or plateaus are better.

(6) When using GPS in regions with questionable surveying and mapping products, operational datum of the local maps must be reconciled with the datum used in navigational and targeting systems. Identify the spheroid and datum information on the pertinent map sheets and then check that the GPS receiver has the compatible datum loaded. If not then you must contact the S2 for updated datum or maps. Otherwise, the GPS will show different locations than those on the map.

(7) Extremely cold temperatures (-4 degrees F and below) and high elevations will adversely affect the operation of the GPS, due to the freezing of the batteries and the LCD screen. Battery life and overall performance can be improved by placing the GPS inside the parka or coat.

d. **Navigation Techniques.** The choice of movement technique often determines the route and navigational technique. For navigation, three techniques can be used: dead reckoning, terrain association, or altimeter navigation. The three are not mutually exclusive and are normally used together, with one chosen as the primary technique. The GPS can be used to supplement these techniques, but due to the problems associated with the restricted line of sight in the mountains, it should not be used as the main technique.

(1) ***Dead Reckoning.*** Because of the complex nature of mountainous terrain, dead reckoning is usually of limited value on most movements. The compass is generally employed more to support terrain association and to orient the map, than as a primary navigational aid. The main exception is during periods of limited visibility on featureless terrain. Heavy fog, snowy or whiteout conditions on a snowfield, glacier, large plateau or valley floor all would call for dead reckoning as a primary navigational technique.

(2) ***Terrain Association.*** The standard terrain association techniques all apply. Handrails, checkpoints, catching features, navigational corridors, boxing-in areas, and attack points are all used. When a small objective lies near or on an easily identifiable feature, that feature becomes an expanded objective. This simplifies the navigational problem by giving a large feature to navigate to first. The altimeter may finalize the search for the objective by identification through elevation. Rough compass headings are used to establish a general direction to the next checkpoint; used when the checkpoint headed toward is a linear feature, and not a precise point. The shape, orientation, size,

elevation, slope (SOSES) strategy is especially valuable in mountain terrain association and should be practiced extensively (FM 3-25.26).

(a) After extensive study of the map and all available sources of information it helps to create a mental image of the route. This will enable the navigator to make the terrain work in his favor. Avoid brush for speed and ease of movement; the military crest of spurs and ridgelines generally provides the best route while providing terrain masking effects. When clear cut, burned-over, or large avalanche slide areas are encountered, it may be necessary to box or contour around them as they may be full of slash or brushy second-growth small trees. Old-growth forest provides the easiest travel.

(b) The following situations will result in objects appearing closer than they actually are:

- When most of the object is visible and offers a clear outline.
- When you are looking across a partially cleared depression.
- When looking down a straight, open road or track.
- When looking over a smooth, uniform surface, such as snow, water, or desert.
- When the light is bright and the sun is shining from behind the observer.
- When the object is in sharp contrast to the background.
- When seen in the clear air of high altitude.
- When looking down from high ground to low ground.

(c) The following situations will result in objects appearing farther away than they actually are:

- When only part of the object is seen or it is small in relation to its surroundings.
- When you are looking across an exposed depression.
- When looking up from low ground to high ground.
- When your vision is narrowly confined.
- When the light is poor, such as dawn, dusk, or low visibility weather; or when the sun is in your eyes, but not behind the object being viewed.
- When the object blends into the background.

(2) ***Altimeter Navigation.*** Altimeters provide assistance to the navigator in several ways. They aid in orientation, in computing rates of ascent or descent, in resection, and in weather prediction.

(a) When moving along any linear feature such as a ridge, watercourse, or trail which is shown on the map, check the altimeter. The point where the indicated elevation contour crosses that feature is your location.

(b) The navigator frequently finds it necessary to determine his position through the use of resection. A modified resection can be performed by shooting an azimuth to a known, clearly visible summit or similar feature and then plotting the back azimuth on the map. By determining your present elevation and finding where that particular contour crosses the back azimuth you should locate your position. This can be difficult when in low ground, as mountain summits can rarely be clearly seen from valley floors. In addition, most mountaintops are so large that there is usually no specific point to shoot at. In this case, the soldier should take multiple azimuths to known features. If he is located on a good linear feature he will have a decent idea of where he is. The altimeter can be used to verify elevation and establish a notional linear feature—a contour line. The point where the resecting back azimuths cross the contour line is the navigator's location.

(c) Using the altimeter to calculate rates of ascent can help in sound decision-making. Rates of travel, along with weather conditions, light conditions (time of day), and the physical condition of the team, are all key variables that can influence the success or failure of the mission.

(d) Altimeters can be used as barometers to assist in weather prediction.

e. **Approach Observations.** Watch the mountain during the approach march, studying it for climbing routes. Distant views can reveal large-scale patterns of ridges, cliffs, snowfields and glaciers. General angles of the large rock masses can be seen from afar.

(1) Closer viewing displays these patterns and angles on a smaller scale. Fault lines, gross bedding planes of rock, cliff bands, and crevasse zones become visible. Snowy or vegetated ledge systems appear. Weaknesses in the mountain walls, such as couloirs or gullies, may present themselves.

(2) Most of these features repeat themselves at increasingly finer levels, as they are generally derived from the overall structure of the particular mountain group. A basic knowledge of mountain geology, combined with the specific geological background of the operational area, pays off in more efficient travel.

f. **Natural Indicators of Direction in the Northern Hemisphere.** Southern slopes are sunnier and drier than northern slopes, with sparser or different types of vegetation. Northern slopes can be snowier and, because of more intense glaciation in past ages, are often steeper.

Note: Opposite rules apply in the Southern Hemisphere.

g. **Winter Route Selection.** The following must be considered when selecting a route in the winter.

(1) Conduct a thorough map reconnaissance considering the weather, individual ski abilities, avalanche danger, vegetation, water features, terrain relief, and the size of the unit.

(2) Weather conditions will affect the chosen route. During calm weather, your rate of movement will be significantly faster than during periods of inclement weather.

(3) Individual ski abilities will affect your rate of movement, constrain your choice of terrain, and impact on your route choices.

(4) Avalanche danger zones must be identified by map review and data gathered during route planning. During movement, snow pits, shovel tests, and ski shear tests must be conducted prior to crossing an avalanche danger zone. Bottom line: avoid avalanche danger areas. If you must cross one, cross above the starting zone or below the run-out zone.

(5) Vegetation can work for you or against you. Thickly forested areas usually have a deep snow pack. For weaker skiers, forested areas are full of potentially dangerous obstacles. On slopes with an angle of 30 to 45 degrees that are sparsely vegetated an avalanche danger is still present. If the weather turns bad, forested areas provide welcome relief from wind and blowing snow.

(6) Water features provide valuable navigation aids. Under deep snow pack small creeks and ponds may be hard to locate. Large frozen lakes and rivers can provide excellent means of increasing your rate of march.

(7) During ski movements, efficient use of terrain will greatly improve morale and reduce fatigue. While traveling in mountainous terrain, do not needlessly give up elevation gained. Maintain a steady climb rate and avoid over exertion. Avoid north, east, and south facing slopes when the avalanche danger is high. Avoid cornices and be aware of their probable and improbable fracture lines. Weather and tactical situation permitting, travel on the windward side of ridgelines. If weak skiers are in the group, stay away from restrictive terrain with sheer drop-offs. When touring use climbing skins to maintain control and lessen lost time per hour due to individuals falling.

(8) The following are additional hints for navigation in snowy conditions:

- Keep the compass warm.
- If no terrain features exist for steering marks, use your back azimuth and tracks to maintain course.
- Limit steering marks to shorter distances since visibility can change quickly.
- Never take azimuths near metallic objects. Hold the compass far enough from your weapon, ice ax, and so on to get accurate readings.
- Make frequent compass checks.
- Preset azimuths on your compass.
- Use a steady, unshifting wind to aid you in maintaining course.

h. **Problems.** The following conditions and characteristics of cold weather and mountainous regions make accurate navigation difficult.

(1) In winter, short hours of daylight, fog, snowfall, blizzards, whiteouts, and drifting snow, especially above tree line, drastically limit visibility. At times, an overcast sky and snow-covered terrain create a phenomenon called flat light, which makes recognition of irregularities in the terrain extremely difficult.

(2) Heavy snow may completely cover existing tracks, trails, outlines of small lakes and similar landmarks. Because the appearance of the terrain is quite different in winter from that in summer, particular attention must be paid to identifying landmarks, both on the ground and from aerial photographs.

(3) Magnetic disturbances, caused by large ore deposits, are frequently encountered and make magnetic compass readings difficult and sometimes unreliable.

(4) Handling maps, compasses, and other navigation instruments in low temperatures with bare hands is difficult. Removing hand wear may only be possible for short periods.

(5) Keeping count of pace is extremely difficult in winter and mountain environments. Thick vegetation and rough, steep slopes hamper attempts at accurate pace counts. The most reliable method is the use of a 50-meter long piece of field wire or rope.

8-5. ROUTE PLANNING

Proper route planning can make the difference between success and failure on long mountain movements. Careful map reconnaissance, knowledge of the enemy situation, terrain analysis of the operational area, and an accurate assessment of the unit's capabilities are all key parts of the planning process.

a. **Map Reconnaissance.** Topographic maps provide the primary source of information concerning the area of operations. A 1:25,000 map depicts greater detail than a 1:50,000 map and should be used whenever possible. Because examination of the micro-terrain is so important for mountain operations, even larger scale maps are desirable. Civilian 1:12,000 maps can be used if available. Aerial, oblique angle, photographs give details not always shown on maps (crags and overhangs). Sketch maps supplement other sources of information but should not be relied on for accuracy since they are seldom drawn to scale. Along with sketch maps, verbal descriptions, documented information gathered from units previously in the area, or published sources such as alpine journals or climbing guides may help. Forest service and logging and mining company maps provide additional information, often showing the most recent changes to logging trails and mining access roads. Standard military topographic maps are generally accurate graphic depictions of the operational area.

(1) When conducting a map reconnaissance, pay close attention to the marginal information. Mountain-specific terrain features may be directly addressed in the legend. In addition, such facilities as ski lifts, cable and tramways are often found. Check the datum descriptor (for foreign maps) to ensure compatibility with entered datum in GPS units. Along with the standard topographic map color scheme, there are some commonly seen applications for mountainous terrain. White with blue contours indicates glaciers or permanent snowfields. The outline of the snow or ice is shown by dashed blue lines while their contour lines are solid blue. High ice cliffs which are equal to or exceed the contour interval will be shown. Low ice cliffs and ice caves may be indicated if they provide local landmarks. Brown contour lines on white mean dry areas without significant forest cover. Areas above tree line, clear cuts, rock or avalanche slide paths and meadows are all possible. Study the surrounding terrain and the legend for other clues. An important point to remember is that thick brush in small gullies and streambeds may not be depicted by green, but should still be expected.

(2) Obstacles, such as rivers and gorges, will require technical equipment to cross if bridges are not present. Fords and river crossing sites should be identified. Due to the potential for hazardous weather conditions, potential bivouac sites are noted on the map. Ruins, barns, sheds and terrain-protected hollows are all possible bivouac sites. Danger areas in the mountains; isolated farms and hamlets, bridges, roads, trails, and large open areas above tree line, are factored in, and plans made to avoid them. Use of terrain-masking becomes essential because of the extended visibility offered by enemy observation points on the dominant high ground.

(3) Helicopter support, weather permitting, requires identification of tentative landing zones for insertions, extractions, resupply and medevac. The confined nature of mountain travel means that crucial passes become significant chokepoints and planners should designate overwatches/surveillance positions beforehand. Alternate routes should be chosen with weather imposed obstacles in mind: spring flood or afternoon snowmelt turns small streams into turbulent, impassable torrents. Avalanche danger prohibits travel on certain slopes or valley floors.

b. **Enemy Situation.** Route selection should only be done after reviewing all available information about the friendly and enemy situation.—Is the enemy force on his own ground? Are they accustomed to the terrain and the weather? Are they trained mountain troops with specialized equipment?—Only after answering these and other questions can an effective route plan be completed. If the enemy force is better prepared to maneuver in the mountains, they have a marked advantage, and route selection must be scrutinized.

c. **Analysis of the Operational Area.** Not all mountainous terrain is created equal and not all movement plans have the same expectation of success. Planners must undertake a thorough analysis of the general terrain to be crossed, including the geology, mountain structure and forms, and ground cover.

(1) Heavily glaciated granite mountains pose different problems than does river-carved terrain. The U-shaped valley bulldozed out by a glacier forces maneuver elements down to the valley floor or up to the ridge tops, while the water-cut V-shape of river valleys allows movement throughout the compartment.

(2) Routes through granite rock (long cracks, good friction; use of pitons, chocks and camming units) will call for different equipment and technique than that used for steep limestone (pockets, smooth rock; bolts, camming units).

(3) Operations above tree line in temperate climates or in the brushy zone of arid mountains means that material for suspension traverse A-frames must be packed. The thick brush and krummholtz mats of subalpine zones and temperate forested mountains can create obstacles that must be bypassed.

(4) Heavy spruce/fir tangles slow progress to a crawl, therefore planners should ensure routes do not blindly traverse these zones.

d. **Unit Assessment.** When assessing unit movement capabilities the key indices are training and conditioning levels. Soldiers who have received basic military mountaineer training, who know how to move through rough terrain, and who have been hardened with training hikes through the mountains, will perform better than troops without this background.

e. **Time-Distance Formulas.** Computing march rates in the mountains is extremely difficult, especially when there is snow cover. The following rates are listed as a guide (Table 8-1). Rates are given for movement over flat or gently rolling terrain for individuals carrying a rifle and loaded rucksack.

	UNBROKEN TRAIL	BROKEN TRAIL
On foot (no snow cover)	2 to 3 kph (cross-country)	3 to 4 kph (trail walking)
On foot (less than 1 foot of snow)	1.6 to 3.2 kph	2 to 3.2 kph
On foot (more than 1 foot of snow)	.4 to 1.2 kph	2 to 3.2 kph
Snowshoeing	1.6 to 3.2 kph	3.2 to 4 kph
Skiing	1. to 5.6 kph	4.8 to 5.6 kph
Skijoring	N/A	3 to 24 kph

Table 8-1. Time-distance formulas.

(1) March distances in mountainous terrain are often measured in time rather than distance units. In order to do this, first measure the map distance. This distance plus 1/3 is a good estimate of actual ground distance. Add one hour for each 1,000 feet of ascent or 2,000 feet of descent to the time required for marching a map distance.

(2) As Table 8-1 indicates, snow cover will significantly affect rates of march. Since snow can be expected in the mountains most months of the year, units should have some experience at basic snow travel.

(3) Individual loads also affect march rates. Combined soldier loads that exceed 50 pounds per man can be expected to slow movement significantly in mountainous terrain. Given the increased weight of extra ammunition for crew-served weapons, basic mountaineering gear, and clothing for mountain travel, it becomes obvious that soldiers will be carrying weights well in excess of that 50-pound limit. Units should conduct cross-country movements in the mountains with the expected rucksack and LCE weights in order to obtain accurate, realistic rates of march.

(4) In the harsh environment of the mountains, helicopter support cannot be relied on. The process of transporting extra equipment and sustainment supplies will result in vastly increased movement times. The heavier loads will exhaust soldiers mentally and physically. Tactical movements, such as patrolling or deliberate assaults, should take this into account.

8-6. ROUTE SELECTION

Many variables affect the selection of the proper route. The following guidelines apply to all situations.

a. **Select a Current Map.** Check the date of the map for an indication of the reliability of the map in depicting vegetation, clearings, roads, and trails accurately. The leader should use all the latest topographic data he can find.

b. **Gather Intelligence Information.** The most important consideration in every leader's mind when plotting a movement is "where is the enemy?" The latest intelligence reports are essential. Additionally, weather reports, snow condition reports, avalanche probability, aerial photos, and any past or recent history of operations in the area may be of help.

c. **Select a Route.** Identify the starting point and determine the movement objective. Plot start and end points. Carefully scrutinize the area in between and begin to select the route. Consider the following:

(1) ***Trafficability.*** This includes degree of slopes, vegetation, width of trails, snow depth, avalanche probability, and the likelihood of crevasses.

(2) ***Time-Distance Formula.*** Time allotted and distance to be covered must be considered.

(3) ***Required Equipment.*** Carry enough equipment to move along the route and to survive if an extended stay becomes necessary. Do not plan a route beyond the means of your equipment.

(4) ***Location of Enemy.*** Plan a route that allows maximum use of the masking effect of the terrain. Avoid danger areas or areas of recent enemy activity unless required by the mission. Use vegetation to mask your movement if possible (especially coniferous forests). Avoid silhouetting on ridgelines.

(5) ***Communications.*** Communications will be severely limited in the mountains. Dead spaces or communications holes are common. Use all available information and plan accordingly.

(6) ***Conditions/Capabilities of Unit.*** The unit must be able to negotiate the route chosen. Take into consideration their present health, as well as their training level when selecting your intended route.

(7) ***Checkpoints/Control Points.*** When plotting a route on the map, utilize prominent terrain features on either side of the route as checkpoints. Ensure that when you select

your checkpoints they are visually significant (elevation) and that they are easily identifiable. Avoid the use of manmade features as checkpoints due to their unreliability and lack of permanence. Select features that are unique to the area.

CHAPTER 9
MOUNTAIN STREAM CROSSINGS

Operations conducted in mountainous terrain may often require the crossing of swift flowing rivers or streams. Such crossings should not be taken lightly. The force of the flowing water may be extremely great and is most often underestimated. All rivers and streams are obstacles to movement. They should be treated as danger areas and avoided whenever possible. When rivers or streams must be crossed, there are a variety of techniques the small-unit leader may choose from, depending upon the type of stream, its width, speed of the current, and depth of the water.

There are limits on the safe use of these techniques. Not all mountain rivers or streams will be fordable with these techniques. If a water obstacle is too wide, swift, or deep, an alternate route should be used, or the crossing will require major bridging by engineers. It may require the use of rafts or boats. Reconnaissance of questionable crossing sites is essential. This chapter covers the techniques for crossing mountain streams that have a depth generally not exceeding waist deep.

9-1. RECONNAISSANCE

Reconnaissance of the route (map, photo, and or aerial) may not always reveal that a water obstacle exists. In a swamp, for example, unfordable sloughs may not show on the map, and they may be concealed from aerial observation by a canopy of vegetation. Whenever it is possible that a unit will be required to cross a water obstacle, its commander must plan some type of crossing capability.

a. Site selection is extremely important once you determine that you must make a crossing (Figure 9-1). Look for a high place from which you can get a good view of the obstacle and possible crossing sites. A distant view, perhaps from a ridge, is sometimes better than a hundred close views from a riverbank. Site selection must be made before the arrival of the main body.

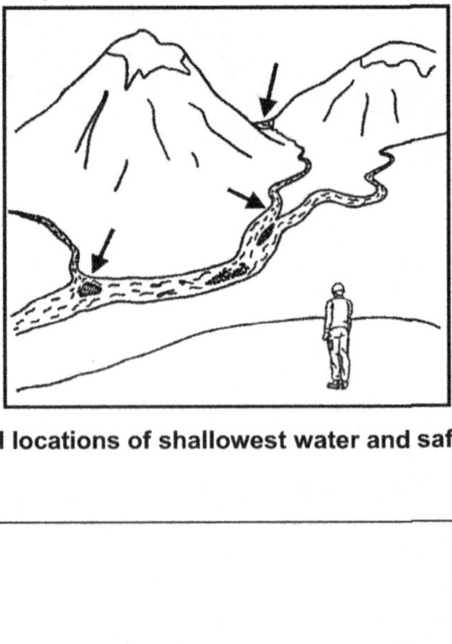

Figure 9-1. Normal locations of shallowest water and safest crossing sites.

b. A dry crossing on fallen timber or log jams is preferable to attempting a wet crossing. Depending upon the time of year, the closer you are to the source, or headwaters, the better your chances are of finding a natural snow or ice bridge for crossing. If a dry crossing is unavailable, the following considerations should be made:

(1) The time of day the crossing can be an important factor. Although early morning is generally best because the water level is normally lower during this period, recent weather is a big factor; there may have been heavy rain in the last eight hours. As glaciers, snow, or ice melt during the day, the rivers rise, reaching their maximum height between mid afternoon and late evening, depending on the distance from the source. Crossings, if made during the early morning, will also allow clothing to dry more quickly during the heat of the day.

(2) A crossing point should normally be chosen at the widest, and thus shallowest, point of the river or stream. Sharp bends in the river should be avoided since the water is likely to be deep and have a strong current on the outside of the bend. Crossings will be easiest on a smooth, firm bottom. Large rocks and boulders provide poor footing and cause a great deal of turbulence in the water.

(3) Many mountain streams, especially those which are fed by glacier run-off, contain sections with numerous channels. It is often easier to select a route through these braided sections rather than trying to cross one main channel. A drawback to crossing these braided channels, however, is the greater distance to the far bank may increase exposure time and often the sand and gravel bars between the channels will offer little cover or concealment, if any.

(4) The crossing site should have low enough banks on the near and far side to allow a man carrying equipment to enter and exit the stream with relative ease. If a handline or rope bridge is to be constructed, the crossing site should have suitable anchors on the near and far bank, along with safe loading and unloading areas. Natural anchors are not a necessity, however the time required to find a site with solid natural anchors will probably be less than the time required to construct artificial anchors. In some areas, above the tree line for example, artificial anchors may have to be constructed. Deadman anchors buried in the ground, or under a large pile of boulders work well.

(5) Log jams and other large obstructions present their own hazards. Logs floating downstream will generally get hung up in shallower sections creating the jam. Once a log jam is formed, however, the water forced to flow around it will erode the stream bottom. Eventually deep drop-offs or holes may develop, especially around the sides and off the downstream end of the log jam. A log jam that totally bridges a section of the stream may be the best way to cross. A wet crossing in the vicinity of a log jam should be performed a good distance below or above it. Some things to consider when crossing near log jams are:

- Cross well to the downstream side when possible.
- Keep a sharp lookout for floating timber that could knock you off your feet.
- If you must cross on the upstream side, stay well upstream from the log jam. If a person is swept off his feet and caught in the debris of the jam, he could easily drown. A handline will greatly increase safety here.

(6) When possible, select a crossing site that has enough natural protection on the near and far banks so that security teams may be placed out and enough cover and concealment is available for the size of the element making the crossing. When cover and

concealment is minimal, as in the higher alpine zones, the crossings must be conducted as efficiently as possible to minimize exposure to enemy observation.

9-2. PREPARATION OF TROOPS AND EQUIPMENT

Prepare men and equipment for a crossing as far in advance as feasible. Final preparation should be completed in a security perimeter on the near side just before crossing. Preparation includes the following.

a. Waterproof water-sensitive items. Wrap radios, binoculars, SOI, papers, maps and any extra clothing in waterproof bags (trash bags also work well), if available. These bags also provide additional buoyancy in case of a fall.

b. Trousers are unbloused and shirts are pulled out of the trousers. All pockets are buttoned. This allows water to escape through the clothing. Otherwise the clothing would fill up and retain water, which would weigh the body down. This is especially critical if an individual must swim to shore. Depending on the circumstances of the crossing (for example, tactical situation, temperature of the air and water), the crossing can be made in minimal clothing so that dry clothing is available after the crossing. Boots should be worn to protect feet from rocks; however, socks and inner soles should be removed. On the far side, the boots can be drained and dry socks replaced.

c. Load-carrying equipment harness and load-bearing vest (LBV) is unbuckled and worn loosely. It is extremely difficult to remove a buckled harness in the water in an emergency.

d. Helmets are normally removed and placed in the rucksack in slow moving streams with sandy or gravel bottoms. If you have to resort to swimming it is easier done without the helmet. However, when crossing swift flowing streams, especially those with large rocks and other debris, the risk of head injury if a person slips is high. In this case the helmet should be worn with the chinstrap fastened.

e. The rucksack should be worn well up on the shoulders and snug enough so it does not flop around and cause you to lose your balance. The waist strap MUST be unbuckled so you can get rid of the pack quickly if you are swept off your feet and have to resort to swimming. If a pack has a chest strap it must also be unbuckled. Secure everything well within your pack. It is easier to find one large pack than to find several smaller items.

f. Individual weapons should be attached to the pack or slung over the shoulder.

9-3. INDIVIDUAL CROSSINGS

Whenever possible, and when the degree of experience permits, streams should be forded individually for a speedier crossing. The average soldier should be able to cross most streams with mild to moderate currents and water depths of not much more than knee deep using proper techniques.

a. The individual should generally face upstream and slightly sideways, leaning slightly into the current to help maintain balance. At times, he may choose to face more sideways as this will reduce the surface area of the body against the current, thus reducing the current's overall force on the individual.

b. The feet should be shuffled along the bottom rather than lifted, with the downstream foot normally in the lead. He should take short, deliberate steps. Lunging steps and crossing the feet result in a momentary loss of balance and greatly increase the chance of a slip.

c. The individual should normally cross at a slight downstream angle so as not to fight the current. There is normally less chance of a slip when stepping off with the current as opposed to stepping off against the current.

d. The individual must constantly feel for obstacles, holes and drop-offs with the lead foot and adjust his route accordingly. If an obstacle is encountered, the feet should be placed on the upstream side of it where the turbulence is less severe and the water normally shallower.

e. To increase balance, and if available, a long ice ax, sturdy tree limb, or other staff can be used to give the individual a third point of contact (Figure 9-2). The staff should be used on the upstream side of the individual and slightly leaned upon for support. The staff should be moved first, then the feet shuffled forward to it. This allows two points of contact to be maintained with the streambed at all times. The individual still moves at a downstream angle with the downstream foot in the lead.

Figure 9-2. Individual crossing with staff.

9-4. TEAM CROSSING

When the water level begins to reach thigh deep, or anytime the current is too swift for personnel to safely perform an individual crossing, a team crossing may be used. For chain crossing, two or more individuals cross arms with each other and lock their hands in front of themselves (Figure 9-3). The line formed faces the far bank. The largest individual should be on the upstream end of the line to break the current for the group. The line formed will then move across the stream using the same principles as for individual crossings, but with the added support of each other. The line should cross parallel to the direction of the current. The team still moves at a slight downstream angle, stepping off with the downstream foot in the lead.

Figure 9-3. Chain method for stream crossing.

9-5. ROPE INSTALLATIONS

When the water level begins to reach waist deep or the current is too swift for even a team crossing, the chosen site must be closely examined. The stream at this point may be impassable. Many times though, a crossing site which may be unsafe for individual or team crossings can be made safe with the installation of a handline or rope bridge. Crossing on a handline will still require each individual to enter the water and get wet. If a one-rope bridge can be constructed, it may require only a couple of individuals to enter the water. Deciding whether to install a handline or a rope bridge will depend on the anchors available, height of the anchors above the water, and the distance from the near and far anchors. The maximum distance a one-rope bridge is capable of spanning is approximately 1/2 to 2/3 the length of the rope in use.

a. **Establishing the Far Anchor.** Whether a handline or rope bridge is to be installed, someone must cross the stream with one end of the rope and anchor it on the far side. This duty should be performed by the most capable and strongest swimmer in the party. The swimmer should be belayed across for his own safety. The belay position should be placed as far above the crossing as possible. In the event that the current is too strong for the individual, he will pendulum back to the near bank. Rescuers should be poised on the near bank at points where the individual will pendulum back, should he fail to reach the far bank. The initial crossing site should be free of obstacles that would snag the rope and prevent the pendulum back to the bank for an easy recovery.

(1) The individual may attach the belay rope to his seat harness or a swami belt with a carabiner. He should NEVER tie directly into the rope when being belayed for a stream crossing. If the swimmer should be swept away and become tangled, he must be able to release himself quickly from the rope and swim to shore as best he can. The individual may also choose to tie a fixed loop into the end of the belay rope and hang on to it, where he can immediately release it in an emergency.

(2) Anytime a crossing site must be used where the swimmer may encounter problems getting to the far bank, he should have on a life vest or other personal flotation

device (PFD). If the swimmer must release the rope at any time, he will have to rely on his own water survival skills and swimming ability to get to shore. A PFD will greatly increase his own personal safety. A PFD may also be used by the last man across, as he will release the rope from the anchor and be belayed across as the first man

b. **Installation of a Handline.** If it is possible to use a rope high enough above the water to enable soldiers to perform a dry crossing, then a rope bridge should be installed as such. If this is impossible, and the rope must be installed to assist in a wet crossing, then it should be installed as a handline (Figure 9-4).

(1) The far anchor should be downstream from the near anchor so that the rope will run at an angle downstream from the near anchor, approximately thirty to forty-five degrees, rather than straight across the stream. Here again, it is easier to move with the current as opposed to directly across or against it.

(2) The rope may be anchored immediately on the far bank, pulled tight, and anchored on the near bank, or it may be installed with a transport tightening system if a tighter rope is required.

(3) Crossing will always be performed on the downstream side of the handline, shuffling the feet with the downstream foot in the lead.

(4) A second climbing rope is used as a belay (Figure 9-5). One end of the belay rope will be on the near bank and the other end on the far bank. It should be sent across with the strong-swimmer. An appropriate knot is tied into the middle of the belay rope to form two fixed loops with each loop being approximately 6 inches long. One loop is connected to the handline with carabiner(s) and the individual crossing connects one loop to himself. The loops are short enough so the individual is always within arms reach of the handline should he slip and let go. The individuals are belayed from both the near and far banks. If a mishap should occur the individual can be retrieved from either shore, whichever appears easiest. The belay on the opposite shore can be released allowing the individual to pendulum to the bank. It is important that the belay rope NOT be anchored or tied to the belayer so that it may be quickly released if necessary.

Figure 9-4. Stream crossing using a handline.

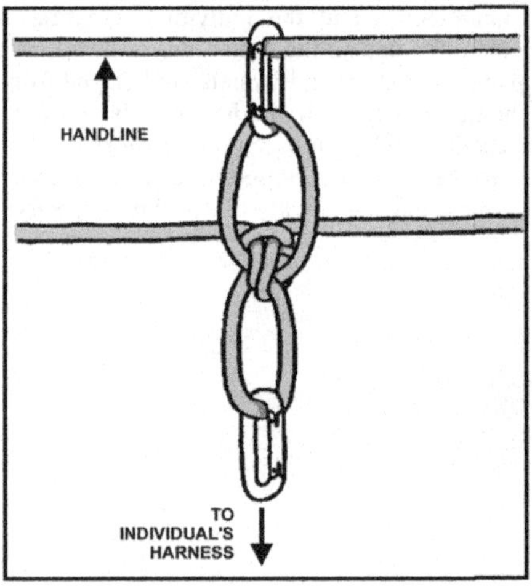

Figure 9-5. Belay rope for crossing using a handline.

(5) Under most circumstances, the handline should be crossed one person at a time. This keeps rope stretch and load on the anchors to a minimum.

(6) Rucksacks can be either carried on the back the same way as for individual crossings, or they can be attached to the handline and pulled along behind the individual.

(7) If a large amount of equipment must be moved across the stream, especially heavier weapons, such as mortars, recoilless rifles, and so on, then a site should be selected to install a rope bridge.

9-6. SAFETY

River and stream crossings present one of the most hazardous situations faced by the military mountaineer. The following safety procedures are minimum guidelines that should be followed when conducting a river or stream crossing.

a. All weak and nonswimmers should be identified before a crossing so that stronger swimmers may give assistance in crossing.

b. Not every river or stream can be crossed safely. It is always possible to cross at a different time or place, use a different technique, or choose another route.

c. The technique used is directly dependent upon water depth, speed of the current, stream bottom configuration, width of the stream, and individual experience.

d. The safest methods of crossing are always with the use of a handline or one-rope bridge.

e. If the installation of a handline or rope bridge becomes too difficult at a given crossing site, then that site should be considered too hazardous and another site selected.

f. A lookout should be posted 50 to 100 meters upstream to watch for any obstacles that may be carried downstream and interfere with the crossing.

g. When conducting individual crossings (those without a handline or rope bridge), lifeguards should be posted downstream with poles or ropes prepared to throw, for assistance or rescue.

h. When the unit knows a rope installation will be required for crossing, at least two life vests or other PFDs should be on hand to provide additional safety for the strong swimmer who must establish the far anchor, and the last man across who retrieves the system.

9-7. SWIMMING

There are times when you might be alone and have no choice but to swim across, or there may be a time that you find yourself suddenly plunged into a swift river or rapids. In either case, the following techniques could save your life.

a. Immediately jettison any equipment or clothing that restricts movement.

b. Do not try to fight the current. Maneuver towards shore in a position with the feet downstream, facing downstream, and fanning the hands alongside the body to add buoyancy and to fend off submerged rocks. Use the feet to protect the rest of the body and to fend off submerged rocks.

c. Keep the head above water to observe for obstacles and attempt to maneuver away from them.

d. Try to avoid backwater eddies and converging currents as they often contain dangerous swirls. Avoid bubbly water under falls as it has little buoyancy. Breath between the wave troughs.

e. If the shore is too difficult to reach, seek out the closest and safest spot, such as a sandbar, to get yourself out of the water as quickly as possible. Hypothermia will set in quickly in colder waters.

CHAPTER 10

MOVEMENT OVER SNOW AND ICE

Movement over snow- and ice-covered slopes presents its own unique problems. Movement on steeper slopes requires an ice ax, crampons, and the necessary training for this equipment. Personnel will also have to learn how to place solid anchors in snow and ice to protect themselves during these movements if roped. Snow-covered glaciers present crevasse fall hazards even when the slope is relatively flat, requiring personnel to learn unique glacier travel and crevasse rescue techniques.

All the principles of rock climbing, anchor placement, belays, and rope usage discussed throughout the previous chapters apply to snow and ice climbing as well. This chapter will focus on the additional skills and techniques required to move safely through snow-covered mountains and over glaciated terrain.

10-1. MOVEMENT OVER SNOW

The military mountaineer must be equally adept on both snow and ice due to route necessity and rapidly changing conditions. On steep slopes in deep snow, the climber may climb straight up facing the slope. The ice ax shaft, driven directly into the snow, provides a quick and effective self-belay in case of a slip—the deeper the shaft penetrates the snow, the better the anchor (Figure 10-1). It is usually best, however, to climb snow-covered slopes in a traversing fashion in order to conserve energy, unless there is significant avalanche danger.

Figure 10-1. Self-belay on snow.

a. The progression from walking on flat terrain to moving on steep terrain is the same as for moving over snow-free terrain. If the snow is packed the sole of the boot will generally hold by kicking steps, even on steep slopes. Where it is difficult to make an effective step with the boot, a cut made with the adze of the ice ax creates an effective step. In these situations crampons should be used for faster and easier movement.

b. When descending on snow, one can usually come straight downhill, even on steep terrain. Movement downhill should be slow and deliberate with the climber using an even pace. The heels should be kicked vigorously into the snow. The body may be kept erect with the aid of an ice ax, which may be jammed into the snow at each step for additional safety. Here again, crampons or step cutting may be necessary. A technique known as glissading may also be used as an easy method of descent and is covered in detail later in this chapter.

10-2. MOVEMENT OVER ICE

Ice is found in many areas of mountains when snow is present, and during the summer months also where perennial snowpack exists. Many times an ice area will be downslope of a snowfield and sometimes the ice pack itself will be lightly covered with snow. Even if using an ice ax and or crampons, movement will still be difficult without proper training.

10-3. USE OF ICE AX AND CRAMPONS

Movement over snow and ice is almost impossible without an ice ax and or crampons.

a. **Ice Ax.** When walking on snow or ice, the ice ax can be used as a third point of contact. When the terrain steepens, there are a number of ways to use the ice ax for snow or ice climbing. Some positions are more effective than others, depending on the intended result. You may find other ways to hold and use the ax, as long the security remains in effect.

(1) ***Cane Position.*** The ice ax can be used on gentle slopes as a walking stick or cane (Figure 10-2). The ax is held by the head with the spike down and the pick facing to the rear in preparation for self-arrest. When moving up or down gentle slopes the ice ax is placed in front as the third point of contact, and the climber moves toward it. When traversing, the ax is held on the uphill side, in preparation for a self-arrest.

Figure 10-2. Using the ice ax in the cane position.

(2) ***Cross Body Position or Port Arms Position.*** On steeper slopes the ax can be used in the port arms position, or cross body position (Figure 10-3). It is carried across the chest, upslope hand on the shaft, spike towards the slope. The head of the ax is held away from the slope with the pick to the rear in preparation for self-arrest. Ensure the leash is connected to the upslope hand, which allows the ax to be used in the hammer position on the upslope side of the climber. The spike, in this case, is used as an aid for maintaining balance.

Figure 10-3. Ice ax in the cross body or port arms position.

(3) ***Anchor Position***. As the slope continues to steepen, the ax may be used in the anchor position (Figure 10-4). The head is held in the upslope hand and the pick is driven into the slope. The spike is held in the downhill hand and pulled slightly away from the slope to increase the "bite" of the pick into the ice. If the climber is wearing a harness, the pick can be deeply inserted in the ice or hard snow and the ax leash could be connected to the tie-in point on the harness for an anchor (ensure the ax is placed for the intended direction of pull).

Figure 10-4. Ice ax in the anchor position.

(4) ***Push-Hold Position.*** Another variation on steep slopes is the push-hold position (Figure 10-5). The hand is placed on the shaft of the ax just below the head with the pick forward. The pick is driven into the slope at shoulder height. The hand is then placed on the top of the ax head for use as a handhold.

Figure 10-5. Ice ax in the push-hold position.

(5) ***Dagger Position.*** The dagger position is used on steep slopes to place a handhold above shoulder height (Figure 10-6). The hand grasps the head of the ax with the pick forward and the shaft hanging down. The ax is driven into the surface in a stabbing action. The hand is then placed on the ax head for use as a handhold.

Figure 10-6. Ice ax in the dagger position.

(6) ***Hammer Position.*** The hammer position will set the pick deepest in any snow or ice condition (Figure 10-7, page 10-6). The ax is used like a hammer with the pick being driven into the slope. On vertical or near-vertical sections, two axes used in the hammer position will often be required.

Figure 10-7. Ice ax in the hammer position.

b. **Crampons.** Walking in crampons is not complicated but it does present difficulties. When walking in crampons, the same principles are used as in mountain walking, except that when a leg is advanced it is swung in a slight arc around the fixed foot to avoid locking the crampons or catching them in clothing or flesh. The trousers should be bloused to prevent catching on crampons. All straps should be secured to prevent stepping on them and, potentially, causing a fall. The buckles should be located on the outside of each foot when the crampons are secured to prevent snagging. Remember, when the crampon snags on the pants or boots, a tear or cut usually results, and sometimes involves the skin on your leg and or a serious fall.

(1) Two methods of ascent are used on slopes: traversing and straight up.

(a) A traverse on ice or snow looks much like any mountain walking traverse, except that the ankles are rolled so that the crampons are placed flat on the surface (Figure 10-8). On snow the points penetrate easily; on ice the foot must be pressed or stamped firmly to obtain maximum penetration. At the turning points of a traverse, direction is changed with the uphill foot as in mountain walking.

Figure 10-8. Correct and incorrect crampon technique.

(b) A straight up method is for relatively short pitches, since it is more tiring than a traverse. The climber faces directly up the slope and walks straight uphill. As the slope steepens, the herringbone step is used to maintain the flatfoot technique. For short steep pitches, the climber may also face downslope, squatting so the legs almost form a

90-degree angle at the knees, driving the spike of the ice ax into the slope at hip level, and then moving the feet up to the ax. By repeating these steps, the ax and crampon combination can be used to climb short, steep pitches without resorting to step cutting. This method can be tiring. The technique is similar to the crab position used for climbing on slab rock and can also be used for short descents.

(2) A technique known as "front-pointing" may be used for moving straight uphill (Figure 10-9). It is especially useful on steep terrain, in combination with the ice ax in the push-hold, dagger, or hammer position. Front-pointing is easiest with the use of more rigid mountain boots and rigid crampons. The technique is similar to doing calf raises on the tips of the toes and is much more tiring than flat-footing.

(a) The technique starts with the feet approximately shoulder width apart. When a step is taken the climber places the front points of the crampons into the ice with the toe of the boot pointing straight into the slope.

(b) When the front points have bitten into the ice the heel of the boot is lowered slightly so that the first set of vertical points can also bite. The body is kept erect, with the weight centered over the feet as in climbing on rock.

Figure 10-9. Front-pointing with crampons.

c. **Vertical Ice.** When a climb on ice reaches the 60- to 70-degree angle, two ice axes may be helpful, and will become necessary as the angle approaches 90 degrees. The same basic climbing techniques described in Chapter 6 should be applied. If leashes of the correct length and fit are attached to both axes, it may be possible to hang completely from the axes while moving the feet.

d. **Descending with Crampons and Ice Ax.** Whenever possible, descend straight down the fall line. As the slope steepens, gradually turn sideways; on steeper slopes, bend at the waist and knees as if sitting, keeping the feet flat to engage all vertical crampon points and keep the weight over the feet as in descending rock slab (Figures 10-10 and 10-11, page 10-8). On steep terrain, assume a cross body or port arms position with the ax, and traverse. The crab position or front-pointing may also be used for descending. Regardless of the technique used, always ensure the points of the crampons are inserted in the snow or ice and take short, deliberate steps to minimize the chance of tripping and falling down the slope.

Figure 10-10. Flat-footing in the crab position.

Figure 10-11. Use of ice ax in descent.

e. **Normal Progression.** The use of the ice ax and crampons follows a simple, logical progression. The techniques can be used in any combination, dictated by the terrain and skill of the individual. A typical progression could be as follows:

(1) ***Crampons.*** Use crampons in the following situations:

- Walking as on flat ground.

- Herringbone step straight up the slope.
- Traverse with feet flat.
- Backing up the slope (crab position).
- Front-pointing.

(2) ***Ice Ax.*** Use the ice ax in these situations:

- Cane position on flat ground.
- Cane position on uphill side as slope steepens.
- Port arms position with spike on uphill side.
- Anchor position with pick on uphill side.
- Push-hold position using front-pointing technique.
- Dagger position using front-pointing technique.
- Hammer position using front-pointing technique.

e. **Climbing Sequence.** Using most of these positions, a single ax can be "climbed" in steps to move upslope on low-angle to near vertical terrain (Figure 10-12). Begin by positioning the feet in a secure stance and placing the ax in the hammer position as high as possible. Slowly and carefully move the feet to higher positions alternately, and move the hand up the ax shaft. Repeat this until the hand is on top of the head of the ax. Remove the ax and place it at a higher position and begin again.

Figure 10-12. Climbing sequence.

f. **Step Cutting.** Step cutting is an extremely valuable technique that is a required skill for any military mountaineer (Figure 10-13). Using cut steps can save valuable time that would be spent in donning crampons for short stretches of ice and can, in some cases, save the weight of the crampons altogether. Steps may also have to be cut by the lead team to enable a unit without proper equipment to negotiate snow- or ice-covered terrain. As units continue to move up areas where steps have been cut they should continue to improve each step. In ascending, steps may be cut straight up the slope, although a traverse will normally be adopted. In descending, a traverse is also the preferred method. When changing direction, a step large enough for both feet and crampons must be made. Once the step is formed, the adze is best used to further shape and clean the step.

(1) ***Snow.*** On slopes of firm snow and soft ice, steps may be cut by swinging the ax in a near-vertical plane, using the inside corner of the adze for cutting. The step should be fashioned so that it slopes slightly inward and is big enough to admit the entire foot. Steps used for resting or for turning must be larger.

(2) ***Ice.*** Hard ice requires that the pick of the ax be used. Begin by directing a line of blows at right angles to the slope to make a fracture line along the base of the intended step. This technique will reduce the chance of an unwanted fracture in the ice breaking out the entire step. Next, chop above the fracture line to fashion the step. When using the pick it should be given an outward jerk as it is placed to prevent it from sticking in the ice.

(3) ***Step Cutting in a Traverse.*** When cutting steps in a traverse, the preferred cutting sequence is to cut one step at an arm's length from the highest step already cut, then cut one between those two. Cutting ahead one step then cutting an intermediate step keeps all of the steps relatively close to one another and maintains a suitable interval that all personnel can use.

(4) ***Handholds.*** If handholds are cut, they should be smaller than footholds, and angled more.

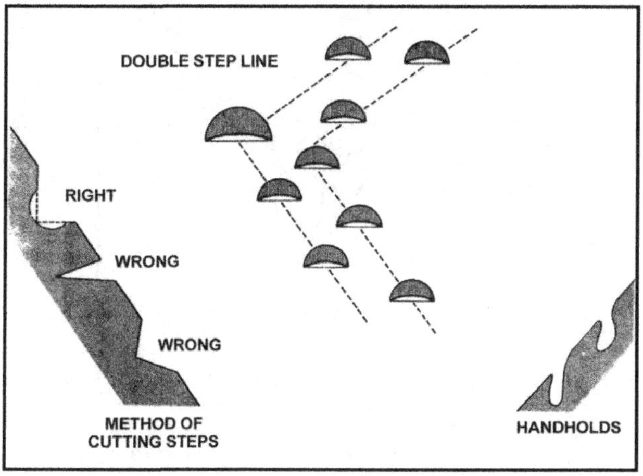

Figure 10-13. Step cutting and handhold cutting.

g. **Self-Arrest.** The large number of climbers injured or killed while climbing on snow and ice can be attributed to two major failings on the part of the climber: climbing unroped, and a lack of knowledge and experience in the techniques necessary to stop, or arrest, a fall (Figure 10-14). A climber should always carry an ice ax when climbing on steep snow or ice; if a fall occurs, he must retain possession and control of his ice ax if he is to successfully arrest the fall. During movement on steep ice, the ax pick will be in the ice solidly before the body is moved, which should prevent a fall of any significance (this is a self belay not a self-arrest).

CAUTION
Self-arrest requires the ax pick to gradually dig in to slow the descent. Self-arrest is difficult on steep ice because the ice ax pick instantly "bites" into the ice, possibly resulting in either arm or shoulder injury, or the ax is deflected immediately upon contact.

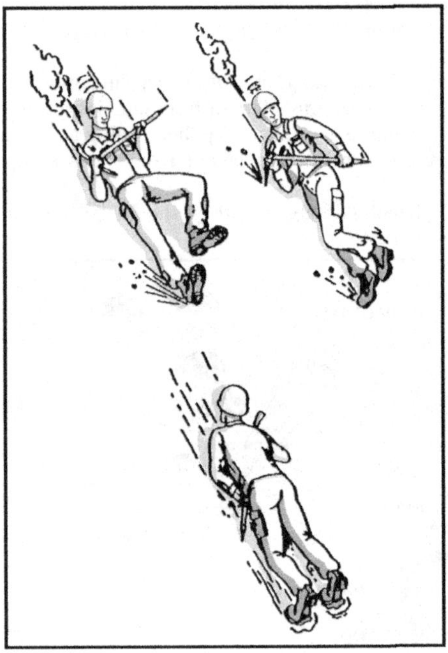

Figure 10-14. Self-arrest technique.

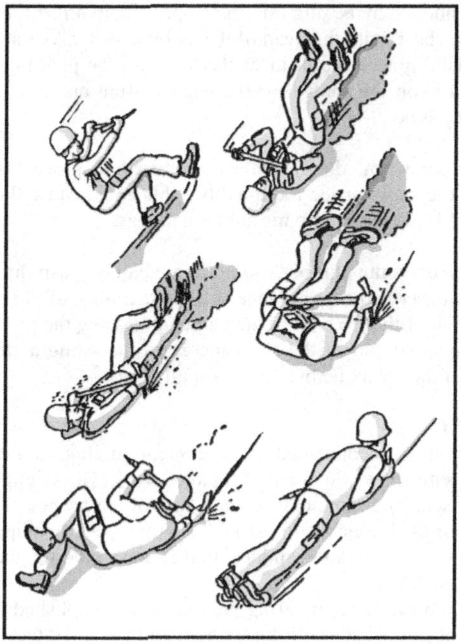

Figure 10-14. Self-arrest technique (continued).

(1) A climber who has fallen may roll or spin; if this happens, the climber must first gain control of his body, whether it is with his ice ax or simply by brute force. Once the roll or spin has been controlled, the climber will find himself in one of four positions.

- Head upslope, stomach on the slope, and feet pointed downslope.
- Head upslope, back to the slope, and feet pointed downslope.
- Head downslope, stomach on the slope, and feet pointed upslope.
- Head downslope, back to the slope, and feet pointed upslope.

(2) To place the body in position to arrest from the four basic fall positions the following must be accomplished.

(a) In the first position, the body is in proper relation to the slope for an arrest.

(b) In the second position, the body must first be rotated from face up to face down on the slope. This is accomplished by rolling the body toward the head of the ax.

(c) In the third position, the pick of the ice ax is placed upslope and used as a pivot to bring the body into proper position.

(d) In the fourth position, the head of the ax must be driven into the snow to the climber's side. This will cause the body to rotate into a head up, stomach down position.

(3) The final position when the arrest of the fall is completed should be with the head upslope, stomach on the slope, with the feet pointed downslope. If crampons are not

worn, the toe of the boots may be dug into the slope to help arrest the fall. The ax is held diagonally across the chest, with the head of the ax by one shoulder and the spike near the opposite hip. One hand grasps the head of the ax, with the pick pointed into the slope, while the other hand is on the shaft near the spike, lifting up on it to prevent the spike from digging into the slope.

Note: If crampons are worn, the feet must be raised to prevent the crampons from digging into the snow or ice too quickly. This could cause the climber to tumble and also, could severely injure his ankles and legs.

(4) When a fall occurs, the climber should immediately grasp the ax with both hands and hold it firmly as described above. Once sufficient control of the body is attained, the climber drives the pick of the ice ax into the slope, increasing the pressure until the fall is arrested. Raising the spike end of the shaft increases the biting action of the pick. It is critical that control of the ice ax be maintained at all times.

10-4. GLISSADING

Glissading is the intentional, controlled, rapid descent, or slide of a mountaineer down a steep slope covered with snow (Figure 10-15, page 10-14). Glissading is similar to skiing, except skis are not used. The same balance and control are necessary, but instead of skis the soles of the feet or the buttocks are used. The only piece of equipment required is the standard ice ax, which serves as the rudder, brake, and guide for the glissade. The two basic methods of glissading are:

a. **Squatting Glissade.** The squatting glissade is accomplished by placing the body in a semi-crouched position with both knees bent and the body weight directly over the feet. The ice ax is grasped with one hand on the head, pick, and adze outboard (away from the body), and the other hand on the shaft. The hand on the shaft grips it firmly in a position that allows control as well as the application of downward pressure on the spike of the ax.

b. **Sitting Glissade.** Using this method the glissader sits on the snow with the legs flat, and the heels and feet raised and pointed downslope. The ice ax is firmly grasped in the same manner as the squatting glissade, with the exception that the hand on the shaft must be locked against the hip for control. The sitting glissade is slower but easier to control than the squatting glissade.

c. **Safety.** A glissade should never be attempted on a slope where the bottom cannot be seen, since drop-offs may exist out of view. Also, a sitting glissade should not be used if the snow cover is thin, as painful injury could result.

Figure 10-15. Glissading techniques.

10-5. SNOW AND ICE ANCHORS

Ice and snow anchors consist of snow pickets, flukes, deadman-type anchors, ice screws, and ice pitons. Deadman anchors can be constructed from snowshoes, skis, backpacks, sleds, or any large items.

a. **Ice Pitons.** The ice piton is used to establish anchor points. The ice piton is not seen in modern ice climbing but may still be available to the military. The standard ice piton is made of tubular steel and is 10 inches in length. Ice pitons installed in pairs are a bombproof anchor; however, ice pitons have no threads for friction to hold them in the ice once placed and are removed easily. Safe use of ice pitons requires placement in pairs. Used singularly, ice pitons are a strong anchor but are easily removed, decreasing the perceived security of the anchor. Follow the instructions below for placing ice pitons in pairs.

(1) Cut a horizontal recess into the ice, and also create a vertical surface (two clean surfaces at right angles to each other).

(2) Drive one piton into the horizontal surface and another into the vertical surface so that the two pitons intersect at the necessary point (Figure 10-16).

(3) Connect the two rings with a single carabiner, ensuring the carabiner is not cross-loaded. Webbing or rope can be used if the rings are turned to the inside of the intersection.

(4) Test the piton pair to ensure it is secure. If it pulls out or appears weak, move to another spot and replace it. The pair of pitons, when placed correctly, are multidirectional.

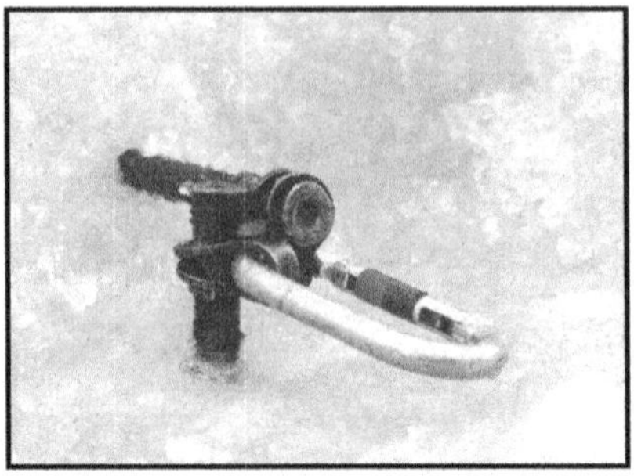

Figure 10-16. Ice piton pair.

(5) The effective time and or strength for an ice piton placement is limited. The piton will heat from solar radiation, or the ice may crack or soften. Solar radiation can be nearly eliminated by covering the pitons with ice chips once they have been placed. If repeated use is necessary for one installation, such as top roping, the pitons should be inspected frequently and relocated when necessary. When an ice piton is removed, the ice that has accumulated in the tube must be removed before it freezes in position, making further use difficult.

c. **Ice Screws**. The ice screw is the most common type of ice protection and has replaced the ice piton for the most part (Figure 10-17). Some screws have longer "hangers" or handles, which allow them to be easily twisted into position by hand. Place ice screws as follows:

(1) Clear away all rotten ice from the surface and make a small hole with the ax pick to start the ice screw in.

(2) Force the ice screw in until the threads catch.

Figure 10-17. Placement of ice screw using the pick.

(3) Turn the screw until the eye or the hanger of the ice screw is flush with the ice and pointing down. The screw should be placed at an angle 90 to 100 degrees from the

lower surface. Use either your hand or the pick of the ice ax to screw it in. If you have a short ax (70 centimeters or less), you may be able to use the spike in the eye or hanger to ease the turning. (Remember that you may only have use of one hand at this point depending on your stance and the angle of the terrain.)

(4) As with ice pitons, melting of the ice around a screw over a period of time must be considered. The effective time and or strength of an ice screw placement is limited. The screw will heat from solar radiation, or the ice may crack or soften. Solar radiation can be nearly eliminated by covering the screw with ice chips once it has been emplaced. If repeated use is necessary for one installation, such as top roping, the screws should be inspected frequently and relocated when necessary. When an ice screw is removed, the ice that has accumulated in the tube, must be removed before further use.

d. **Horseshoe or Bollard Anchor.** This is an artificial anchor shaped generally like a horseshoe (Figure 10-18). It is formed from either ice or snow and constructed by either cutting with the ice ax or stamping with the boots. When constructed of snow, the width should not be less than 10 feet. In ice, this width may be narrowed to 2 feet, depending on the strength of the ice. The length of the bollard should be at least twice the width. The trench around the horseshoe should be stamped as deeply as possible in the snow and should be cut not less than 6 inches into the ice after all rotten ice is removed. The backside of the anchor must always be undercut to prevent the rope from sliding off and over the anchor.

(1) This type of anchor is usually available and may be used for fixed ropes or rappels. It must be inspected frequently to ensure that the rope is not cutting through the snow or ice more than one-third the length of the anchor; if it is, a new anchor must be constructed in a different location.

(2) A horseshoe anchor constructed in snow is always precarious, its strength depending upon the prevailing texture of the snow. For dry or wind-packed snow, the reliability of the anchor should always be suspect. The backside of the bollard can be reinforced with ice axes, pickets, or other equipment for added strength.

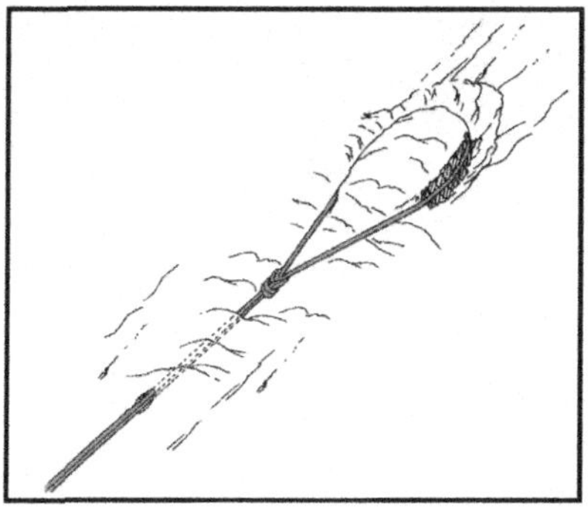

Figure 10-18. Horseshoe or bollard anchor.

e. **Pickets and Ice Axes.** Pickets and ice axes may be used as snow anchors as follows.

(1) The picket should be driven into the snow at 5 to 15 degrees off perpendicular from the lower surface. If the picket cannot be driven in all the way to the top hole, the carabiner should be placed in the hole closest to the snow surface to reduce leverage. The picket may also be tied off with a short loop of webbing or rope as in tying off pitons.

(2) An ice ax can be used in place of a picket. When using an ice ax as a snow anchor, it should be inserted with the widest portion of the ax shaft facing the direction of pull. The simplest connection to the ax is to use a sling or rope directly around the shaft just under the head. If using the leash ensure it is not worn, frayed, or cut from general use; is strong enough; and does not twist the ax when loaded. A carabiner can be clipped through the hole in the head, also.

(3) Whenever the strength of the snow anchor is suspect, especially when a picket or ax cannot be driven in all the way, the anchor may be buried in the snow and used as a "dead man" anchor. Other items suitable for dead man anchor construction are backpacks, skis, snowshoes, ski poles, or any other item large enough or shaped correctly to achieve the design. A similar anchor, sometimes referred to as a "dead guy," can be made with a large sack either stuffed with noncompressible items or filled with snow and buried. Ensure the attaching point is accessible before burying. The direction of pull on long items, such as a picket or ax, should be at a right angle to its length. The construction is identical to that of the dead man anchor used in earth.

f. **Equalized Anchors.** Snow and ice anchors must be constantly checked due to melting and changing snow or ice conditions.

(1) Whenever possible, two or more anchors should be used. While this is not always practical for intermediate anchor points on lead climbs or fixed ropes, it should be mandatory for main anchors at all belay positions, rappel points, or other fixed rope installations. (Figure 10-19, page 10-18, shows an example of three snow pickets configured to an equalized anchor.)

(2) As with multipoint anchors on rock, two or more snow or ice anchors can be joined together with a sling rope or webbing to construct one solid, equalized anchor. A bowline on a bight tied into the climbing rope can also be used instead of sling ropes or webbing.

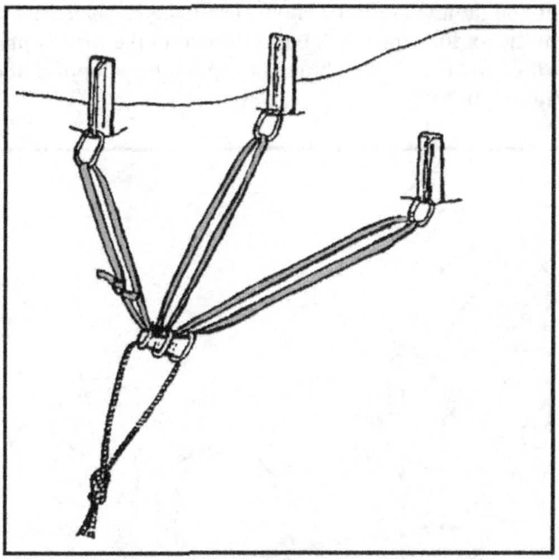

Figure 10-19. Equalized anchor using pickets.

10-6. ROPED CLIMBING ON ICE AND SNOW

When climbing on ice or snow team members tie into a climbing rope the same as when they climb on rock. When crevasses are expected, a three-man rope team is recommended.

a. **Tie-In Method.** For climbing on snow and ice, the tie-in procedure is normally the same as for rock climbing; however, when moving over snow-covered glaciers, the tie-in is modified slightly. (See paragraph 10-7, Movement on Glaciers, for more information).

b. **Movement.** For movement on gentle or moderate slopes where there is little chance of a serious fall, all climbers move simultaneously. Normally the climbers move in single file using the steps created by the lead climber and improving them when necessary. The rope between the climbers should be fully extended and kept reasonably tight. Should any member fall, he immediately yells "FALLING." The other rope team members immediately drop into a self-arrest position. The fallen climber also applies the self-arrest procedure. By using this method, called the "team arrest," the entire team as a whole arrests the fall of one member. On steeper slopes, and when crossing snow-covered crevasses where the snow bridges appear weak, the climbers use belayed climbing techniques as in rock climbing.

c. **Belaying on Snow and Ice.** The principles of belaying on ice and snow are the same as on rock. Generally, the high-force falls found in rock climbing are not present on snow and ice unless the pitch being climbed is extremely steep.

(1) ***Boot-Ax Belay.*** This belay can be useful in areas where the full length of the ice ax can penetrate the snow. The holding strength of the boot-ax belay is directly related to the firmness of the snow and to the strength of the ice ax shaft. The shaft of the ax is

tilted slightly uphill and jammed into the snow. The belayer places his uphill foot against the downhill side of the ax for support. A bight formed in the rope is placed over the boot and around the shaft of the ice ax. The brake is applied by wrapping the rope around the heel of the boot (Figure 10-20).

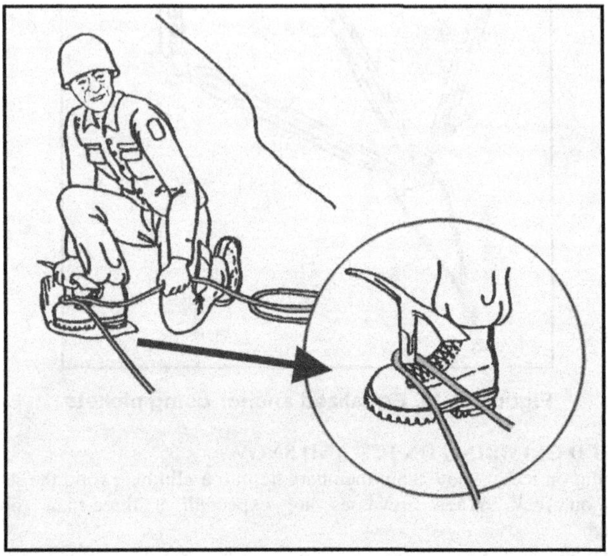

Figure 10-20. Boot-ax belay.

(2) ***Body Belay.*** The body belay can be used on snow and ice, also. The principles are the same as for belays on rock—solid anchors must be used and a well-braced position assumed. The position can be improved by digging depressions into the snow or ice for a seat and footholds. A strong platform should be constructed for the standing body belay.

(3) ***Munter Hitch.*** This belay technique is also used on snow and ice. When using the hitch off of the anchor, a two-point equalized anchor should be constructed as a minimum.

d. **Fixed Ropes.** The use of fixed ropes on ice is recommended for moving units through icefall areas on glaciers or other steep ice conditions. The procedures for emplacing fixed ropes on ice are basically the same as on rock with the exception that anchors need more attention, both in initial placement and in subsequent inspection, and steps may have to be cut to assist personnel.

10-7. MOVEMENT ON GLACIERS

Movement in mountainous terrain may require travel on glaciers. An understanding of glacier formation and characteristics is necessary to plan safe routes. A glacier is formed by the perennial accumulation of snow and other precipitation in a valley or draw. The accumulated snow eventually turns to ice due to metamorphosis. The "flow" or movement of glaciers is caused by gravity. There are a few different types of glaciers identifiable primarily by their location or activity.

- Valley glacier—resides and flows in a valley.
- Cirque glacier—forms and resides in a bowl.
- Hanging glacier—these are a result of valley or cirque glaciers flowing and or deteriorating. As the movement continues, portions separate and are sometimes left hanging on mountains, ridgelines, or cliffs.
- Piedmont glacier—formed by one or more valley glaciers; spreads out into a large area.
- Retreation glacier—a deteriorating glacier; annual melt of entire glacier exceeds the flow of the ice.
- Surging glacier—annual flow of the ice exceeds the melt; the movement is measurable over a period of time.

a. **Characteristics and Definitions.** This paragraph describes the common characteristics of glaciers, and defines common terminology used in reference to glaciers. (Figure 10-21 shows a cross section of a glacier, and Figure 10-22 depicts common glacier features.)

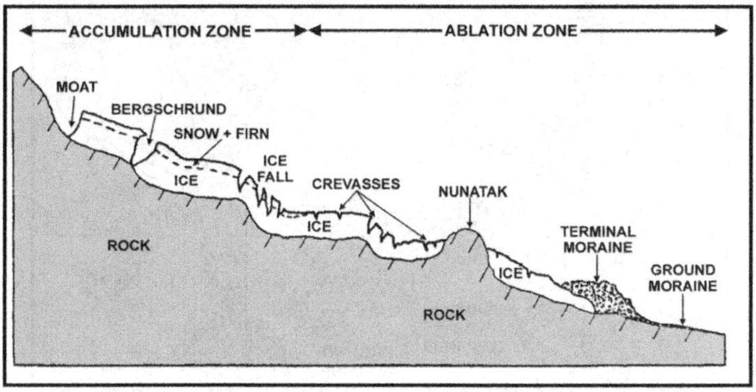

Figure 10-21. Glacier cross section.

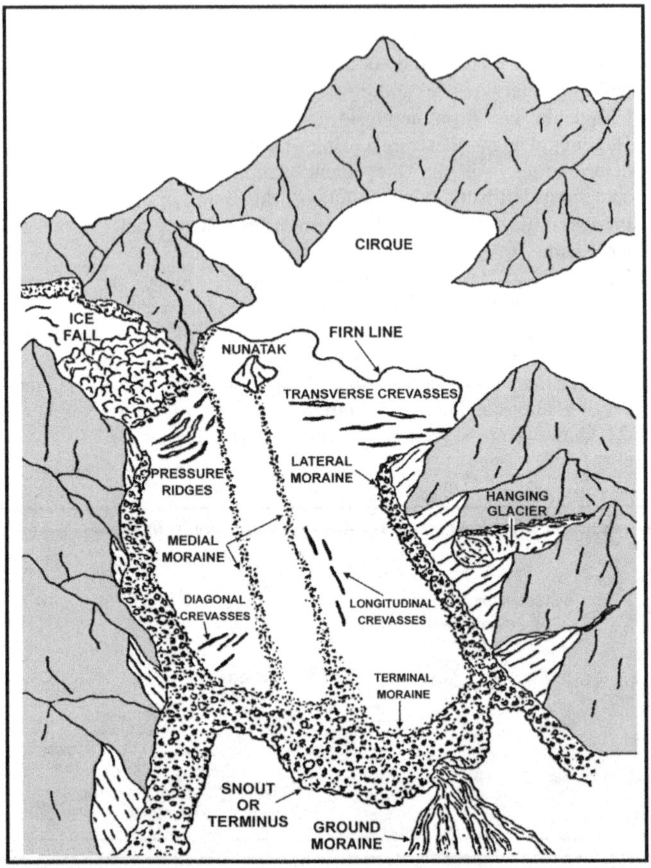

Figure 10-22. Glacier features.

(1) Firn is compacted granular snow that has been on the glacier at least one year. Firn is the building blocks of the ice that makes the glacier.

(2) The accumulation zone is the area that remains snow-covered throughout the year because of year-round snowfall. The snowfall exceeds melt.

(3) The ablation zone is the area where the snow melts off the ice in summer. Melt equals or exceeds snowfall.

(4) The firn line separates the accumulation and ablation zones. As you approach this area, you may see "strips" of snow in the ice. Be cautious, as these could be *snow bridges* remaining over crevasses. Remember that snow bridges will be weakest lower on the glacier as you enter the accumulation zone. The firn line can change annually.

(5) A bergschrund is a large crevasse at the head of a glacier caused by separation of active (flowing) and inactive (stationary) ice. These will usually be seen at the base of a major incline and can make an ascent on that area difficult.

(6) A moat is a wall formed at the head (start) of the glacier. These are formed by heat reflected from valley wall.

(7) A crevasse is a split or crack in the glacier surface. These are formed when the glacier moves over an irregularity in the bed surface.

(8) A transverse crevasse forms perpendicular to the flow of a glacier. These are normally found where a glacier flows over a slope with a gradient change of 30 degrees or more.

(9) Longitudinal crevasses form parallel to the flow of a glacier. These are normally found where a glacier widens.

(10) Diagonal crevasses form at an angle to the flow of a glacier. These are normally found along the edges where a glacier makes a bend.

(11) A snow bridge is a somewhat supportive structure of snow that covers a crevasse. Most of these are formed by the wind. The strength of a snow bridge depends on the snow itself.

(12) Icefalls are a jumble of crisscross crevasses and large ice towers that are normally found where a glacier flows over a slope with a gradient change of 25 degrees or more.

(13) Seracs are large pinnacles or columns of ice that are normally found in icefalls or on hanging glaciers.

(14) Ice avalanches are falling chunks of ice normally occurring near icefalls or hanging glaciers.

(15) The moraine is an accumulation of rock or debris on a glacier caused by rockfall or avalanche of valley walls.

(16) The lateral moraine is formed on sides of glacier.

(17) The medial moraine is in the middle of the glacier. This is also formed as two glaciers come together or as a glacier moves around a central peak.

(18) The terminal moraine is at the base of a glacier and is formed as moraines meet at the snout or terminus of a glacier.

(19) The ground moraine is the rocky debris extending out from the terminus of a glacier. This is formed by the scraping of earth as the glacier grew or surged and exposed as the glacier retreats.

(20) A Nunatak is a rock projection protruding through the glacier as the glacier flows around it.

(21) An ice mill is a hole in the glacier formed by swirling water on the surface. These can be large enough for a human to slip into.

(22) Pressure ridges are wavelike ridges that form on glacier normally after a glacier has flowed over icefalls.

(23) A glacier window is an opening at the snout of the glacier where water runs out of the glacier.

b. **Dangers and Obstacles.** The principle dangers and obstacles to movement in glacial areas are crevasses, icefalls, and ice avalanches. Snow-covered crevasses make movement on a glacier extremely treacherous. In winter, when visibility is poor, the difficulty of recognizing them is increased. Toward the end of the summer, crevasses are widest and covered by the least snow. Crossing snow bridges constitutes the greatest potential danger in movement over glaciers in the summer. On the steep pitch of a glacier, ice flowing over irregularities and cliffs in the underlying valley floor cause the ice to break up into ice blocks and towers, criss-crossed with crevasses. This jumbled cliff of ice is known as an icefall. Icefalls present a major obstacle to safe movement of troops on glaciers.

(1) Moving on glaciers brings about the hazard of falling into a crevasse. Although the crevasses are visible in the ablation zone in the summer (Figure 10-23), the accumulation zone will still have hidden crevasses. The risk of traveling in the accumulation zone can be managed to an acceptable level when ropes are used for connecting the team members (Figure 10-24, page 10-24). Crampons and an ice ax are all that is required to safely travel in the ablation zone in the summer.

Figure 10-23. Ablation zone of glacier in summer.

Figure 10-24. Rope teams moving in the accumulation zone of a glacier.

(2) When conditions warrant, three to four people will tie in to one rope at equal distances from each other. To locate the positions, if three people are on a team, double the rope and one ties into the middle and the other two at the ends. If four people are on a team, form a "z" with the rope and expand the "z" fully, keeping the end and the bight on each "side" of the "z" even. Tie in to the bights and the ends.

(3) Connect to the rope with the appropriate method and attach the Prusik as required. The rope should be kept relatively tight either by Prusik belay or positioning of each person. If the team members need to assemble in one area, use the Prusik to belay each other in.

(4) If a team member falls into a crevasse, the remaining members go into team arrest, assess the situation, and use the necessary technique to remove the person from the crevasse. The simplest and most common method for getting someone out of a crevasse is for the person to climb out while being belayed.

(5) All items should be secured to either the climber or the rope/harness to prevent inadvertent release and loss of necessary items or equipment. Packs should be secured to the rope/harness with webbing or rope. If traveling with a sled in tow, secure it not only to a climber to pull it, but connect it to the rope with webbing or rope also.

(6) If marking the route on the glacier is necessary for backtracking or to prevent disorientation in storms or flat-light conditions, use markers that will be noticeable against the white conditions. The first team member can place a new marker when the last team member reaches the previous marker.

c. **Roped Movement.** The first rule for movement on glaciers is to rope up (Figure 10-25). A roped team of two, while ideal for rock climbing, is at a disadvantage on a snow-covered glacier. The best combination is a three-man rope team. Generally, the rope team members will move at the same time with the rope fully extended and reasonably tight between individuals, their security being the team arrest. If an individual should break through a snow bridge and fall into a crevasse, the other members immediately perform self-arrest, halting the fall. At points of obvious weakness in the snow bridges, the members may decide to belay each other across the crevasse using one of the established belay techniques.

Figure 10-25. Preparation for roped movement.

(1) Even with proper training in crevasse rescue techniques, the probability exists that an individual may remain suspended in a crevasse for a fairly lengthy amount of time while trying to get himself out or while awaiting help from his rope team members. Because of this, it is strongly recommended that all personnel wear a seat/chest combination harness, whether improvised or premanufactured.

(2) Rope team members must be able to quickly remove the climbing rope from the harness(es) during a crevasse rescue. The standard practice for connecting to the rope for glacier travel is with a locking carabiner on a figure-eight loop to the harness. This allows quick detachment of the rope for rescue purposes. The appropriate standing part of the rope is then clipped to the chest harness carabiner.

(3) If a rope team consists of only two people, the rope should be divided into thirds, as for a four-person team. The team members tie into the middle positions on the rope, leaving a third of the rope between each team member and a third on each end of the rope. The remaining "thirds" of the rope should be coiled and either carried in the rucksack, attached to the rucksack, or carried over the head and shoulder. This gives each

climber an additional length of rope that can be used for crevasse rescue, should one of the men fall through and require another rope. If necessary, this excess end rope can be used to connect to another rope team for safer travel.

Note: The self-arrest technique used by one individual will work to halt the fall of his partner on a two-man rope team; however, the chance of it failing is much greater. Crevasse rescue procedures performed by a two-man rope team, by itself, may be extremely difficult. For safety reasons, movement over a snow-covered glacier by a single two-man team should be avoided wherever possible.

d. **Use of Prusik Knots.** Prusik knots are attached to the climbing rope for all glacier travel. The Prusiks are used as a self-belay technique to maintain a tight rope between individuals, to anchor the climbing rope for crevasse rescue, and for self-rescue in a crevasse fall. The Prusik slings are made from the 7-millimeter by 6-foot and 7-millimeter by 12-foot ropes. The ends of the ropes are tied together, forming endless loops or slings, with double fisherman's knots. Form the Prusik knot on the rope in front of the climber. An overhand knot can be tied into the sling just below the Prusik to keep equal tension on all the Prusik wraps. Attach this sling to the locking carabiner at the tie in point on the harness.

Note: An ascender can replace a Prusik sling in most situations. However, the weight of an ascender hanging on the rope during movement will become annoying, and it could be stepped on during movement and or climbing.

e. **Securing the Backpack/Rucksack.** If an individual should fall into a crevasse, it is essential that he be able to rid himself of his backpack. The weight of the average pack will be enough to hinder the climber during crevasse rescue, or possibly force him into an upside down position while suspended in the crevasse. Before movement, the pack should be attached to the climbing rope with a sling rope or webbing and a carabiner. A fallen climber can immediately drop the pack without losing it. The drop cord length should be minimal to allow the fallen individual to reach the pack after releasing it, if warm clothing is needed. When hanging from the drop cord, the pack should be oriented just as when wearing it (ensure the cord pulls from the top of the pack).

f. **Routes.** An individual operating in the mountains must appreciate certain limitations in glacier movement imposed by nature.

(1) Additional obstacles in getting onto a glacier may be swift glacier streams, steep terminal or lateral moraines, and difficult mountain terrain bordering the glacier ice. The same obstacles may also have to be overcome in getting on and off a valley glacier at any place along its course.

(2) Further considerations to movement on a glacier are steep sections, heavily crevassed portions, and icefalls, which may be major obstacles to progress. The use of current aerial photographs in conjunction with aerial reconnaissance is a valuable means of gathering advance information about a particular glacier. However, they only supplement, and do not take the place of, on-the-ground reconnaissance conducted from available vantage points.

g. **Crossing Crevasses.** Open crevasses are obvious, and their presence is an inconvenience rather than a danger to movement. Narrow cracks can be jumped, provided the take off and landing spots are firm and offer good footing. Wider cracks will have to be circumvented unless a solid piece of ice joins into an ice bridge strong enough to support at least the weight of one member of the team. Such ice bridges are often formed in the lower portion of a crevasse, connecting both sides of it.

(1) In the area of the firn line, the zone that divides seasonal melting from permanent falls of snow, large crevasses remain open, though their depths may be clogged with masses of snow. Narrow cracks may be covered. In this zone, the snow, which covers glacier ice, melts more rapidly than that which covers crevasses. The difference between glacier ice and narrow snow-covered cracks is immediately apparent; the covering snow is white, whereas the glacier ice is gray.

(2) Usually the upper part of a glacier is permanently snow covered. The snow surface here will vary in consistency from dry powder to consolidated snow. Below this surface cover are found other snow layers that become more crystalline in texture with depth, and gradually turn into glacier ice. It is in this snow-covered upper part of a glacier that crevasses are most difficult to detect, for even wide crevasses may be completely concealed by snow bridges.

h. **Snow Bridges.** Snow bridges are formed by windblown snow that builds a cornice over the empty interior of the crevasse. As the cornice grows from the windward side, a counter drift is formed on the leeward side. The growth of the leeward portion will be slower than that to the windward so that the juncture of the cornices occurs over the middle of the crevasse only when the contributing winds blow equally from each side. Bridges can also be formed without wind, especially during heavy falls of dry snow. Since cohesion of dry snow depends only on an interlocking of the branches of delicate crystals, such bridges are particularly dangerous during the winter. When warmer weather prevails the snow becomes settled and more compacted, and may form firmer bridges.

(1) Once a crevasse has been completely bridged, its detection is difficult. Bridges are generally slightly concave because of the settling of the snow. This concavity is perceptible in sunshine, but difficult to detect in flat light. If the presence of hidden crevasses is suspected, the leader of a roped team must probe the snow in front of him with the shaft of his ice ax. As long as a firm foundation is encountered, the team may proceed, but should the shaft meet no opposition from an underlying layer of snow, a crevasse is probably present. In such a situation, the prober should probe closer to his position to make sure that he is not standing on the bridge itself. If he is, he should retreat gently from the bridge and determine the width and direction of the crevasse. He should then follow and probe the margin until a more resistant portion of the bridge is reached. When moving parallel to a crevasse, all members of the team should keep well back from the edge and follow parallel but offset courses.

(2) A crevasse should be crossed at right angles to its length. When crossing a bridge that seems sufficiently strong enough to hold a member of the team, the team will generally move at the same time on a tight rope, with each individual prepared to go into self-arrest. If the stability of the snow bridge is under question, they should proceed as follows for a team of three glacier travelers:

(a) The leader and second take up a position at least 10 feet back from the edge. The third goes into a self-belay behind the second and remains on a tight rope.

(b) The second belays the leader across using one of the established belay techniques. The boot-ax belay should be used only if the snow is deep enough for the ax to be inserted up to the head and firm enough to support the possible load. A quick ice ax anchor should be placed for the other belays. Deadman or equalizing anchors should be used when necessary.

(c) The leader should move forward, carefully probing the snow and evaluating the strength of the bridge, until he reaches firm snow on the far side of the crevasse. He then continues as far across as possible so number two will have room to get across without number one having to move.

(d) The third assumes the middle person's belay position. The middle can be belayed across by both the first and last. Once the second is across, he assumes the belay position. Number one moves out on a tight rope and anchors in to a self-belay. Number two belays number three across.

(3) In crossing crevasses, distribute the weight over as wide an area as possible. Do not stamp the snow. Many fragile bridges can be crossed by lying down and crawling to the other side. Skis or snowshoes help distribute the weight nicely.

i. **Arresting and Securing a Fallen Climber.** The simplest and most common method for getting someone out of a crevasse is for the person to climb out while being belayed. Most crevasse falls will be no more than body height into the opening if the rope is kept snug between each person.

(1) To provide a quick means of holding an unexpected breakthrough, the rope is always kept taut. When the leader unexpectedly breaks through, the second and third immediately go into a self-arrest position to arrest the fall. A fall through a snow bridge results either in the person becoming jammed in the surface hole, or in being suspended in the crevasse by the rope. If the leader has fallen only partially through the snow bridge, he is supported by the snow forming the bridge and should not thrash about as this will only enlarge the hole and result in deeper suspension. All movements should be slow and aimed at rolling out of the hole and distributing the weight over the remainder of the bridge. The rope should remain tight at all times and the team arrest positions adjusted to do so. It generally is safer to retain the rucksack, as its bulk often prevents a deeper fall. Should a team member other than the leader experience a partial fall, the rescue procedure will be same as for the leader, only complicated slightly by the position on the rope.

(2) When the person falls into a crevasse, the length of the fall depends upon how quickly the fall is arrested and where in the bridge the break takes place. If the fall occurs close to the near edge of the crevasse, it usually can be checked before the climber has fallen more than 6 feet. However, if the person was almost across, the fall will cause the rope to cut through the bridge, and then even an instantaneous check by the other members will not prevent a deeper fall. The following scenario is an example of the sequence of events that take place after a fall by the leader in a three-person team. (This scenario is for a team of three, each person referred to by position; the leader is number 1.)

(a) Once the fall has been halted by the team arrest, the entire load must be placed on number 2 to allow number 3 to move forward and anchor the rope. Number 3 slowly releases his portion of the load onto number 2, always being prepared to go back into self-arrest should number 2's position begin to fail.

(b) Once number 2 is confident that he can hold the load, number 3 will proceed to number 2's position, using the Prusik as a self belay, to anchor the rope. In this way the rope remains reasonably tight between number 2 and number 3. Number 3 must always be prepared to go back into self-arrest should number 2's position begin to fail.

(c) When number 3 reaches number 2's position he will establish a bombproof anchor 3 to 10 feet in front of number 2 (on the load side), depending on how close number 2 is to the lip of the crevasse. This could be either a deadman or a two-point equalized anchor, as a minimum.

(d) Number 3 connects the rope to the anchor by tying a Prusik with his long Prusik sling onto the rope leading to number 1. An overhand knot should be tied into the long Prusik sling to shorten the distance to the anchor, and attached to the anchor with a carabiner. The Prusik knot is adjusted toward the load.

(e) Number 2 can then release the load of number 1 onto the anchor. Number 2 remains connected to the anchor and monitors the anchor.

(f) A fixed loop can be tied into the slack part of the rope, close to number 2, and attached to the anchor (to back up the Prusik knot).

(g) Number 3 remains tied in, but continues forward using a short Prusik as a self-belay. He must now quickly check on the condition of number 1 and decide which rescue technique will be required to retrieve him.

(3) These preliminary procedures must be performed before retrieving the fallen climber. If number 3 should fall through a crevasse, the procedure is the same except that number 1 assumes the role of number 3. Normally, if the middle person should fall through, number 1 would anchor the rope by himself. Number 3 would place the load on number 1's anchor, then anchor his rope and move forward with a Prusik self-belay to determine the condition of number 2.

j. **Crevasse Rescue Techniques.** Snow bridges are usually strongest at the edge of the crevasse, and a fall is most likely to occur some distance away from the edge. In some situations, a crevasse fall will occur at the edge of the snow bridge, on the edge of the ice. If a fall occurs away from the edge, the rope usually cuts deeply into the snow, thus greatly increasing friction for those pulling from above. In order to reduce friction, place padding, such as an ice ax, ski, ski pole, or backpack/rucksack, under the rope and at right angles to the stress. Push the padding forward as far as possible toward the edge of the crevasse, thus relieving the strain on the snow. Ensure the padding is anchored from falling into the crevasse for safety of the fallen climber.

(1) ***Use of Additional Rope Teams.*** Another rope team can move forward and assist in pulling the victim out of a crevasse. The assisting rope team should move to a point between the fallen climber and the remaining rope team members. The assisting team can attach to the arresting team's rope with a Prusik or ascender and both rope teams' members can all pull simultaneously. If necessary, a belay can be initiated by the fallen climber's team while the assisting team pulls. The arresting team member closest to the fallen climber should attach the long Prusik to themselves and the rope leading to the fallen climber, and the assisting team can attach their Prusik or ascender between this long Prusik and the arresting team member. As the assisting team pulls, the Prusik belay will be managed by the arresting team member at the long Prusik.

Note: Safety in numbers is obvious for efficient crevasse rescue techniques. Additional rope teams have the necessary equipment to improve the main anchor or establish new ones and the strength to pull a person out even if he is deep in the crevasse. Strength of other rope teams should always be used before establishing more time-consuming and elaborate rescue techniques.

(2) ***Fixed Rope.*** If the fallen climber is not injured, he may be able to climb out on a fixed rope. Number 1 clips number 3's rope to himself. He then climbs out using number 3's rope as a simple fixed line while number 2 takes up the slack in number 1's rope through the anchor Prusik for a belay.

(3) ***Prusik Ascending Technique.*** There may be times when the remaining members of a rope team can render little assistance to the person in the crevasse. If poor snow conditions make it impossible to construct a strong anchor, the rope team members on top may have to remain in self-arrest. Other times, it may just be easier for the fallen climber to perform a self-rescue. (Figure 10-26 shows the proper rope configuration.) The technique is performed as follows:

(a) The fallen climber removes his pack and lets it hang below from the drop cord.

(b) The individual slides their short Prusik up the climbing rope as far as possible.

(c) The long Prusik is attached to the rope just below the short Prusik. The double fisherman's knot is spread apart to create a loop large enough for one or both feet. The fallen climber inserts his foot/feet into the loop formed allowing the knot to cinch itself down.

(d) The individual stands in the foot loop, or "stirrup," of the long sling.

(e) With his weight removed from the short Prusik, it is slid up the rope as far as it will go. The individual then hangs from the short Prusik while he moves the long Prusik up underneath the short Prusik again.

(f) The procedure is repeated, alternately moving the Prusiks up the rope, to ascend the rope. Once the crevasse lip is reached, the individual can simply grasp the rope and pull himself over the edge and out of the hole.

(g) Besides being one of the simplest rope ascending techniques, the short Prusik acts as a self-belay and allows the climber to take as long a rest as he wants when sitting in the harness. The rope should be detached from the chest harness carabiner to make the movements less cumbersome. However, it is sometimes desirable to keep the chest harness connected to the rope for additional support. In this case the Prusik knots must be "on top" of the chest harness carabiner so they can be easily slid up the rope without interference from the carabiner. The long Prusik sling can be routed through the chest harness carabiner for additional support when standing up in the stirrup.

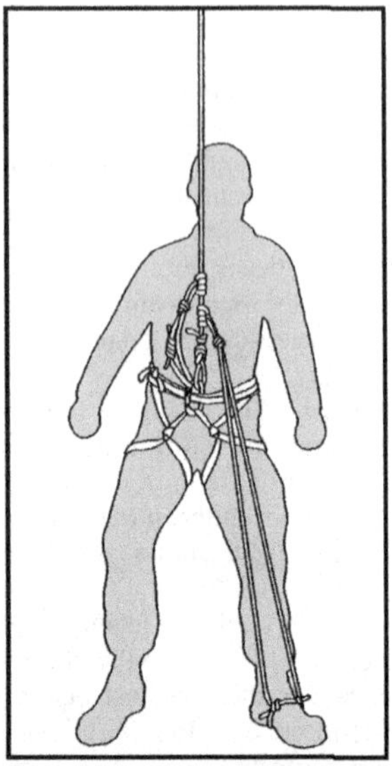

Figure 10-26. Prusik ascending technique.

(4) ***Z-Pulley Hauling System.*** If a fallen climber is injured or unconscious, he will not be able to offer any assistance in the rescue. If additional rope teams are not immediately available, a simple raising system can be rigged to haul the victim out of the crevasse. The Z-pulley hauling system is one of the simplest methods and the one most commonly used in crevasse rescue (Figure 10-27, page 10-32). The basic Z rig is a "3-to-1" system, providing mechanical advantage to reduce the workload on the individuals operating the haul line. In theory, it would only take about 33 pounds of pull on the haul rope to raise a 100-pound load with this system. In actual field use, some of this mechanical advantage is lost to friction as the rope bends sharply around carabiners and over the crevasse lip. The use of mechanical rescue pulleys can help reduce this friction in the system. The following describes rigging of the system. (This scenario is for a team of three, each person referred to by position; the leader is number 1.)

(a) After the rope team members have arrested and secured number 1 to the anchor, and they have decided to install the Z rig, number 2 will attach himself to the anchor without using the rope and clear the connecting knot used. Number 3 remains connected to the rope.

(b) The slack rope exiting the anchor Prusik is clipped into a separate carabiner attached to the anchor. A pulley can be used here if available.

(c) Number 3 will use number 2's short Prusik to rig the haul Prusik. He moves toward the crevasse lip (still on his own self-belay) and ties number 2's short Prusik onto number 1's rope (load rope) as close to the edge as possible.

(d) Another carabiner (and pulley if available) is clipped into the loop of the haul Prusik and the rope between number 3's belay Prusik and the anchor is clipped (or attached through the pulley). Number 3's rope becomes the haul rope.

(e) Number 3 then moves towards the anchor and number 2. Number 2 could help pull if necessary but first would connect to the haul rope with a Prusik just as number 3. If the haul Prusik reaches the anchor before the victim reaches the top, the load is simply placed back on the anchor Prusik and number 3 moves the haul Prusik back toward the edge. The system is now ready for another haul.

CAUTION
The force applied to the fallen climber through use of the Z-pulley system can be enough to destroy the harness-to-rope connection or injure the fallen climber if excess force is applied to the pulling rope.

Notes: 1. The Z-pulley adds more load on the anchor due to the mechanical advantage. The anchor should be monitored for the duration of the rescue.

2. With the "3-to-1" system, the load (fallen climber) will be raised 1 foot for every 3 feet of rope taken up during the haul.

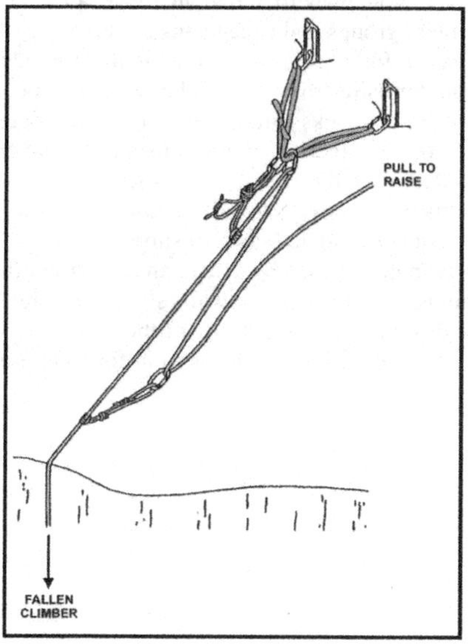

Figure 10-27. Z-pulley hauling system.

10-8. GLACIER BIVOUAC PROCEDURES

When locating a bivouac site or a gathering area where the team might need or want to unrope, at least one person will need to "probe" the area for hidden crevasses. The best type of probe will be the manufactured collapsing probe pole, at least eight feet in length. Other items could be used but the length and strength of the probe is most important. Other rope team members will belay the probers. The prober is "feeling" for a solid platform to place the tent by pushing the probe as hard and deep as possible into the surface. Probing should be in 2-foot intervals in all directions within the site.

a. If the probe suddenly has no resistance while pushing down, a crevasse is present. Attempts to outline the crevasse can be futile if the crevasse is large. Normally, the best decision is to relocate the proposed bivouac area far enough away to avoid that crevasse. (Sometimes only a few feet one way or the other is all that's needed to reach a good platform.) Probe the tent site again after digging to the desired surface. Mark boundaries with wands or other items such as skis, poles, and so on.

b. Occasionally while probing, increased pressure will be noticed without reaching a solid platform. The amount of snowfall may be such that even after digging into the snow, the probe still doesn't contact a hard surface. Try to find a solid platform.

c. There should be no unroped movement outside the probed/marked areas. If a latrine area is needed, probe a route away from the bivouac site and probe the latrine area also. If a dugout latrine is necessary, probe again after digging.

d. Multiple tent sites can be connected, which keeps tents closer together. Probe all areas between the tents if you plan to move in those areas. Closer tents will make communicating between tent groups and rope teams easier.

e. If there is a chance for severe storms with high winds, snow walls may be constructed to protect the tent site from wind. The walls can be constructed from loose snow piled on the perimeter, or blocks can be cut from consolidated snow layers. In deep soft snow, digging three or four feet to find a consolidated layer will result in enough snow moved to build up decent walls around the tent site.

(1) For block construction, move the soft snow from the surface into the wall foundation areas (down to a consolidated layer of snow).

(2) Cut blocks approximately 1 by 1 by 2 feet, and construct the walls by interlocking the blocks with overlapping placements. The walls should be slightly higher than the tent. At a minimum, build walls on the windward side of the tent site.

(3) Snow walls can also provide shelter from wind for food preparation.

CHAPTER 11

MOUNTAIN RESCUE AND EVACUATION

Steep terrain and adverse weather are common in mountainous environments. Under these conditions, relatively minor injuries may require evacuation. The evacuation technique chosen is determined by the type of injury, distance to be moved, terrain, and existing installations. Air evacuation is preferred; however, the weather, tactical situation, or operational ceiling of the aircraft may make this impossible. It is, therefore, imperative that all personnel are trained in mountain evacuation techniques and are self-sufficient. Casualties should be triaged before evacuation. Triage is performed by the most experienced medical personnel available (physician, physician's assistant, medic).

Performing a rescue operation can be a significant emotional event. Rescue scenarios must be practiced and rehearsed until rescue party members are proficient in the many tasks required to execute a rescue. To perform most of the high-angle rescues, Level I and Level II mountaineers are required with a Level III supervising.

11-1. CONSIDERATIONS

The techniques of evacuation are proven techniques. They are, however, all subject to improvement and should be discarded or modified as better methods of handling victims are developed.

a. When evacuating a victim from mountainous areas keep in mind that the purpose of a rescue operation is to save a life, and physical risk to the rescuers must be weighed against this purpose. However, there is no excuse for failing to make the maximum effort within this limitation. Work and expense should be no deterrent when a life is at stake.

b. Rescues will be unplanned (improvised) or planned rescue operations. For a planned rescue, equipment that is especially suited and designed for rescue should be used. For training missions always have a medical plan developed before an emergency arises (plan for the worst and hope for the best). Ensure that the MEDEVAC plan is a comprehensive plan and must be thought out and understood by all that may be involved in a potential rescue.

c. The following actions will be done immediately at the rescue scene.

(1) Assume command. One person, and one person only, is overall in charge at all times.

(2) Prevent further injuries to the victim and to others. Use reasonable care in reaching the victim.

(3) Immediately ensure the victim has an open airway, resume victim's breathing, control serious bleeding, and maintain moderate body warmth. If the victim is unconscious, continually monitor pulse. Protect the patient from environmental hazards.

(4) Do not move the victim until you have ascertained the extent of injuries, unless it is necessary to prevent further injuries or the victim is located in a dangerous location (for example, avalanche run-out zone, hanging glacier, possibility of falling rocks).

(5) Do nothing more until you have thoroughly considered the situation. Resist the urge for action. Speed is less important than correct action.

(6) Decide whether to evacuate with available facilities or to send for help. Speed in getting to a hospital must be balanced against the probability of further injury if working with inexperienced people, lack of equipment or wrong equipment, and terrain at hand.

(7) When the evacuation route is long and arduous, a series of litter relay points or stations should be established. These stations must be staffed with the minimum medical personnel to provide proper emergency treatment. When a victim develops signs of shock or worsens while being evacuated, he should be treated and retained at one of these stations until his condition allows evacuation.

(8) Helicopters or heated vehicles, if available, should be used for evacuation. While the use of aircraft or vehicles is preferred and can expedite a rescue operation, evacuation of a seriously wounded soldier should never be delayed to await aircraft, vehicle, or a change in weather.

11-2. PLANNING RESCUE OPERATIONS

Every commander should have a medical evacuation plan before undertaking an operation. This plan should have contingencies included so as not to rely on a single asset.

a. When rescuing a casualty (victim) threatened by hostile action, environmental hazard, or any other immediate hazard, the rescuer should not take action without first determining the extent of the hazard and his ability to handle the situation. THE RESCUER MUST NOT BECOME A CASUALTY.

b. The rescue team leader must evaluate the situation and analyze the factors involved. This evaluation can be divided into three major steps:

- Identify the task.
- Evaluate the circumstances of the rescue.
- Plan the action.

c. The task must be identified. In planning a rescue, the rescuer tries to obtain the following information:

- Who, what, where, when, why, and how the situation happened.
- Number of casualties by precedence (urgent, priority, routine, tactical immediate),
- number of casualties by type (litter or ambulatory), and the nature of their injuries.
- Terrain features and location of the casualties.
- Tactical situation.
- If adequate assistance is available to aid in security, rescue, treatment, and evacuation.
- If treatment can be provided at the scene; if the victims require movement to a safer location.
- Equipment required for the rescue operation.

d. Circumstances of the rescue are as follows:

(1) After identifying the task, relate it to the circumstances of the situation.

- Are additional personnel, security, medical, or special rescue equipment needed?
- Are there circumstances, such as aircraft accidents (mass casualties), that may require specialized skills?

- What is the weather condition?
- Is the terrain hazardous?
- How much time is available?

(2) The time element may cause a rescuer to compromise planning stages or treatment (beyond first aid). Make a realistic estimate of time available as quickly as possible to determine the action time remaining. The key elements are the casualty's condition and environment.

(3) Mass casualties are to be expected on the modern battlefield. All problems or complexities of rescue are now multiplied by the number of casualties. Time becomes the critical element.

(4) Considerations for the main rescue group for a planned rescue are as follows:

(a) Carry all needed equipment, hot food and drinks, stove, sleeping bags, tents, bivouac sacks, warm clothes, ropes, and stretchers.

(b) Prepare the evacuation route (ground transport to hospital, walking trails, fixed lines, lowering lines, anchor points, and rescue belay points). If the victim is airlifted out, attach a paper with the medical actions that were performed on the ground (for example, blood pressure, pulse rate, drugs started, and so on).

(c) When performing all rescues, the rescuers are always tied in for safety. With all rescue techniques, remember to think things through logically for safety and to prevent the rescuer from accidentally untying himself or the fallen climber.

(d) Constantly inform the casualty (if they are conscious) as to what you are doing and what he must do.

e. The rescue plan should proceed as follows:

(1) In estimating time available, the casualties' ability to endure is of primary importance. Age and physical condition may vary. Time available is a balance of the endurance time of the casualty, the situation, and the personnel and equipment available.

(2) Consider altitude and visibility. Maximum use of secure, reliable trails or roads is essential.

(3) Ensure that blankets and rain gear are available. Even a mild rain can complicate a normally simple rescue. In high altitudes, extreme cold, or gusting winds, available time is drastically reduced.

(4) High altitudes and gusting winds reduce the ability of fixed-wing or rotary-wing aircraft to assist in operations. Rotary-wing aircraft may be available to remove casualties from cliffs or inaccessible sites, and to quickly transport casualties to a medical treatment facility. Relying on aircraft or specialized equipment is a poor substitute for careful planning.

11-3. MASS CASUALTIES

When there are mass casualties, an orderly rescue may involve further planning.

a. To manage a mass casualty rescue or evacuation, separate stages are taken.

- FIRST STAGE: Remove personnel who are not trapped among debris or who can be easily evacuated.
- SECOND STAGE: Remove personnel who may be trapped by debris, but whose extraction only requires the equipment on hand and little time.

- THIRD STAGE: Remove the remaining personnel who are trapped in extremely difficult or time-consuming situations, such as moving large amounts of debris or cutting through a wall.
- FOURTH STAGE: Remove dead personnel.

b. Evacuation of wounded personnel is based on the victim's condition and is prioritized as follows:

- PRIORITY ONE: Personnel with life-threatening injuries that require immediate emergency care to survive; first aid and stabilization are accomplished before evacuation.
- PRIORITY TWO: Personnel with injuries that require medical care but speed of evacuation is not essential.
- PRIORITY THREE: Injured personnel who can evacuate themselves with minimal assistance.
- PRIORITY FOUR: The logistics removal of dead personnel.

11-4. SPECIAL TRAINING

Before receiving training in basic mountain evacuation, litter teams should receive instruction in military mountaineering and basic first aid. Litter bearers and medics must know the use and care of rope as an item of equipment. The members of litter teams must be proficient in the techniques of belaying and choosing belay points. Proper support and protection must be given to victims and litter bearers when evacuating over steep, difficult terrain.

11-5. PREPARATION FOR EVACUATION

Although the wounded soldier's life may have been saved by applying first aid, it can be lost through carelessness, rough handling, or inadequate protection from the elements. Therefore, before trying to move the wounded soldier, the type and extent of his injury must be evaluated. Dressings over wounds must be reinforced, and fractured bones must be properly immobilized and supported. Based upon the evaluation of the type and extent of the soldier's injury, the best method of manual transportation is selected.

11-6. MANUAL CARRIES

Personnel who are not seriously injured but cannot evacuate themselves may be assisted by fellow soldiers. Personnel who are injured and require prompt evacuation should not be forced to wait for mobile evacuation or special equipment.

a. **One-Man Carries.** The basic carries taught in the Soldier's Manual of Common Tasks (fireman's carry, two-hand, four-hand, saddleback, piggyback, pistol belt, and poncho litter) are viable means of transporting injured personnel; however, the mountainous terrain lends itself to several other techniques. One-man carries include the sling-rope carry and the rope coil carry.

(1) ***Sling-Rope Carry.*** The sling-rope carry (Figure 11-1) requires a 4.5-meter sling rope and two men—one as the bearer and the other as an assistant to help secure the casualty to the bearer's back. Conscious or unconscious casualties may be transported this way.

(a) The bearer kneels on all fours.

(b) The assistant places the casualty face down on the bearer's back ensuring the casualty's armpits are even with the bearer's shoulders.

(c) The assistant then finds the middle of the sling rope and places it between the casualty's shoulders.

(d) The assistant runs the ends of the sling rope under the casualty's armpits, crosses the ends, and runs the ends over the bearer's shoulders and back under the bearer's arms.

(e) The assistant runs the ends of the rope between the casualty's legs, around the casualty's thighs, and back around to the front of the bearer. The rope is tied with a square knot with two overhand knots just above the bearer's belt buckle.

(f) The rope must be tight. Padding, when available, should be placed where the rope passes over the bearer's shoulders and under the casualty's thighs.

Figure 11-1. Sling-rope carry.

(2) ***Rope Coil Carry.*** The rope coil carry requires a bearer and a 36 1/2-meter coiled rope. It can be used to transport a conscious or unconscious victim.

(a) Place the casualty on his back.

(b) Separate the loops on one end of the coil, forming two almost equal groups.

(c) Slide one group of loops over the casualty's left leg and the other group over the right leg. The wraps holding the coil should be in the casualty's crotch with the loops on the other end of the coil extending upward toward the armpits.

(d) The bearer lies on his back between the casualty's legs and slides his arms through the loops. He then moves forward until the coil is extended.

(e) Grasping the casualty's arm, the bearer rolls over (toward the casualty's uninjured side), pulling the casualty onto his back.

(f) Holding the casualty's wrists, the bearer carefully stands, using his legs to lift up and keeping his back as straight as possible.

(g) A sling rope around both the casualty and bearer, tied with a joining knot at chest level, aids in keeping an unconscious victim upright. This also prevents the coils from slipping off the carrier's chest.

Note: The length of the coils on the rope coil and the height of the bearer must be considered. If the coils are too long and the bearer is shorter, the rope must be uncoiled and recoiled with smaller coils. If this is not done, the casualty will hang too low on the bearer's back and make it a cumbersome evacuation. A sling-rope harness can be used around the victim's back and bearer's chest, which frees the bearer's hands.

b. **Buddy Rappel.** The carrier can also conduct a seat-hip rappel with a victim secured to his back. In this case, the rappeller faces the cliff and assumes a modified L-shape body position to compensate for the weight of the victim on his back. The victim is top-rope belayed from above, which provides the victim with a point of attachment to a secured rope. The methods for securing a victim to a rappeller's back are described below.

(1) To secure the victim to the carrier's back with a rope, the carrier ties a standard rappel seat (brake hand of choice, depending on the injury) and rests his hands on his knees while the victim straddles his back.

(2) A 4.2-meter sling rope is used. A 45-centimeter tail of the sling is placed on the victim's left hip. (This method describes the procedure for a seat-hip rappel with right-hand brake.)

(3) The remaining long end of the sling rope is routed under the victim's buttocks, and passed over the victim's and carrier's right hip. The rope is run diagonally, from right to left, across the carrier's chest, over his left shoulder, and back under the victim's left armpit.

(4) The rope is then run horizontally, from left to right, across the victim's back. The rope is passed under the victim's right armpit and over the carrier's right shoulder.

(5) The rope is run diagonally, from right to left, across the carrier's chest and back across the carrier's and victim's left hip.

(6) The two rope ends should now meet. The two ends are tied together with a square knot and overhand knots.

(7) The knot is positioned on the victim's left hip. The carrier's shoulders may need to be padded to prevent cutting by the rope.

(8) An alternate method is to use two pistol belts hooked together and draped over the carrier's shoulders. The victim straddles the carrier, and the belay man secures the loose ends of the pistol belts under the victim's buttocks. Slack in the pistol belt sling should be avoided, since the carrier is most comfortable when the victim rests high on his back (see FM 8-35).

(9) A large rucksack can be slit on the sides near the bottom so that the victim can step into it. The victim is belayed from the top with the carrier conducting a standard rappel. The carrier wears the rucksack with the victim inside.

(10) A casualty secured to a carrier, as described above, can be rappelled down a steep cliff using a seat-shoulder or seat-hip rappel. The casualty's and rappeller's shoulders should be padded where the sling rope and rappel lines cross if a seat-shoulder

rappel is used. The buddy team should be belayed from above with a bowline tied around the victim's chest under his armpits. The belay rope must run over the rappeller's guide hand shoulder.

11-7. LITTERS

Many types of litters are available for evacuating casualties in rough mountain terrain. Casualties may be secured to litters in many different ways, depending on the terrain, nature of injuries, and equipment available. **All casualties must be secured.** This should be done under medical supervision after stabilization. It is also important to render psychological support to any victim awaiting evacuation.

If the litter must be carried, belayed, and then carried again, a sling rope should be wound around the litter end and tied off in a l-meter-long loop. This enables the carriers to hook and unhook the litter from the belay. Slings are available to aid the soldiers with litter carrying. Utility rope or webbing 6 meters long may be used. The rope is folded in half, and the loose ends are tied together with an overhand knot. These slings are attached to the litter rails (two or three to a side, depending on the number of litter bearers) by a girth hitch, and then routed up along the handling arm, over the shoulder, behind the neck, and then down along the other arm. The knot can be adjusted to help the outside arm grip the webbing. These slings help distribute the load more evenly, which is important if a great distance must be traveled.

a. **Manufactured Litters.** The following litters are readily available to mountaineering units.

(1) The poleless, nonrigid litter (NSN 6530-00-783-7510) is best issued for company medics since it is lightweight, easy to carry, and readily available. Casualties should be secured with the chest strap and pelvic straps, which are sewn on one side. This litter may be used when rappelling, on traverse lines, and on hauling lines in the vertical or horizontal position. It can be improvised with poles.

(2) The poleless semi-rigid litter (NSN 6530-00-783-7600) may be used the same as the nonrigid litter. It offers more victim protection and back support because of the wooden slats sewn into it.

(3) The mountain basket-type rigid litter (NSN 6530-00-181-7767) is best suited for areas where several casualties are to be transported. All other litters may be placed inside this litter basket and transported across traverse lines. This litter is rectangular and has no vertical leg divider so that it will accommodate other litters. It is also known as a modified Stokes litter.

(4) The Stokes metal litter (NSN 6530-00-042-8131) is suited for situations as above; however, the casualty must be moved in and out of the litter since no other litter will fit inside it. Some Stokes litter frames have a central weld on the frame end, which is a potential breaking point. Winding the rope around the frame end will distribute the force over a wider area and stabilize the system. (See FM 8-10-6 or USAF TO 00-75-5 for additional information on the Stokes litter.)

(5) The standard collapsible litter (NSN 6530-00-783-7905) (rigid pole folding litter) is most readily available in all units and, although heavy and unsuited to forward deployment, may be rigged for movement over rough or mountainous terrain. The folding aluminum litter (NSN 6530-00-783-7205) is a compact version of the pole litter and is better suited for forward deployment.

(6) The UT 2000 is manufactured in Austria and is specifically designed for mountaineering operations. The litter consists of two parts that join together to form a rigid litter. Each part has shoulder and waist straps that can be used to man-pack the litter making it extremely light and portable. When joined together the shoulder and waist straps are used to secure the casualty to the litter. Strapping is also provided to make a secure hoist point for aircraft extraction and high-angle rescues. Wheel sets are another accessory to the UT 2000 litter (either two wheels or one); they attach to the litter for use during a low-angle rescue.

(7) The patient rescue and recovery system (NSN 6530-01-260-1222) provides excellent patient support and protection (Figure 11-2). However, it is not a spinal immobilization device. A backboard must be used with this system for patients who have injuries to the shoulder area. This system will accommodate long and short backboards, scoop stretchers, and most other immobilization equipment.

Figure 11-2. Rescue and recovery system (NSN 6530-01-260-1222).

b. **Field-Expedient Litters.** A litter can be improvised from many different things. Most flat-surface objects of suitable size can be used as litters. Such objects include boards, doors, window shutters, benches, ladders, cots, and poles. Some may need to be tied together to obtain the required size. If possible, these objects should be padded.

(1) Litters can also be made by securing poles inside blankets, ponchos, shelter halves, tarpaulin, jackets, shirts, sacks, bags, or mattress covers. Poles can be improvised from strong branches, tent supports, skis, and other similar items.

(2) If poles cannot be found, a large item, such as a blanket, can be rolled from both sides toward the center. Then the rolls can be used to obtain a firm grip to carry the victim. If a poncho is used, the hood must be up and under the victim, not dragging on the ground.

(3) A rope litter is prepared using one rope (Figure 11-3). It requires 20 to 30 minutes to prepare and should be used only when other materials are not available. Four to six bearers are required to carry the litter. The rope litter is the most commonly used field-expedient litter.

Note: Above the tree line, little material exists to construct litters.

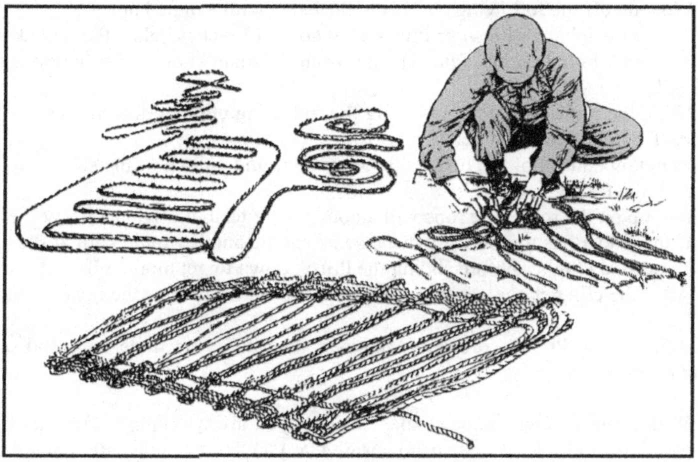

Figure 11-3. Rope litter.

(a) Make 24 bights about 45 to 61 centimeters long, starting in the middle of the rope so that two people can work on the litter at one time.

(b) With the remainder of the rope, make a clove hitch over each bight. Each clove hitch should be about 15 centimeters from the closed end of the bight when the litter is complete.

(c) Pass the remainder of the rope through the bights outside of the clove hitches. Dress the clove hitches down toward the closed end of the bight to secure the litter and tie off the ends of the rope with clove hitches.

(d) Line the litter with padding such as clothing, sleeping bags, empty boxes.

(e) Make the rope litter more stable by making it about 6 inches wider. After placing the clove hitches over the bights, slide them in (away from the closed end) about 15 centimeters. Take two 3- to 4-meter poles, 8 centimeters in diameter at the butt ends, and slide each pole down through the bights on each side. Dress down the clove hitches against the poles. Take two 1-meter poles, and tie them off across the head and foot of the litter with the remaining tails of the climbing rope.

Note: The above measurements may have to be altered to suit the overall length of rope available.

11-8. RESCUE SYSTEMS

Rescue systems are indispensable when conducting rescue operations. A large number of soldiers will not always be available to help with a rescue. Using a mechanical advantage rescue system allows a minimal amount of rescuers to perform tasks that would take a larger number of people without it.

a. **Belay Assist.** This system is used to bring a climber over a section that he is unable to climb, but will continue climbing once he is past the difficult section.

(1) First, tie off the following climber at the belay with a mule knot.

(2) Tie a Prusik knot with short Prusik cord about 12 inches below the mule knot, and place a carabiner into the loop. Place the tail from the mule knot into the carabiner in the Prusik cord.

(3) Untie the mule knot without letting the following climber descend any more than necessary. Do not remove the belay.

(4) Maintain control of the brake side of the rope and pull all of the slack through the carabiner in the Prusik cord.

(5) Pull up on the rope. The rope will automatically feed through the belay.

(6) If the leader has to pull for a distance he can tie another mule knot at the belay to secure the second climber before sliding the Prusik down to get more pulling distance.

(7) After the climber can continue climbing, the leader secures the belay with a mule knot.

(8) Remove the Prusik cord and carabiner, then untie the mule knot and continue belaying normally.

Notes: With all rescue techniques make sure that you always think everything through, and double check all systems to ensure that you don't accidentally untie the fallen climber or find yourself without back-up safety. Do not compound the problem! When doing any rescue work the rescuers will always be tied in for safety.

b. **Belay Escape.** The belay escape is used when a climber has taken a serious fall and cannot continue. The belayer is anchored and is performing an indirect belay, and must assist the injured climber or go for assistance. To accomplish this he must escape the belay system. The belayer will remain secured to an anchor at all times.

(1) After a climber has been injured, tie off the belay device on your body using a mule knot. To improve this system, clip a nonlocking carabiner through the loop in the overhand knot and clip it over the rope.

(2) Attach a short Prusik cord to the load rope and secure it to the anchor with a releasable knot.

(3) Using a guard knot or Munter mule, attach the climbing rope from the belay device.

(4) Untie the mule knot in the belay device attached to the harness and slowly lower the climber, transferring the weight of the climber onto the Prusik.

(5) Remove the climbing rope from the belay device attached to the harness.

(6) Release the mule knot securing the Prusik, transferring the weight to the anchor.

(7) At this point the climber is secured by the rope to the anchor system and the belayer can now assist the injured climber.

11-9. LOW-ANGLE EVACUATION

Cliffs and ridges, which must be surmounted, are often encountered along the evacuation path. Raising operations place a greater load on all elements of the system than do lowering operations. Since all means of raising a victim (pulley systems, hand winches, and power winches) depend on mechanical advantage, it becomes easy to overstress and break anchors and hand ropes. Using mechanical raising systems tends to reduce the soldier's sensitivity to the size of the load. It becomes important to monitor the system and to understand the forces involved.

a. **Raising Systems, Belays, and Backup Safeties.** Raising systems, belays, and backup safeties are of special importance in any raising operation. The primary raising system used is the Z-pulley system, which theoretically gives three pounds of lift for each pound of force expended. In practice, these numbers decrease due to rope-pulley friction, rope-edge friction, and other variables. A separate belay rope is attached to the litter and belayed from a separate anchor. Backup Prusik safeties should be installed in case any part of the pulley system fails.

(1) ***Raising System***. When considering a raising system for evacuations, the Z-pulley system is the most adaptable. It can be rigged with the equipment on hand, and can be modified and augmented to handle heavier loads. Although the vertical or horizontal hauling lines can also be used, the Z-pulley system offers a mechanical advantage that requires less exertion by the transport team.

(2) ***Belays***. Whenever ropes are used for an evacuation operation, the overriding safety concern is damage to the ropes. This is the main reason for two-rope raising systems (raising rope and belay rope).

(3) ***Backup Safeties.*** Because the stresses generated by the Z-pulley system can cause anchors to fail, backup safety Prusiks are used to safeguard the system. These should be attached to alternate anchor points, if possible.

b. **Raising the Litter.** The litter is prepared as already described.

(1) The raising ropes and belay ropes are secured to top anchors and are thrown down to the litter crew.

(2) Padding is placed at the cliff edge and over any protrusions to protect the ropes from abrasion.

(3) The litter attendants secure the ropes to the litter.

(4) The raising crew sets up the Z-pulley system.

(5) One member of the crew secures himself to an anchor and moves to the edge of the cliff to transmit signals and directions. (This is the signalman or relay man.)

Note: If the load is too heavy at this time, another pulley is added to the system to increase the mechanical advantage.

(6) Attendants guide the litter around obstacles while the crew continues to raise the system.

(7) As the litter nears the cliff edge, the signalman assists the attendant in moving the litter over the edge and onto the loading platform, taking care not to jar the casualty.

c. **Descending Slopes.** When descending a moderately steep slope that can be down-climbed, the litter and victim are prepared as described earlier (Figure 11-4, page 11-12).

(1) One man serves as the belay man and another takes his position on the rope in front of the belay man, assisting him in lowering the litter. The litter bearers take their positions and move the litter down with the speed of descent controlled by the belay man.

Figure 11-4. Low-angle evacuation—descending.

(2) The extra man may assist with the litter or precede the team to select a trail, clearing away shrubs and vines. He reconnoiters so that the team need not retrace its steps if a cliff is encountered.

(3) The most direct, practical passage should be taken utilizing available natural anchors as belay positions.

11-10. HIGH-ANGLE EVACUATION

Evacuation down cliffs should be used only when absolutely necessary and only by experienced personnel. The cliffs with the smoothest faces are chosen for the route. Site selection should have the following features: suitable anchor points, good loading and unloading platforms, clearance for the casualty along the route, and anchor points for the A-frame, if used. There are many ways to lower a casualty down a steep slope. As long as safety principals are followed, many different techniques can be used. One of the easiest and safest techniques is as follows (Figure 11-5):

a. Use multiple anchors for the litter and litter tenders.

b. Secure the litter to the lowering rope with a minimum of four tie-in points (one at each corner of the litter). Lengths of sling rope or 7-millimeter cordage work best. Make the attached ropes adjustable with Prussik knots so that each corner of the litter can be raised or lowered to keep the litter stable during descent. Tie the top of the ropes with loops and attach to the lowering rope with a pear shaped locking carabiner.

c. Two litter tenders will descend with the litter to control the descent and to monitor the casualty. They can be attached to separate anchors and either self-belay themselves or be lowered by belayers.

d. Once the steep slope has been negotiated, continue the rescue with a low-angle evacuation.

Figure 11-5. Cliff evacuation descent.

APPENDIX A

LEVELS OF MILITARY MOUNTAINEERING

Military mountaineering training provides tactical mobility in mountainous terrain that would otherwise be inaccessible. Soldiers with specialized training who are skilled in using mountain climbing equipment and techniques can overcome the difficulties of obstructing terrain. Highly motivated soldiers who are in superior physical condition should be selected for move advanced military mountaineering training (Levels 2 and 3) conducted at appropriate facilities. Soldiers who have completed advanced mountaineering training should be used as trainers, guides, and lead climbers during collective training. They may also serve as supervisors of installation teams and evacuation teams. Properly used these soldiers can drastically improve mobility and have a positive impact disproportionate to their numbers. Units anticipating mountain operations should strive to have approximately ten percent of their force achieve advanced mountaineering skills.

A-1. LEVEL 1: BASIC MOUNTAINEER

The basic mountaineer should be a graduate of a basic mountaineering course and have the fundamental travel and climbing skills necessary to move safely and efficiently in mountainous terrain. These soldiers should be comfortable functioning in this environment and, under the supervision of qualified mountain leaders or assault climbers, can assist in the rigging and use of all basic rope installations.

a. On technically difficult terrain, the basic mountaineer should be capable of performing duties as the "follower" or "second" on a roped climbing team, and should be well trained in using all basic rope systems. These soldiers may provide limited assistance to soldiers unskilled in mountaineering techniques.

b. Particularly adept soldiers may be selected as members of special purpose teams led and supervised by advanced mountaineers. Figure A-1 lists the minimum knowledge and skills required of basic mountaineers.

c. In a unit training program, Level 1 qualified soldiers should be identified and prepared to serve as assistant instructors to train unqualified soldiers in basic mountaineering skills. All high-risk training, however, must be conducted under the supervision of qualified Level 2 or 3 personnel.

• Characteristics of the mountain environment (summer and winter). • Mountaineering safety. • Use, care, and packing of individual cold weather clothing and equipment. • Care and use of basic mountaineering equipment. • Mountain bivouac techniques. • Mountain communications. • Mountain travel and walking techniques. • Hazard recognition and route selection. • Mountain navigation. • Basic medical evacuation.	• Rope management and knots. • Natural anchors. • Familiarization with artificial anchors. • Belay and rappel techniques. • Use of fixed ropes (lines). • Rock climbing fundamentals. • Rope bridges and lowering systems. • Individual movement on snow and ice. • Mountain stream crossings (to include water survival techniques). • First aid for mountain illnesses and injuries.

Figure A-1. Level 1, basic mountaineering tasks.

A-2. LEVEL 2: ASSAULT CLIMBER

Assault climbers are responsible for the rigging, inspection, use, and operation of all basic rope systems. They are trained in additional rope management skills, knot tying, and belay and rappel techniques, as well as using specialized mountaineering equipment. Assault climbers are capable of rigging complex, multipoint anchors, and high-angle raising/lowering systems. Level 2 qualification is required to supervise all high-risk training associated with Level 1. At a minimum, assault climbers should possess the additional knowledge and skills shown in Figure A-2.

• Use specialized mountaineering equipment.
• Perform multipitch climbing:
  – Free climbing and aid climbing.
  – Leading on class 4 and 5 terrain.
• Conduct multipitch rappelling.
• Establish and operate hauling systems.
• Establish fixed ropes with intermediate anchors.
	• Move on moderate angle snow and ice.
	• Establish evacuation systems and perform high-angle rescue.
	• Perform avalanche hazard evaluation and rescue techniques.
	• Be familiar with movement on glaciers.

Figure A-2. Level 2, assault climber tasks.

A-3. LEVEL 3: MOUNTAIN LEADER

Mountain leaders possess all the skills of the assault climber and have extensive practical experience in a variety of mountain environments in both winter and summer conditions. Level 3 mountaineers should have well-developed hazard evaluation and safe route finding skills over all types of mountainous terrain. Mountain leaders are best qualified to advise commanders on all aspects of mountain operations, particularly the preparation and leadership required to move units over technically difficult, hazardous, or exposed terrain. The mountain leader is the highest level of qualification and is the principle trainer for conducting mountain operations. Instructor experience at a military mountaineering center or as a member of a special operations forces (SOF) mountain team is critical to acquiring Level 3 qualification. Figure A-3 outlines the additional knowledge and skills expected of mountain leaders. Depending on the specific AO, mountain leaders may need additional skills such as snowshoeing and all-terrain skiing.

• **Recognizing and evaluating peculiar terrain, weather, and hazards.** • **Preparing route, movement, bivouac, and risk management plans for all conditions and elevation levels.** • **Using roped movement techniques on steep snow and ice.** • **Performing multipitch climbing on mixed terrain (rock, snow, and ice).**	• **Performing glacier travel and crevice rescue.** • **Establishing and operating technical high-angle, multipitch rescue and evacuation systems.** • **Usiing winter shelters and survival techniques.** • **Leading units over technically difficult, hazardous, or exposed terrain in both winter and summer conditions.**

Figure A-3. Level 3, mountaineer leader tasks.

APPENDIX B

MEASUREMENT CONVERSION FACTORS

MULTIPLY	BY	TO OBTAIN
Millimeters	.03937	Inches
Centimeters	.3937	Inches
Centimeters	.03281	Feet
Meters	39.37	Inches
Meters	3.281	Feet
Meters	1.0936	Yards
Kilometers	.62137	Miles
Knots	1.1516	MPH

APPENDIX C

AVALANCHE SEARCH AND RESCUE TECHNIQUES

The effect of an avalanche can be disastrous. Chances of survival after burial by an avalanche are approximately 90 percent if the victim is located within the first 15 minutes. Probability of survival drops rapidly and, after two hours, chances of survival are remote. Suffocation accounts for 65 percent of avalanche fatalities, collision with obstacles such as rocks and trees accounts for 25 percent, and hypothermia and shock accounts for 10 percent.

The best chance of survival in snow country is to avoid an avalanche; but, if a member of your group ***is*** *in an avalanche, they are depending on you for rescue!*

C-1. IMMEDIATE ACTION

Survivors at the avalanche site organize into the first rescue team and immediately start rescue operations. If any indication of the location of the victim is found, random probing starts in that vicinity. The tip and edges of the slide are also likely areas to search. A human body is bulky and is apt to be thrown toward the surface or the sides.

C-2. GENERAL PROCEDURES

Establish from witnesses where the victim was located just before the avalanche to determine the point where the victim disappeared—the "last seen" point. Using this and any other information, establish a probable victim trajectory line leading to high priority search areas. Make a quick but systematic check of the slide area and the deposition area, and mark all clues. Look for skis, poles, ice axes, packs, gloves, hats, goggles, boots, or any other article the person may have been carrying—it might still be attached to the victim.

a. Organize initial searchers and probers. If using avalanche beacons, immediately select personnel to begin a beacon search. Ensure all other beacons are turned off or to receive to eliminate erroneous signals. All personnel should have a shovel or other tool for digging or, if enough personnel are available, a digger can be standing by to assist when needed. If the initial search reveals items from the victim, make an initial probe search in that area. This probing should take only a few seconds.

b. Make a coarse probe of all likely areas of burial, and repeat it as long as a live rescue remains possible. Resort to the fine probe only when the possibility of a live rescue is highly improbable. Unless otherwise indicated, start the coarse probe at the deposition area.

C-3. ESTABLISHING THE VICTIM'S MOST PROBABLE LOCATION

In many respects, a moving avalanche resembles a liquid. A human body, with a higher density than the flowing snow, would be expected to sink deeper and deeper into the avalanche; however, several factors influence the body's location. Turbulence, terrain, and the victim's own efforts to extricate himself all interact to determine the final burial position. Study of a large number of case histories leads to the following conclusions.

- The majority of buried victims are carried to the place of greatest deposition, usually the toe of the slide.
- If two points of the victim's trajectory can be established, a high probability exists that the victim will be near the downhill flow line passing through these two points.
- Any terrain features that catch and hold avalanche debris are also apt to catch a victim.
- If an avalanche follows a wandering gully, all debris deposit areas are likely burial spots. The likelihood of a victim being buried in a particular bend is proportional to the amount of debris deposited there.
- Vegetation, rocks, and other obstacles act as snares. The victim tends to be retained above the obstacle. An obstacle may simply delay the victim's motion, leading to final burial down flow from the obstacle.
- Maximum speed of the flowing snow occurs at the avalanche center. Friction reduces flow velocity along the edges. The closer the victim's trajectory is to the center of the slide, the greater will be his burial depth.
- Efforts of the victim to extricate himself by vigorous motion and "swimming" definitely minimize burial depth. Conversely, the limp body of an unconscious victim is likely to be buried deeply.
- An occasional exception to the above is emphasized. The victim may not be buried but may have been hurled away from the avalanche by wind blast. In the case of large and violent avalanches, a search of the surrounding terrain is advisable. Victims have been located in tree tops outside the slide area.

Use of avalanche transceivers is the most efficient method of searching for an avalanche victim, but only if the victim is wearing an active transceiver. Many models of transceivers are available, each with its own manufacturer's instructions for proper use and care. All currently available transceivers are compatible, although they may operate differently in the search mode.

C-4. PROBING FOR AVALANCHE VICTIMS

Probing offers the advantage of requiring simple equipment that can be operated by personnel with no previous training. Although the probers do not need previous training the search leader must be familiar with the technique to ensure proper execution of the probe line.

a. **Probe Poles.** Rigid steel tubing approximately 3/4-inch in diameter and approximately 10 feet long is recommended for the primary probe pole. Longer poles are difficult to manage, especially in a high wind. Although this type of pole performs best, it is difficult to transport to the avalanche site because of its length and weight.

(1) Each person operating in avalanche areas should carry folding sectional poles. These poles are similar to folding tent poles, but are stronger and are connected with cable instead of bungee cord. These poles should be carried on the outside of the pack for immediate access.

(2) If no probing poles are available, initial probing attempts can be started using ski poles in one of two ways: the ski pole can be reversed, probing with the wrist strap down; or the basket can be removed so that the point is down (the preferred method), which allows the ski pole to penetrate the snow more easily.

b. **Probing Lines.** For the probing operation to be effective, probing lines must be orderly and properly spaced. To ensure systematic and orderly probing, the number of personnel per line should be limited. Twenty per line is satisfactory, while thirty is normally the upper limit. The number of probers in the line will be dictated by not only the width of the area to be probed but the number of personnel available. A string may be used to keep the probe lines aligned, but will require added time to maintain.

(1) The probe line maintains a steady advance upslope. Advancing uphill automatically helps set the pace and permits easy probing to the full length of the probe. Probing does not come to a halt when a possible contact is made. The probe is left in contact and the line continues. A shovel crew follows up on the strike by digging down along the pole. Extra probes are carried by the shovel crew to replace those left in contact. Such a plan of operation is especially important when more than one victim is buried.

(2) Striking a body gives a distinct feel to the probe, which is easily recognizable in soft snow but less recognizable in hard compacted snow. A common problem is encountering debris within the snow that can be mistaken for the victim. The only sure check is by digging.

c. **Probing Techniques.** Two distinct probing methods are recognized: coarse probe and fine probe. As evidenced by their names, coarse probing implies a wider spacing of probe pole insertions with emphasis on speed. Fine probing involves close-spaced probing with emphasis on thoroughness. Coarse probing is used during initial phases of the search when live recovery is anticipated. Fine probing is the concluding measure, which almost guarantees finding the body. The coarse probe technique has a 70-percent chance of locating the victim on a given pass, while the fine probe has, essentially, a 100-percent chance of locating the body.

(1) The coarse probe functions as follows:

(a) Probers are spaced along a line 30 inches center to center, with feet about 15 inches apart.

(b) A single probe pole insertion is made at the center of the straddle span.

(c) On command of the probe line commander, the group advances 20 inches and repeats the single probe.

(d) Three commands are used for the complete sequence:

- "DOWN PROBE."
- "UP PROBE."
- "STEP FORWARD."

By using these commands, the leader can maintain closer control of the advancing probe line. It is important that the commands be adjusted to a rhythm that enforces the maximum reasonable pace. A string may also be used along the probe line to keep the probers dressed, although this requires the use of two soldiers to control the string. Strict discipline and firm, clear commands are essential for efficient probing. The probers themselves work silently.

(2) The fine probe functions as follows:

(a) Probers are spaced the same as for the coarse probe. Each man probes in front of his left foot, then in the center of his straddled position, and finally in front of his right foot.

(b) On command, the line advances 1 foot and repeats the probing sequence. Each probe is made 10 inches from the adjacent one.

(c) The commands for the fine probe are:

- "LEFT PROBE."
- "UP PROBE."
- "CENTER PROBE."
- "UP PROBE."
- "RIGHT PROBE."
- "UP PROBE."
- "STEP FORWARD."

(d) Good discipline and coordinated probing is even more important in fine probing than with the coarse probe. Careless or irregular probing can negate the advantages of fine probing. Use of a string to align the probers is especially important with the fine probe. The three insertions are made along the line established by the string, which is then moved ahead 1 foot.

GLOSSARY

AI	alpine ice
ALICE	all-purpose, lightweight, individual carrying equipment
AMS	acute mountain sickness
APS	anchor pulley system
ATC	air traffic controller
BDU	battle dress uniform
BTC	bridge team commander
C	Celsius
CE	community Europe
CEN	combined European norm
cm	centimeter
CMS	chronic mountain sickness
CPR	cardio-pulmonary resuscitation
CTA	common table of allowance
DA	Department of the Army
ECWCS	extreme cold weather clothing system
EN	European norm
F	Fahrenheit
FM	field manual
ft	foot; feet
GPS	Global Positioning System
GTA	graphic training aid
HACE	high-altitude cerebral edema
HAPE	high-altitude pulmonary edema
IAW	in accordance with
LBV	load-bearing vest
LCD	liquid crystal diode
LCE	load-carrying equipment
MEDEVAC	medical evacuation
MPH	miles per hour
MPS	moveable pulley system
MRE	meal ready-to-eat

NSN	national stock number
OP	observation post
PABA	para-amino benzoic acid
PCD	progress capture device
PFD	personal flotation device
RURP	realized ultimate reality piton
SE	southeast
SLCD	spring-loaded camming device
SOF	special operations forces
SOP	standard operating procedure
SOSES	shape, orientation, size, elevation, slope
SPF	sun protection factor
TOE	table of organization and equipment
UIAA	Union des International Alpine Association
U.S.	United States
USAF	United States Air Force
USN	United States Navy
UV	ultraviolet
UVA	ultraviolet A (radiation wavelengths between 320 and 400 nanometers)
UVB	ultraviolet B (radiation wavelengths between 295 and 320 nanometers)
WI	waterfall ice
YDS	Yosemite Decimal System

INDEX

ROPE LOG *(USAGE AND HISTORY)* For use of this form, seeFM 3-97.61; the proponent agency is TRADOC			UNIT ID MARKING
NSN	DOCUMENT NUMBER	SERIAL NUMBER	MFR LOT NUMBER
DATE OF MFR	ISSUE DATE	DATE IN SERVICE	LENGTH
DIAMETER	FIBER	COLOR	CONSTRUCTION

INSPECT ROPE FOR DAMAGE OR EXCESSIVE WEAR EACH TIME IT IS DEPLOYED AND AGAIN AFTER EACH USE. IMMEDIATELY RETIRE ALL SUSPECT ROPES.

DATE USED	LOCATION	TYPE OF USE	ROPE EXPOSURE	INSPECTOR'S INITIAL/DATE	ROPE CONDITION AND COMMENTS

DA FORM 5752-R, MAY 89

DATE USED	LOCATION	TYPE OF USE	ROPE EXPOSURE	INSPECTOR'S INITIAL/DATE	ROPE CONDITION AND COMMENTS

MANUFACTURER ______________________

____

REVERSE OF DA FORM 5752-R, MAY 89

Made in United States
North Haven, CT
05 November 2026

21795741R00300